Adaptive Individuals
IN
Evolving Populations

MODELS AND

BIOLOGY

PSYCHOLOGY

COMPUTER
SCIENCE

ALGORITHMS

Adaptive Individuals in Evolving Populations: Models and Algorithms

Editors

Richard K. Belew
CSE Department
University of California at San Diego

Melanie Mitchell
Santa Fe Institute
Santa Fe, New Mexico

Proceedings Volume XXVI

Santa Fe Institute
Studies in the Sciences of Complexity

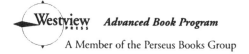 **Westview** *Advanced Book Program*
A Member of the Perseus Books Group

Publisher: *David Goehring*
Executive Editor: *Jeff Robbins*
Production Manager: *Michael Cirone*
Production Supervisor: *Lynne Reed*
Production Assistant: *Amy Raufman Hannum*

Director of Publications, Santa Fe Institute: *Ronda K. Butler-Villa*
Production, Santa Fe Institute: *Della L. Ulibarri*
Indexer, Santa Fe Institute: *Marylee Thomson*

Library of Congress Cataloging-in-Publication Data

Adaptive individuals in evolving populations : models and algorithms /
    editors, Richard K. Belew, Melanie Mitchell.
        p.   cm. — (Santa Fe Institute studies in the sciences of
    complexity. Proceedings volume ; 26)
    Includes bibliographical references and index.
    ISBN 0-201-48364-5 (hardcover). — ISBN 0-201-48369-6 (p)
    1. Animal behavior. 2. Animals—Adaptation. 3. Animal
populations. 4. Animal behavior—Computer simulation. 5. Animals—
Adaptation—Computer simulation. 6. Animal populations—Computer
simulation.   I. Belew, Richard K.   II. Mitchell, Melanie.
III. Series: Proceedings volume in the Santa Fe Institute studies in
the sciences of complexity ; v. 26.
QL751.A44   1996
591.5—dc20   96-13591
                                                                   CIP

This volume was typeset using TEXtures on a Macintosh IIsi computer. Camera-ready
output from a Hewlett Packard Laser Jet 4M Printer.

ISBN 0-201-48364-5 (Hardcover)
ISBN 0-201-48369-6 (Paperback)

    5 6 7 8 9

About the Santa Fe Institute

The *Santa Fe Institute* (SFI) is a private, independent, multidisciplinary research and education center, founded in 1986. Since it's founding SFI has devoted itself to creating a new kind of scientific research community, pursuing emerging science. Operating as a small, visiting institution, SFI seeks to catalyze new collaborative, multidisciplinary projects that break down the barriers between the traditional disciplines, to spread its ideas and methodologies to other individuals and encourage the practical applications of its results.

All titles from the *Santa Fe Institute Studies in the Sciences of Complexity* series will carry this imprint which is based on a Mimbres pottery design (circa A.D.950–1150), drawn by Betsy Jones. The design was selected because the radiating feathers are evocative of the outreach of the Santa Fe Institute Program to many disciplines and institutions.

Santa Fe Institute
Studies in the Sciences of Complexity

Lectures Volumes

Vol.	Editor	Title
I	D. L. Stein	Lectures in the Sciences of Complexity, 1989
II	E. Jen	1989 Lectures in Complex Systems, 1990
III	L. Nadel & D. L. Stein	1990 Lectures in Complex Systems, 1991
IV	L. Nadel & D. L. Stein	1991 Lectures in Complex Systems, 1992
V	L. Nadel & D. L. Stein	1992 Lectures in Complex Systems, 1993
VI	L. Nadel & D. L. Stein	1993 Lectures in Complex Systems, 1995

Lecture Notes Volumes

Vol.	Author	Title
I	J. Hertz, A. Krogh, & R. Palmer	Introduction to the Theory of Neural Computation, 1990
II	G. Weisbuch	Complex Systems Dynamics, 1990
III	W. D. Stein & F. J. Varela	Thinking About Biology, 1993

Reference Volumes

Vol.	Author	Title
I	A. Wuensche & M. Lesser	The Global Dynamics of Cellular Automata: Attraction Fields of One-Dimensional Cellular Automata, 1992

Contributors to This Volume

Ackley, David H.
Department of Computer Science, University of New Mexico, Albuquerque, NM 87131 <E-mail: ackley@cs.unm.edu>

Belew, Richard K.
Department of Computer Science and Engineering, University of California at San Diego, La Jolla, CA 92093 <E-mail: rik@cs.ucsd.edu>

Bergman, Aviv
Interval Research Corporation, 1801 Page Mill Rd., Building C, Palo Alto, CA 94304 <E-mail: bergman@interval.com>

Feldman, Marcus W.
Department of Biological Sciences, Stanford University Stanford, CA 94305 <E-mail: marc@kimura.stanford.edu>

Forrest, Stephanie
Department of Computer Science, University of New Mexico, Albuquerque, NM 87131 <E-mail: forrest@cs.unm.edu>

Godfrey-Smith, Peter G.
Center for the Study of Language and Information, Stanford University, Stanford, CA 94305 <E-mail: pgsmith@csli.stanford.edu>

Hart, William E.
Sandia National Labs, P. O. Box 5800 - MS 1110, Albuquerque, NM 87185 <E-mail: wehart@cs.sandia.gov>

Hightower, Ron
Department of Computer Science, University of New Mexico, Albuquerque, NM 87131 <E-mail: high@cs.unm.edu>

Littman, Michael L.
Department of Computer Science, Brown University, Providence, RI 02912 <E-mail: mlittman@cs.brown.edu>

Menczer, Filipo
Department of Computer Science and Engineering, University of California, San Diego, La Jolla, CA 92093 <E-mail: fil@cs.ucsd.edu>

Miglino, Orazio
Department of Psychology, University of Palermo, Palermo, Italy <E-mail: orazio@caio.irmkant.rm.cnr.it>

Mitchell, Melanie
Santa Fe Institute, 1399 Hyde Park Road, Santa Fe, NM 87501 <E-mail: mm@santafe.edu>

Nolfi, Stefano
Institute of Psychology, National Research Council, 15, Viale Marx, 00137, Rome, Italy <E-mail: stefano@kant.irmkant.rm.cnr.it>

Parisi, Domenico
Institute of Psychology, National Research Council, 15, Viale Marx, 00137, Rome, Italy <E-mail: domenico@gracco.irmkant.rm.cnr.it>

Perelson, Alan S.
   Theoretical Division, Los Alamos National Laboratory, Los Alamos, NM 87545
   <Email: asp@receptor.lanl.gov>
Roughgarden, Jonathan
   Department of Biological Sciences, Stanford University Stanford, CA 94305
   <E-mail: rough@pangea.stanford.edu>
Schull, Jonathan
   Downtown Anywhere and SoftLock Services, 36 Brunswick St., Rochester, NY
   14607
   <E-mail: schull@awa.com>
Shafir, Sharoni
   Department of Biological Sciences, Stanford University Stanford, CA 94305
   <E-mail: sharoni@ecology.stanford.edu>
Taylor, Charles
   Department of Biology, University of California at Los Angeles, Los Angeles,
   CA 90024 <E-mail: taylor@cognet.ucla.edu>
Todd, Peter, M.
   Max Planck Institute for Psychological Research, Leopoldstrasse 24, 80802 Munich Germany <Email: ptodd@mpipf-muenchen.mpg.de>
Zhivotovsky, Lev A.
   Institute of General Genetics, Russian Acadamy of Sciences, 3 Gubkin St.,
   B-333, Moscow, 117809, Russia

Authors of Classic Papers

Baldwin, J. M.
Bateson, G.
Hinton, G. E.
James, W.
Johnston, T. D.
Lamarck, J. B.
Lloyd Morgan, C.
Nowlan, S. J.
Piaget, J.
Simpson, G. G.
Skinner, B. F.
Spencer, H.
Waddington, C. H.

Contents

PSYCHOLOGY

Contents

COMPUTER SCIENCE

Acknowledgments

This book arose from a workshop held in July, 1993, sponsored by the Santa Fe Institute's Adaptive Computation Program. We are very grateful to Beth and Charles Miller for providing the setting for this workshop at their beautiful Santa Fe home, Sol y Sombra, and to Andi Sutherland of SFI and Fred Mims of Sol y Sombra for their superlative organizational support. Thanks also to the contributing authors for their continued energy, patience and forebearance during the two years between our meeting and this published record. RB also wishes to thank members of the Cognitive Science 200 seminar (Winter, 1995 term) at UC San Diego, especially Deborah Forster and Bill Grundy, for their helpful discussion of an early version of this manuscript and contributions towards the Glossary. Finally, thanks to Ronda Butler-Villa, Della Ulibarri, and Marylee Thomson for their editorial wizardry.

Adaptive Individuals in Evolving Populations: Models and Algorithms

Chapter 1:
Introduction

1. BOOK OVERVIEW: FORM AND CONTENT

Even the simplest creature is marvelous to observe as it transforms itself to better match the environment in which it finds itself. How is such adaptation accomplished? How much of this capability should be attributed to the particular individual we happen to be observing, how much to its species, and how much to the inclusive evolutionary processes wedding all life to this planet? How did the elaborate individual learning we find in more complex organisms evolve? Once in place, does an increased individual capacity for adaptation alter the selective pressures causing the species to adapt to its niche?

This book has grown out of a workshop organized to address questions like these. The meeting was sponsored by the Santa Fe Institute and held at Sol y Sombra in Santa Fe, New Mexico, during July, 1993. It brought together a group of about 20 scientists from the disciplines of biology, psychology, and computer science, all studying interactions between the evolution of populations and individuals' adaptations in those populations, and all of whom make some use of computational tools in their work.

The questions that brought us together have been addressed by scientists and scholars for centuries. An excellent example is the "Baldwin effect" (a phenomenon identified by the psychologist J. Mark Baldwin exactly a century ago) that arises repeatedly in many modern computer simulations. Despite the potentially rich historical overtones offered by literature like this, it too often appears that the modern enterprise of science requires successful scientists to become a sort of "write-only memory," writing too much and not reading enough. To combat this tendency and acknowledge the scholarship that precedes our own, a collection of "classic" papers addressing these questions was circulated to all participants as preparation for our discussion.

Here we have reprinted eleven of these seminal classic readings, all addressing interactions between individual adaptation and population evolution. These texts include Baldwin's original paper (Chapter 5), for example, and range from a chapter from Lamarck's *Zoological Philosophy* of 1809 (Chapter 4) to the very brief 1988 paper by Hinton and Nowlan, "How Learning Can Guide Evolution" (Chapter 25), which served to awaken computer scientists to the Baldwin effect. Our selection was further refined by restricting it to papers that are less widely accessible than they deserve. It is for this reason only that centrally relevant papers (e.g., Dennett's [1981] "Why the Law of Effect Won't Go Away," or Plotkin's [1988] "Learning and Evolution") are not included. For many of the reprinted classics, we provide a specially written preface that helps to place this work in the context of the modern research agenda.[1]

In the face of this prior scholarship, the issue then becomes how these classic questions have been changed by knowledge gained in the interim, and what new insights we might gain from modern techniques, particularly computational ones. The other papers in this volume are based on presentations made during the meeting and the discussions that followed them.

Much of this discussion concerned the interpretation of results derived from one discipline by those outside the discipline. For example, what lesson can an ecologist draw from a computational analysis of function optimization? One consequence of this interdisciplinary dynamic[2] is that each paper was reviewed by one or more workshop participants, typically from disciplines different than that of the author. These papers are also preceded by short prefaces, written by the reviewer(s). The goal of the preface to each "modern" (versus classic) paper is to explain the interdisciplinary relevance of the ideas and results of the paper. We hope these gentle

---

[1] There are two minor variations on the basic Preface + Classic pattern. In Chapter 3, J. Schull has provided a more holistic introduction to the classics of Lamarck, Baldwin, Lloyd Morgan, Waddington, Simpson, and Bateson (Chapters 4, 5, 6, 7, 8, and 9, respectively). Also, since the relevant works of William James fall across a number of his writings, Schull has intermixed extended quotations from James with his own commentary on these in Chapter 17.

[2] Other issues surrounding interdisciplinary research are considered below in Section 4.

introductions will provide readers outside the authors' discipline with an appreciation of each contribution's value.

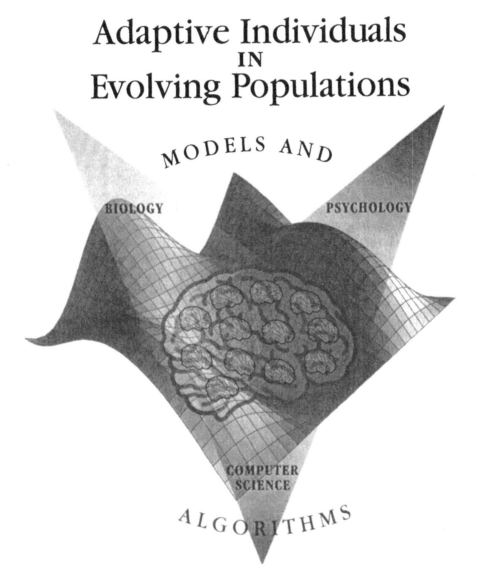

FIGURE 1   AI in EP logo.

The book is divided into three parts corresponding to primarily biological, psychological, or computational themes. Each part begins with an overview chapter, and readers with backgrounds in one of these three areas may find their respective overview chapter the best place to begin.

With overviews, classic reprints, and new work, target readings and prefaces, this book's form is admittedly complex; the issues it attempts to address are no less so. Obeying the maxim that a picture is worth an entire chapter of words, we have attempted to capture the themes of the workshop and of this volume in a single diagram or "logo" that appears on the book's frontispiece and in Figure 1. First, this graphic reflects a sort of self-similarity between adaptive systems at two different levels: individuals (the multiple small blobs) capable of certain malleable deformations and the population (single large blob) in which they are contained are both capable of being plastic or adaptive.

Next, the diagram captures our fundamentally interdisciplinary approach. As mentioned above, the core issues we address are not new, but our attempt to attack them with a "pincer movement" using three distinct "armies" corresponding to the biological, psychological, and computational perspectives may be. We imagine these three perspectives as green, red, and blue spotlights, respectively, each highlighting salient aspects of the phenomena. The biologists at the workshop were especially focused on adaptation at the *population* level and on the *environment* in which this population finds itself. The psychologists were especially concerned with cognitive activity and plasticity in *individuals*. The computer scientists were most concerned with commonalties across these two levels—commonalties that will help in developing adaptive and evolving computer systems.

Given our (RB and MM) backgrounds as computer scientists, it is no surprise that the workshop's participants shared interest in and experience with computation. The purposes toward which the computational tools are applied can be quite different, however. The final theme captured by the logo (and also part of our title), then, is the distinction between *models* and *algorithms*. Briefly, we view "models" as tools operating as an extension of the enterprise of biology, psychology, and other natural sciences. "Algorithms," on the other hand, are tools designed to do some job for somebody; they are engineered artifacts whose success is measured in terms of effectiveness and efficiency in performing the given task. This dichotomy is far from pure; in Chapter 24 we address a number of complications to this simple picture, especially as they suggest research issues for computer science.

2. MAJOR THEMES OF THIS BOOK

The readings in this book have a number of themes in common. All of them deal in some way with questions concerning *adaptation*: what is it, at what levels in living

systems does it operate (e.g., genetic, phenotypic, and population levels), ¿
does adaptation at these various levels affect adaptation at the other levels

Herbert Simon would have us take adaptation to be a *sine qua non* for any
cognitive system:

> "Cognitive science is...a fundamental set of common concerns shared by...
> disciplines concerned with systems that are adaptive—that are what they
> are from being ground between the nether millstone of their physiology
> or hardware, as the case may be, and the upper millstone of a complex
> environment in which they exist" (Simon, 1980).

As used by Simon and others, "adaptation" connotes not only the capacity for
change, but the additional requirement that this change represents an improvement
in "fit." As will become clear in a number of the readings, establishing a clear
criterion by which such improvements can be measured becomes a difficult issue
on its own, and so the term "plasticity," which refers more simply to "flexibility;
capacity for change" (Gordon, 1992, p.255), is sometimes appropriate.[3] With this
relaxation, we immediately make contact with evolutionary concerns, as Darwin
himself noted: "I speculated whether a species very liable to repeated and great
changes of conditions might not assume a fluctuating condition ready to be adapted
to either condition" (Darwin, 1881, from Gordon, ibid.).

Note the important move, however, from considering plasticity to be a trait
of individuals to one of populations. For populations, plasticity becomes "substantial individual variation among members of the same species" (Brauth, Hall, and
Dooling, 1991, p.1). Common to both individual and population plasticity is that a
capacity for change is implied. The comparison for individuals considers the same
individual at two points in time; a plastic individual is one that exhibits a wide
range of variability during its lifetime. A population's plasticity, on the other hand,
requires comparison across members of the same species; a plastic species is one
that exhibits wide phenotypic variation.

The methodological difficulties involved in assessing the range of phenotypic
variation, particularly as it may be sensitive to *environmental* variation, has been
a fundamental obstacle to the investigation of the role of plasticity in evolutionary theory (Mayr, 1982). The concept of "norms of reaction" (illustrated in Figure 2) has proven an important conceptual device in understanding this relationship. Imagine some range of environmental variation, and a genotype that exhibits
a range of phenotypic variations in the face of these environmental variations. The
genotype can then be viewed as a function, transforming environmental variables

---

[3]The word "plastic" was in fact the defining word of our workshop: *"Plastic* individuals in evolving
populations." Among this group of self-selected psychologists, computer scientists, and biologists
the appropriate sense of "plastic" was understood. Our publishers have convinced us, however, that
the typical book buyer is more likely to be reminded of recycling and perhaps "The Graduate,"
not cognition!

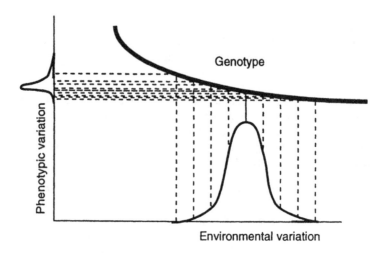

FIGURE 2   Norm of reaction.

to phenotypic traits. This perspective helps to suggest a relationship between genotypic characteristics and environmental variation that together give rise to a particular phenotypic form, as well as variants in form that we might observe. The sticky,
sticky issues of just how much of the variability we see across individuals should be
ascribed to *heritable* traits that we associate with genetic factors, and how much
to shared environmental commonalties will, in natural situations, be very hard to
determine. In some carefully designed laboratory experiments we may be able to
distinguish between these factors. In our computational simulations making these
distinctions is trivial: by virtue of the program's construction we can know precisely
the variables corresponding to the "genotypic" and "environmental" factors.

Under this general rubric of adaptation and plasticity in individuals and in
populations, some common threads tying the readings together are:

- How direct are the interactions coupling genetic and phenotypic adaptations—
  i.e., interactions between adaptations that occur as a result of changes to the
  genome over evolutionary time, and those that occur via plasticity during an
  organism's lifetime?
- In what ways should the *behavior* of an organism (as contrasted with its morphology or physiology) be viewed as a key phenotypic trait that is acted on by
  evolution?
- More specifically, in what ways does an individual's *learning* behaviors during
  its lifetime interact with the evolution of its species?
- To what extent do behaving/learning organisms effect a kind of "natural selection" on their environments, thereby producing a symmetric complement to

the (more conventionally considered) selective forces that their environments apply to them?

■ To what extent are learning (of an individual over its lifetime) and evolution (of a species over many generations) two sides of the same coin—i.e., two manifestations (at different time scales) of a similar process?

In the following we give an overview of these questions and the chapters of this book that address them.

2.1 DIRECT VERSUS INDIRECT INTERACTIONS BETWEEN GENOTYPIC AND PHENOTYPIC ADAPTATIONS

Many of the readings in this book concern the types of effects, both direct and indirect, that genotypic adaptations can have on phenotypic adaptations, and vice versa. The excerpt we reprint from Lamarck's *Zoological Philosophy* (Chapter 4) is one of the earliest discussions of this theme. As is well known, Lamarck proposed (long before the mechanisms of heredity were discovered) that phenotypic traits acquired by an organism during its lifetime can be passed on—*directly*—to the organism's offspring. One of the hallmarks of modern genetics is its rejection of this hypothesis. It is almost universally accepted among biologists that "reverse transcription" of acquired phenotypic traits back into the heritable genotype does not occur.

Figure 3 presents a simple, conventional perspective of the relationship between genotype and phenotype. We begin with a particular genotype at some time $G(t_0)$. According to Weismann's doctrine (e.g., see Weismann, 1893), sequestration of the germ line insulates subsequent phenotypic changes from evolution of the genotype. Following Elman et al. (in press), we will use the term *development* to refer to all types of within-lifetime change in individuals, but then distinguish between two important subcategories. The first type, *maturation*, refers to the process of transforming genotype into phenotype, the cause of which we primarily attribute to the genetic code. This gives rise to a neonate phenotype at a somewhat later time t_1. *Learning* is the second type of individual change, the cause of which we attribute primarily to interactions with the environment. This gives rise to an "adapted" phenotype that we consider at a still later time t_3. It is important to note that "maturation" and "learning" are idealized analytic categories, acting as poles for the full spectrum of developmental changes due to mixtures of genetic and environmental influence; most biological examples of developmental change will be extraordinarily complicated mixtures of the two. The final component of this simple diagram is an explicit blockage of the direct inheritance Lamarck proposed.[4] That is, our theories must disallow the genetic inheritance of characteristics acquired during the lifetime of an individual.

---

[4] Note that Bateson (Chapter 9) makes this blockage his very first premise of theory building.

If only the story were as straightforward as this figure makes it seem! As our brief excerpt of Lamarck (Chapter 4) documents, this distinguished biologist and keen observer had many parts of the evolutionary story right 50 years before Darwin, and good reasons for the confusion that henceforth became his primary legacy. As suggested by Figure 4, the fact is that a *common environment mediates* the direct causal linkages captured in Figure 3. A range of other *indirect* interactions between genotype and phenotype then become possible. These include: environmental influences on the maturational process transforming genotype to phenotype; the learning changes that shape a phenotype during its lifetime; the effects of its behaviors on

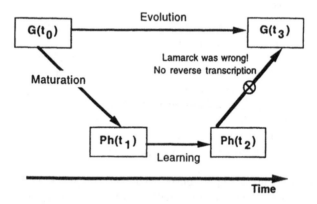

FIGURE 3 Genotype/phenotype separation: standard view.

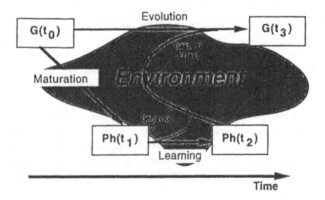

FIGURE 4 Genotype/phenotype interactions.

the environment; and long-term (indirect!) effects these actions by individuals can have on populations.

A good example of a phenomenon requiring more sophisticated analysis is provided by the recent experiments of Cairns Smith et al. (Cairns Smith, Overbaugh, and Miller, 1988; Foster, 1993). They have shown that *E. coli* are capable of unusual adaptations to characteristics of their lactose growth medium. A fundamental tenet of the neo-Darwinian synthesis is the independence of the sources of genetic variation from the selective forces subsequently applying to phenotypes. Yet recent laboratory data suggests that, at least in some simple bacteria, the genotype seems able to "adaptively mutate" and be responsive to environmental variability. The potential impact of mechanisms of adaptive mutation on a more complete theory of the interacting mechanisms of variation and selection are therefore profound:

> "The discovery that cells use biochemical systems to change their DNA in response to physiological inputs moves mutation beyond the realm of "blind" stochastic events and provides a mechanistic basis for understanding how biological requirements can feed back onto genome structure...there is no unicorn in the genomic garden. But we have found a genetic engineer there, and she has an impressive toolbox full of sophisticated molecular devices for reorganizing DNA molecules" (Shapiro, 1995, p.374).

It is important to be clear that any such interactions must be *indirect*, and not of the sort proposed by Lamarck. We do not consider ourselves Lamarckian, nor to our knowledge do any of our contributors. All the explanations we consider are in basic accordance with the fundamental components of the neo-Darwinian explanation suggested by Figure 3; no new magical mechanism is proposed that might directly communicate information gained by the phenotype directly to the genotype. Much of the work in this book is motivated by a belief in the existence of plausible, non-Lamarckian mechanisms by which information acquired by phenotypic adaptations can eventually become genetically encoded. The Baldwin effect is but one example.

It is unfortunately true, however, that many of the phenomena that will most interest us are often confused with Lamarckianism. Consider:

> "Piaget was greatly influenced by Waddington, but he went well beyond support for Waddington's notion of the exploitative system; he championed the 'Baldwin Effect' and *other rather Lamarckian notions* by which behavior might *directly* affect the genotype—and for which no known mechanisms exist" (Plotkin, 1988, p.143, emphasis added).

It is not clear what notions Plotkin intends to preclude as "rather Lamarckian"; guilt by association seems sufficient! In his discussion of the Baldwin effect (Chapter 8 in this volume) Simpson makes a similarly confusing reference to "neo-Lamarckian," apparently as a parallel to the very well defined and useful "neo-Darwinian." Even Waddington seemed confused by the Baldwin effect, despite the fact that his mechanism of genetic assimilation shares a number of features with it.

We hope re-publication of some of the seminal papers at the heart of this confusion will help to clarify these critically important issues. Here we reprint readings by Baldwin, Lloyd Morgan, Waddington, Piaget, Simpson, and Bateson (Chapters 5, 6, 7, 19, 8, and 9, respectively) on this topic. Bateson (Chapter 9) perhaps goes furthest, arguing that; setting aside the impossibility of reverse transcription, the complexity and "economics of somatic flexibility" in organisms makes pure Lamarckianism unworkable, but gives an advantage to "simulated Lamarckianism" of the sort described by the Baldwin effect. From a computational perspective, Hinton and Nowlan's paper (Chapter 25) presents a highly simplified model of the Baldwin effect in order to demonstrate its plausibility. Among the contributed chapters, Parisi and Nolfi (Chapter 23) also demonstrate the Baldwin effect investigated, though in a richer, computational setting, and give a more extensive analysis of the underlying mechanisms. Hightower, Forrest, and Perelson (Chapter 11) examine the possibility of Baldwin-like effects in the immune system—another biological system in which adaptations occur both on evolutionary and within-lifetime time scales.

A number of these authors will be seen to be championing the Baldwin effect or its relatives, and all without recourse to Lamarckian magic. The critical issue seems to be just how strong the causal but indirect role environmentally mediated mechanisms can play in the evolutionary process. A central goal of this book is to get beyond a simple knee-jerk rejection of Lamarck's *direct* inheritance, so as to consider *indirect* ways by which phenotypic change can influence evolution.

2.2 BEHAVIOR AS A PHENOTYPIC TRAIT

There seems to be an inevitable bias in evolutionary theory toward things that fossilize well—morphological structures. A message sounded loudly by ethologists, however, is that *behaviors* too count as first-class phenotypic traits. Moreover, behaviors "close the loop" between an organism and its environment. It may be possible to describe a behavior in terms of physiological changes to an organism (e.g., range of limb movement), but without also describing the action's effect on the world in which the organism naturally finds itself, these physiological characteristics are meaningless. As several recent critiques of early artificial intelligence methods have made clear, this is as true of high-level, cognitive behaviors performed by humans as it is of amoebic locomotion. Planning that is "reactive" to environmental characteristics is simpler and more effective than forcing an agent to maintain an internal, completely consistent world model (Agre, 1995). Most intelligent behaviors are "situated": they must be understood in terms of the environmental and cultural context in which they occur (Suchman, 1987; Hutchins, 1995). In short, behavior weds organisms to their environments in a way that is critical to a full understanding of what it means to be adaptive (see Section 2.4).

Behavior will prove a particularly important class of traits to us in this book because behaviors are especially plastic, and because they play a particularly powerful evolutionary role. Their range of variability, relative to morphological changes, is pronounced. Lorenz (1973) describes this flexibility in terms of "gaps" in behavioral chains, places in the program controlling behavior in which a range of alternatives can be filled in by individuals of particular environmental situations. Mayr is often identified with a particularly strong account of behavior's evolutionary role:

> "Many if not most acquisitions of new structures in the course of evolution can be ascribed to selection forces exerted by newly acquired behaviors. Behavior thus plays an important role as the pacemaker of evolutionary change" (Mayr, 1982, p.612).

Because behavioral traits are so hard to quantify, biologists typically consider behaviors with especially direct connections to selective fitness, such as foraging. The papers by Shafir and Roughgarden; Menczer and Belew; Parisi and Nolfi; and Miglino, Nolfi, and Parisi (Chapters 12, 13, 23, and 22, respectively) fall into this category. This last paper highlights several distinct, hierarchically nested levels—genotype, structural neural network, functional neural network, potential behaviors, actual behaviors—at which evolution may operate in a behaving organism, even when selection acts upon only the last.

Psychologists typically consider much more elaborate behaviors with less direct connection to selective pressures. One goal for our interdisciplinary discussion is to relate the more sophisticated computational models coming from cognitive science to the evolutionary issues of central concern to biologists. The ways in which this can be done are discussed in detail in the psychology overview, Chapter 14. The critical cognitive ability to *predict* future events is a topic of special concern in the papers of Zhivitovsky, Bergman, and Feldman (Chapter 10), and Parisi and Nolfi (Chapter 23).

2.3 INTERACTIONS BETWEEN LEARNING AND EVOLUTION

The benefits of individual plasticity to evolutionary adaptation seem self-evident: in any sufficiently complex, nonstationary environment, an individual able to adapt to changes in the environment during its lifetime is surely more fit than one that cannot. There are evolutionary costs, however, that must also be associated with this same plasticity, and a full analysis must consider both the costs and benefits as they apply to any particular species to understand the net evolutionary impact of individual adaptation. Here we will mention only some of the most important costs and benefits of learning with respect to evolution (Johnston's paper, Chapter 20, provides a much more extensive analysis), and then discuss the even more complicated interactions related to the evolutionary origins of learning.

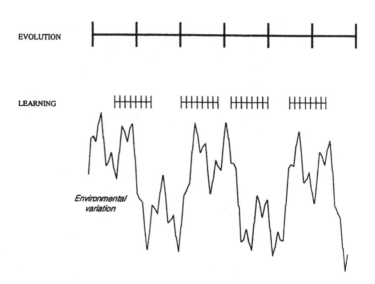

EVOLUTION

LEARNING

Environmental
variation

FIGURE 5   Environmental regularities.

LEARNING PROVIDES A SELECTIVE ADVANTAGE. If we do as Herbert Simon says (above), and treat the problem of adaptation as one of identifying and exploiting environmental regularities, we can model a very simple version of this in terms of a one-dimensional time series (e.g., daily temperature values). Dynamical systems analysis tells us that these regularities can occur at many different time scales (e.g., days, years, millenia), and that their pattern may be arbitrarily complex. We might hope that an adaptive system is capable of finding regularities in such time series and exploiting them (via, for example, seasonal variations in coat thickness, burrowing behaviors, or migration patterns).

An abstracted time series and a simplified version of the relationship between evolutionary and learning systems is shown in Figure 5. First consider evolution, plodding along. Its "sampling" of the environment (denoted in the figure by short vertical bars) occur in the form of generations, and track those glacial changes occurring slowly enough that a species' gene pool can follow.[5] Within each generation, many individuals each experience spatially and temporally localized environments (denoted by more closely spaced but disjoint sets of vertical bars). We can easily imagine that, as in the figure, some frequencies of environmental change occur too quickly for evolutionary adaptation. In such cases learning may become another

---

[5]Of course, we must be careful as human observers not to impose *our* chauvinistic notion of generational time onto this adaptation; evolution can be a fairly rapid process in some species. The important comparison is between the relative rates of evolutionary and individual change.

adaptive process by which individuals are transformed in response to regularities of the local environment they experience. For example, if temperature changes too rapidly for evolution to track, individuals can *learn* to predict short-term temperature changes during their lifetime and perhaps control their environments, by learning to burrow or hibernate when it is too cold outside. Such individual adaptations are *in addition to* any evolutionary changes their species may make.

In Piaget's words, useful behavioral adaptations are those:

> "...that characterize varieties of behavior which naturally facilitate survival...
> [and] serve to increase the powers of the individual of a species by putting
> greater means at their disposal" (Piaget, 1978, p.xix).

Waddington has argued more broadly that since learning can be expected to increase the variance of individuals, and since variance across populations is evolution's primary grist, individual plasticity should *speed up* evolution. Ways in which individuals' adaptations can help to "inform" the evolutionary process are addressed by many of the papers in this collection, most centrally in the papers by Hart and Belew; Littman; and Zhivitovsky, Bergman, and Feldman (Chapters 27, 26, and 10, respectively).

LEARNING CAN ALSO CONSTITUTE A SELECTIVE DISADVANTAGE. There are equally good arguments as to why learning, and plasticity in general, might slow down the evolutionary processes shaping populations. Sewall Wright (1980) observed that individual plasticity can effectively hide desirable phenotypic traits by allowing them to become temporarily achieved. That is, desirable phenotypic traits can be effectively masked because they can be achieved temporarily by some individuals via learning. Evolution gets less information concerning selective pressures in a particular environment. Gordon summarizes the argument:

> "Of a range of phenotypes, only some are optimal. If the production of
> phenotypes were irreversible, not subject to change, then selection could
> choose the best ones and eliminate the others. But plasticity allows less
> than optimal organisms to slip out from under selection's heel, by taking a
> more optimal form temporarily" (Gordon, 1992, p.262).

It is also possible to point to fairly specific reproductive costs associated with learning. Most concrete are the additional metabolic costs associated with developing and maintaining the machinery for learning; the vertebrate central nervous system is an excellent example of both an advanced substrate for learning and its attending metabolic demands. A learning juvenile is also less well prepared for the world, and hence creates an additional "load" on the parents that must rear it (Cecconi, Menczer, and Belew, 1996). The same environmental and experiential factors to which a plastic individual must be sensitive exposes it to concomitant risk that the necessary environmental conditions for successful development will not prevail. The paradoxical balance of such risks with the equally obvious benefits of plasticity with which we began, sets the stage for the more refined analysis this book attempts.

THE EVOLUTION OF LEARNING. Prerequisite to an understanding of the potential effects, facilitory or inhibitory, of learning on evolution, it is useful to consider the effects of evolution on learning, i.e., the evolution of the original learners, and the subsequent evolution of increasingly powerful learning capabilities. Bateson's article (Chapter 9) is one of the few to address such questions. If we view genetics as one adaptive "circuit" in the interior of a second adaptive system comprised of learning individuals, Bateson argues that there exist both "centrifugal" and "centripetal" forces shifting a "locus of control" to and fro between genetic and learning systems. Further, he suggests that the long-term tendency will be in favor of the centripetal force "affirming" learned behaviors via "genetic assimilation" (Waddington's term; see Chapter 7) in exactly those cases where environmental conditions remain constant long enough for this to occur. While Bateson is the first to acknowledge that "To speculate about problems so vast is perhaps romantic" (p.123), it may be that hypotheses such as these which seem impossible to test biologically can be tested computationally.

2.4 "COEVOLUTION" BETWEEN ORGANISMS AND THEIR ENVIRONMENTS

It is easy to envision evolution as a process by which passive individuals are actively selected (for or against) by their environments. In this view, genetic alternatives are generated and the environment culls some and perpetuates others. A related simplification is to assume that all the information determining the outcome of these evolutionary experiments resides in the individual's genes.

As the discussion of reaction norms above makes clear, a wide range of "epigenetic" variation can be elicited from the same genome by the environment. Waddington (1975) is often credited with going the next step, elevating the environment to a full partner in a coevolutionary process. Certainly environments selects individuals, but through a process sometimes called "niche selection" individuals— especially *behaving* individuals—also select their environments (a popular song refers to this as "finding a place where the weather suits my clothes"). Waddington (1960) identified this capacity of individuals as a third "exploitative system," operating together with genetics and natural selection.

In *The Ontogeny of Information*, Oyama (1985) argues that a more balanced view of genetic and environmental roles can also help us move beyond polarized debates of "nature" versus "nurture." Common metaphors for the genes (e.g., programs, blueprints, etc.) as the *information*-carrying component in evolution all rest on an inappropriate "preformationist" view of information, as if information "...exists before its utilization or expression":

"Instead, it is ontogenesis, the inherently orderly but contingent coming into being, that is essential about the emergence of pattern and form..." (p.3).

"Nativism and empiricism require each other as do warp and weft. What they share is the belief that information can preexist the processes that give rise to it. Yet information 'in the genes' or 'in the environment' is not biologically relevant until it participates in the phenotypic processes" (p.13).

"[Thus there is] causal symmetry...whereby genetic and nongenetic factors alike can be sources of variation in form and whereby constancy of form generally requires constancy in developmentally relevant aspects of both genome and environment" (p.14).

This theme—causal symmetry between organisms and their environments—is taken up in several of the readings. Schull's exposition of James' work (Chapter 17) makes it clear that this was one of James' central ideas; James extended notions of selection and organism/environment causal symmetry to psychology, social systems, and cultural evolution. This issue is also touched on in Bateson's paper (Chapter 9), with his mention of "extra regulators"—mechanisms by which organisms are able to control their environments and thus keep environmental variables within homeostatic limits. Even Skinner, who might be expected to focus exclusively on environmental characteristics, expresses an even-handed balance between factors[6]:

"Early behaviorists, impressed by the importance of newly discovered environmental variables, found it particularly reinforcing to explain what appeared to be an instinct by showing that it could have been learned, just as ethnologists have found it reinforcing to show that behavior attributed to the environment is still exhibited when environmental variables have been ruled out" (Skinner, this volume, p.274).

In this book, Todd's paper on sexual selection (Chapter 21) considers organism/environment interactions from another angle. Under sexual selection, organisms act as selectors on *each other*, yielding a coevolutionary process that can run amok as "runaway sexual selection." Finally, Menczer and Belew's paper (Chapter 13) on "Latent Energy Environments" reports on LEE, a computer model specifically designed to investigate organism/environment interactions.

---

[6]Though note the bias expressed by his invocation of the language of learning, "...found it particularly reinforcing," as an explanation for the behaviors of the two classes of scientists!

2.5 ARE LEARNING AND EVOLUTION TWO FACES OF SAME COIN?

It is quite evident that learning individuals have evolved, and so we have immediate reason to be interested in situations that involve both learning and evolution. A less obvious connection between learning and evolution are arguments that the two forms of adaptation are in fact variations on a single, common theme. If true, our explanations of each will share critical features, for deep reasons.

A typical starting point is Broadbent's (1961) Law of Effect: actions followed by rewards are repeated. This seems to fit both forms of adaptation quite well. Dennett explains "Why the Law of Effect Won't Go Away":

> "The Law of Effect and the principle of natural selection are not just analogies; they are designed to work together.... If creatures with some plasticity in their input-output relations were to appear, some of them might have an advantage over even the most sophisticated of their trophistic cousins. Which ones? Those that were able to distinguish good results of plasticity from bad, and preserve the good. The problem of selection reappears and points to its own solution: let some class of events in the organisms be genetically endowed with the capacity to increase the likelihood of the recurrence of behavior-controlling events upon which they act" (Dennett, 1981, p.75).

Over a century earlier, Spencer attempted to make a similar parallel between learning by individuals and evolution by populations. This attempt is reflected here in two readings from Spencer's books (Chapters 15–16). The reader is warned that in support of his arguments, Spencer sometimes expresses alarmingly racist and unscientific views about differences in human intelligence. For example:

> "...thus it happens that the European comes to have from twenty to thirty cubic inches more brain than the Papuan. Thus it happens that faculties, as that of music, which scarcely exist in the inferior human races, become congenital in the superior ones. Thus it happens that out of savages unable to count even up to the number of their fingers, and speaking a language containing only nouns and verbs, come at length our Newtons and Shakspeares [sic]" (p.242).

Spencer was certainly not alone. Even William James, acknowledged by many as one of the greatest thinkers of the twentieth century, believed that:

> "[the]...absence of prompt tendency in [the male] brain to set into particular modes is the very condition which insures that it shall ultimately become so much more efficient than the woman's.... [T]he masculine brain deals with new and complex matter...in a manner which the feminine method of direct intuition, admirably and rapidly as it performs within its limits, can vainly hope to cope with" (James, 1927/1955, pp.691–692).

Rather than attempting to exorcise such offenses from the historical readings, we encourage the modern student of evolution, behavior, and intelligence to remember the role tacit preconceptions and social agendas have played in the history of these subjects. As publications as recent as *The Bell Curve* (Herrnstein and Murray, 1994) demonstrate, these dangers remain with us.

Such political issues aside, the parallel between learning and evolution that Spencer identifies takes a more modern form with Karl Popper. His famous phrase, "Let our hypotheses die in our stead," is made quite explicit by models like that of Hinton and Nowlan (Chapter 25). In such models the tradeoff between a population's use of mortal individuals and an individual's use of repeated trials is as direct as can be. A bit of trial-and-error experience in a lifetime is viewed as a "cheap trial" that is easier to test as an individual's hypothesis than going to the bother of constructing an entirely new genetic experiment.

While the tie between a life and a bit of individual learning can be effected in a computer model by simply renaming appropriate variables (e.g., `Allele` → `Guess`), it is important to remember that in real biological systems the two forms of adaptation rely upon entirely different mechanisms. The exact biochemistry of learning in the nervous system remains an area of intense investigation, but many features—phosphate channels, neurotransmitters, long-term potentiation, etc.—are known. Similarly, many details of population and molecular genetics—nucleotide substitution patterns, polymorphism's, etc.—are also well known. What we do know of the *biological* processes of learning and evolution suggests that however similar the adaptive problems these processes face, and even if they both do in fact share a common Law of Effect-style learning rule, these similarities exist at what a computer scientist would call the "logical" level only. And just as every logical computer design must be transformed into a physical design, incorporating implementation constraints imposed by features of the hardware, we should expect that the biological phenomena of learning and evolution will also be subject to their respective biological "implementation constraints." To use Broadbent's language, the generation of an "action," the perception of "reward," the "memory" required to allow the same action to be repeated, etc., will involve very different representations and modes of access to these representations in the individual and population cases. While it is provocative to think of learning and evolution as two examples of a similar adaptive technique, their differences are equally provocative.

---

3. CAVEATS

Almost as important as the themes these chapters share may be warnings about the problems we have encountered, and the steps we have taken to avoid them. Two require special attention.

The first is a closely related topic we purposefully avoid: cultural evolution. If we use "culture," (or perhaps "proto-culture," Lumsden and Wilson, 1981) to refer to all of the many ways conspecifics affect one another via regularized modifications to their shared environment, the connections with our themes are obvious and many. Culture is a prime determinant of individuals' experience. If the environment shaping and shaped by adaptive populations and individuals is as central as we argue above, then certainly cultural systems which ritualize the environmental experience of individuals across generations must be hugely important (Hutchins, 1995). If the parallels between learning and evolution just mentioned are striking, those relating processes of cultural change to evolution are no less so. Dawkins (1976) has popularized the notion of cultural "memes" (which may perhaps take the form of myths, hypotheses, theories, books, etc.) that are to be considered as the analog for genes in the evolutionary system; others have advanced very similar arguments (Popper, 1972; Campbell, 1974). In a similar vein, Bateson's article (Chapter 9) draws an intriguing parallel between the Baldwin effect and the actions of a legislative body, "affirming in law that which has already become the custom of the people" (p.117, this volume). Culture certainly seems—and can be modeled as (e.g., see Belew, 1990)—a third, intermediate form of adaptation, operating between learning and evolution. The time scale of the environmental regularities to which culture "attends" (cf. Section 2.3 above) falls squarely between the paces of evolutionary change and individual lifetimes. In all these ways, processes of cultural adaptation seem natural bedfellows with learning and evolution.

Nevertheless, in this book we skirt the issue almost completely, for several reasons. First, as the brief list of issues above and the chapters to follow hopefully illustrate, it seems quite hard enough to consider interactions among individual plasticity and specic evolution without complicating our topic still further. Second, these issues appear prior to those involving culture, and their independent consideration therefore seems justified. In Section 2.3 above, we briefly considered the evolution of the first species with significant learning abilities, in comparison to prior species whose adaptation depended exclusively on evolutionary forces. In the same way, we propose to restrict our attention to those earlier, more "primitive" situations in which learning and evolution interact, without an appreciable role for culture.

Our third reason is that a full analysis of culture's role as yet another, equally powerful adaptive force seems hard indeed. This is quite apparent already from studies of bird song, canine play, etc. The enormous complexity, and danger of facile simplification, that face us as we attempt to address *human* cultural phenomena is staggering. The fourth and final reason for avoiding culture as much as possible is just how attractive the parallels between cultural adaptation and evolution are. Viewing myths or legislation as a culture's genetic material is clearly provocative. But whether such images are useful merely as metaphors or instead point toward fundamental characteristics of adaptive systems depends on better understanding of the phenomena of adaptation and more concrete data concerning cultural change.

For now, then, we must satisfy ourselves with the hope that a better understanding of learning and evolution will prepare us for this next critical stage of investigation.

The second caveat concerns the fine line we must walk between believing in adaptation's power and succumbing to the "adaptationist fallacy." That fallacy, that evolution inexorably improves its designs until reaching perfection, has dogged evolutionary theory for much of its history. Modern evolutionary analysis, certainly since Gould and Lewontin's (1979) devastating critique of the "Panglossian paradigm," has made it clear that the biological process is much more stochastic and much less driven to perfection than Pangloss might desire; being good enough to get by is typically just fine. Our cause for concern about adaptationism, then, is motivated first by the fact that this well-known message of the biologists may not have filtered fully into the other disciplines.

But the issue is complicated further by the participation of computer scientists in this discussion. Typically these scientists work as part of engineering schools, within which the search for good, and preferably optimal, designs is almost axiomatic. Our initial distinction between computational models and algorithms gains real teeth here. To the extent that we are attempting to build veridical models of biological evolution, we should remain wary of adaptationist tendencies toward optimal outcomes. But as engineers of computational algorithms, we will be expected by our peers to achieve just this optimality. It therefore becomes especially important to be clear as to the scientific purpose of any one piece of work.

4. THE BENEFITS OF INTERDISCIPLINARY RESEARCH

As we said above, a distinguishing characteristic of this book—one it shares with much of the work associated with the Santa Fe Institute—is that it depends critically on the shared perspective provided by the involvement of scientists from multiple disciplines. As anyone who has participated in interdisciplinary work can attest, such conversations can often be unproductive and even painful. Beyond our central focus on the interactions between two forms of adaptation, we believe this book can also provide a case study of a successful interdisciplinary interaction. We will mention several characteristics that seem important to that success here.

First, we did not attempt to blur the distinctions among our participants' background disciplines. Said positively, we believe that the best way to be interdisciplinary is to respect disciplines. We relied, for example, on the biologists attending our meeting to tell us about evolution and the environment; they necessarily did this from the paradigmatic perspective of biology, something they had spent their careers absorbing. As organizers and editors, our respect for each of the three disciplinary perspectives of our workshop (and of this book) has caused us to, if anything, artificially highlight the distinctions between them. For example, the readings that follow have been grouped into Biological, Psychological and Computational sections, when in fact almost every reading has at least some aspects of the other two. These groupings are therefore in some cases fairly arbitrary, sometimes reflecting the discipline with which the author is primarily associated (e.g., the chapter by Piaget (Chapter 19) is in the Psychology section, although it largely deals with intersections between biology and psychology), and sometimes reflecting the discipline in which the reading has had the most influence (e.g., Hinton and Nowlan's model of the Baldwin effect is in the Computer Science section, Chapter 25). The unifying views of each of the three areas are captured in the "overview" chapters, summarizing the meetings of the three disciplinary subgroups.

The subtleties of each paradigm's perspective, and their differences as they address a shared set of issues, are impossible to adequately summarize here. Some very interesting telltale signs are provided, however, by quite simple evidence: the words each of these participants use. Logical positivists aside, current philosophy of science acknowledges the tremendous burden carried by the "keywords" of every science. This point has been made especially clear in biology by Fox Keller and Lloyd's (1992) *Keywords in Biology:*

"[Words] serve as conduits for unacknowledged, unbidden, and often unwelcome traffic between worlds. Words also have memories; they can insinuate theoretical or cultural past into the present. Finally, they have force. Upon examination, their multiple shadows and memories can be seen to perform real conceptual work, in science as in ordinary language" (p.2).

"Indeed, it is precisely because of the large overlap between forms of scientific thought and forms of societal thought that "keywords"—terms whose meanings chronically and insistently traverse the boundaries between ordinary and technical discourse—can serve not simply as indicators of either social meanings and social change *or* scientific meaning and scientific change, but as indicators of the ongoing traffic *between* social and scientific meaning and, accordingly, between social and scientific change" (pp.4–5, emphasis in original).

The confusion's caused by "simple" misinterpretation of critical words in our interdisciplinary conversations are therefore far from simple. In this and in other Santa Fe Institute workshops, we have found that one of the hardest parts of interdisciplinary discussions is reaching a consensus on the meaning of terms central to the discussion. In this spirit, we have collected a set of the most troublesome keywords that arose in a glossary at the end of this volume. As with the *Keywords* book, these characterizations of the words' meanings are:

"...not intended to provide definitive or correct definitions...[but] to provide a rough map of some of the territory of dispute and change" (*Keywords,* p.6).

Unlike *Keywords*, however, we have restricted our attempts to paragraphs rather than multiple pages. The "definitions" have come from our authors and workshop participants, a group of students at the University of California at San Diego who were early readers of this manuscript, and the dictionary. In some cases this has resulted in multiple definitions, reflecting some of the terminological debate that arose in our meeting.

In many ways, this glossary forms a fitting conclusion for our volume. It reflects the variety of perspectives of our interdisciplinary group, and captures some of the dynamic of the meeting from which this printed record has sprung. It highlights the role artifacts of the scientific process, from words to computational models, play in shaping the science they convey. And while these words' characterizations, and the rest of this book's elements, are far from definitive, we hope that they may put subsequent discussion of these important issues on firmer ground.

REFERENCES

Agre, P. E. "Computational Research on Interaction and Agency." *Artificial Intelligence* **72** (1995): 1–52.

Belew, R. K. "Evolution, Learning, and Culture: Computational Metaphors for Adaptive Search." *Complex Systems* **4(1)** (1990): 11–49.

Brauth, S. E., W. S. Hall, and R. J. Dooling, eds. *Plasticity of Development.* Cambridge, MA: MIT Press, 1991.

Broadbent, D. E. *Behavior.* New York: Basic Books, 1961.

Cairns Smith, J., J. Overbaugh, and S. Miller. "The Origins of Mutants." *Nature* **335** (1988): 142–145.

Campbell, D. T. "Evolutionary Epistemology." In *The Philosophy of Karl Popper*, edited by P. A. Schlipp. LaSalle, IL: Open Court, 1974.

Cecconi F., F. Menczer, and R. K. Belew. "Maturation and the Evolution of Imitative Learning in Artificial Organisms." *Adaptive Behavior* **4(1)** (1996): in press.

Dawkins, R. *The Selfish Gene.* New York: Oxford University Press, 1976.

Dennett, D. C. "Why the Law of Effect Won't Go Away." In *Brainstorms.* Cambridge, MA: MIT Press, 1981.

Elman, J. L., E. A. Bates, M. H. Johnson, A. Karmiloff-Smith, D. Parisi, and K. Plunkett. *Rethinking Innateness: A Connectionist Perspective on Development.* Cambridge, MA: MIT Press (in press).

Foster, P. L. "Adaptive Mutation: The Uses of Adversity." *Annual Review of Microbiology* **47** (1993): 467–504.

Fox Keller, E., and E. A. Lloyd, eds. *Keywords in Evolutionary Biology.* Cambridge, MA: Harvard University Press, 1992.

Gordon, D. M. "Phenotypic Plasticity." In *Keywords in Evolutionary Biology*, edited by E. Fox Keller and E. A. Lloyd. Cambridge, MA: Harvard University Press, 1992.

Gould, S. J., and R. C. Lewontin. "The Spandrels of San Marco and the Panglossian Paradigm: A Critique of the Adaptationist Programme." *Proceedings of the Royal Society of London* **B205** (1979): 581–598.

Herrnstein, R. J., and C. Murray. *The Bell Curve: Intelligence and Class Structure In American Life*. New York: Free Press, 1994.

Hutchins, E. *Cognition in the Wild*. Cambridge, MA: MIT Press, 1995.

James, W. *The Principles of Psychology*. Great Books of the Western World (vol. 53), Encyclopedia Brittanica, Inc., (1927/1955). (First published in 1927 by H. Holt and company, New York.)

Lorenz, K. *The Foundations of Ethnology*. New York: Springer-Verlag, 1981.

Lorenz, K. *Behind the Mirror: A Search for the Natural History of Knowledge*. (R. Taylor, translator). New York: Harvest/HBJ, 1973.

Lumsden, C. J., and E. O. Wilson. *Genes, Minds, and Culture: The Coevolutionary Process*. Cambridge, MA: Harvard University Press, 1981.

Mayr, E. *The Growth of Biological Thought*. Cambridge, MA: Harvard University Press, 1982.

Oyama, S. *The Ontogeny of Information: Developmental Systems and Evolution*. Cambridge: Cambridge University Press, 1985.

Piaget, J. *Behavior and Evolution*. New York: Random House, 1978.

Plotkin, H. C. "Learning and Evolution." In *The Role of Behavior in Evolution*, edited by H. C. Plotkin. Cambridge MA: MIT Press, 1988.

Popper, K. R. *Objective Knowledge: An Evolutionary Approach*. Oxford: Clarendon Press, 1972.

Shapiro, J. A. "Adaptive Mutation: Who's Really in the Garden?" *Science* **268** (1995): 373–374.

Simon, H. A. "Cognitive Science: The Newest Science of the Artificial. *Cognitive Science* 4 (1980): 33–46.

Suchman, L. A. *Plans and Situated Actions: The Problem of Human-Machine Communication*. Cambridge, MA: Cambridge University Press, 1987.

Waddington, C. H. "Evolutionary Adaptation." In *Evolution of Life*, edited by S. Tax. Chicago, IL: University of Chicago Press, 1960.

Waddington, C. H. *Evolution of an Evolutionist*. Ithaca, NY: Cornell University Press, 1975.

Weismann, A. *The Germ-Plasm: A Theory of Heredity*. New York: Scribners, 1893.

Wright, S. "Genic and Organismic Selection." *Evolution* **34** (1980): 825–843.

BIOLOGY

Jonathan Roughgarden, Aviv Bergman, Sharoni Shafir, and Charles Taylor

Chapter 2:
Adaptive Computation in Ecology and Evolution: A Guide for Future Research

The relatively new subject of adaptive computation in computer science introduces concepts and methods attractive to ecologists and evolutionary biologists. Models in ecology and evolutionary biology consist mostly of systems of differential equations. The state variables are quantities such as the number of individuals in a population, the frequency of an allele in a population's gene pool, or the amount of mass or energy in an aggregate of populations (e.g., a trophic level). The differential equations for these quantities lead to complexity through nonlinear interactions, and the terms in such equations involve mass action expressions (e.g., total rate of predation is proportional to the number of predators times the number of prey). Models stated as differential equations naturally lead to phase space analysis, often with an emphasis on equilibrium points or other limit sets. In contrast, adaptive computation is object oriented, which means for biologists that the focus is on individual organisms. The interactions between the organisms are discrete events (a particular hawk captures a particular snake at a particular time), and are specified as behavioral rules. In adaptive computation models, complexity "emerges" as the objects and their interactions develop during a computer simulation. The approach of adaptive computation is not automatically better than the mass action differential equation approach—for this depends on the scale of the organisms relative to habitat size. A barnacle is about 1 cm^2 and the number along a one-kilometer stretch of coast is of order, say, 10^8. It is perfectly sensible to model barnacles as a population of particles because of their tiny size alone, and even more so when we acknowledge that little is known about how an individual barnacle or other tiny creature lives its daily life. The territory of a kestrel (a kind of falcon) is about 1 km^2 and their population in some region of interest is usually small and readily countable. Each bird can be known and studied as an individual, and interactions between them directly witnessed. When an individual organism is the natural object of interest, adaptive computation is particularly useful from a biological standpoint, and no alternative theoretical methodology is available.

Given the importance for biology of adaptive computation in ecology and evolutionary biology, and the evident interest of computer scientists in making contributions to biology, our group has assembled a brief guide to promote collaborations between computer scientists and biologists. We focus on the role of models in biology, what are they intended to accomplish and how useful information is extracted from them, and list theoretical topics likely to be of interest to ecologists and evolutionists. Before beginning though, we wish to appreciate the perspective of computer scientists that there is nothing special about the real world relative to other possible worlds that may emerge through computer simulation. The phrase "reality is a computation" refers to the causal processes that have produced our world as one possible instance. One may imagine that these processes can be formulated as rules, as a large algorithm, with the world we live in emerging as one possible realization of a Monte Carlo simulation using that algorithm. One might further suppose that the aim of science is to elucidate the grand algorithm that underlies reality, rather than to focus on reality itself. Still, biologists have in fact made reality their central mission, and this report is about promoting collaboration with the existing biological community of scientists, and not about reforming the self-perception of their disciplinary goals.

1. WHAT ECOLOGICAL AND EVOLUTIONARY MODELS DO

Ecology and evolutionary biology possess three kinds of models: minimal models for ideas, minimal models for systems, and synthetic models of systems, and these differ fundamentally in purpose.

A minimal model for an idea is intended to explore a concept without reference to a particular species or place. An example is the evolution of sex. Biologists do not understand why sexual reproduction has evolved so ubiquitously in nature. Many conjectures have been stated about the costs and benefits of sexual reproduction as compared with asexual reproduction. These conjectures have been explored theoretically to see simply if they make sense. Does a hypothetical population with recombination, mutation, and selection really evolve faster than a hypothetical asexual population subject to the same mutation and selection but lacking in recombination? The answer turns out to depend on the initial condition, and on the statistical nature of environmental change. Thus, models about the evolution of sex explore an idea; they are not really intended for testing, nor to apply to specific systems. Most of the early textbook models in ecology and evolutionary biology are of this type.

A minimal model for a system offers a simplified view of a particular kind of system. Models for optimal foragers, for example, embody a view of a bird or a bee moving from one bunch of flowers to another, and spending a certain time within each bunch (or patch) before moving on. The models are not well posed for animals that do not forage in this way, but that may be filter feeders, or may be territorial, because different kinds of choices must be made. These models offer simplifications of reality because aspects of the biology are deliberately omitted in the hope that they are extraneous. Of course, omitted information may turn out to be vital, but if not and the simplification is valid, then a very satisfying sense of understanding results. Today, most theoretical work in ecology and evolutionary biology is of this type.

A synthetic model for a system, often called a systems model, is a synthesis of relatively detailed descriptions of all the component parts and processes in some system. These models are often aids to organize the work of a large team of scientists. These models offer the greatest promise of realism. But this promise is often not attained because the complexity of the model precludes diagnosis of problems when something goes wrong and inhibits understanding of why interesting results emerge when they do. Also, team work among the contributing scientists is rarely perfect, and some parts of the model are usually well specified while other parts are just guesses. Nonetheless, synthetic models are invaluable. They are increasingly demanded of ecologists for inclusion in global climate and ocean circulation models to predict ecological aspects of global change.

Computer scientists are more than welcome to contribute to all three types of modeling efforts, but adaptive computation seems particularly useful in the first and second types of models, the minimal models for ideas or systems. To be effective, computer scientists should be especially sensitive to how they extract information from a computer simulation. The biological appeal of Tom Ray's simulations (Ray, 1992), for example, is that the output was analyzed as though an expedition had been taken to an island where a new group of species was found. One could see biology emerging because of how the data were analyzed. The danger for a computer scientist is that output can be presented as a core dump of graphs. Look, Ma, here's a neat parasite emerging. So, what should a computer scientist do? Well, a step in the right direction is to "know thy market." If biologists are going to buy the product, know what they are interested in. Consider some big unsolved questions. Possible answers to these will automatically attract interest. So we move now to some questions for which there is a known market. Another point to discuss is how the biological interest will be expressed. Computer scientists should not expect to see a model intended for testing actually to be tested very soon. Testing a model is time-consuming and expensive, and a significant fraction of the scientific community must appreciate its significance before a grant to fund the testing will be awarded.

Model testing is usually five years behind model formulation, and thus acceptance of a model is about ten years away from model formulation. So don't hold your breath awaiting instant acclaim; this is academia's development cycle.

2. ISSUES IN ECOLOGY

2.1 TOPICS FOR IDEA MODELS

Few, if any, organisms develop from egg to adult under complete direction of the genome. Instead, the phenotype results from a genotype-environment interaction. In plants this is called phenotypic plasticity and is exemplified by vines that grow different shapes of leaves depending on whether the leaf is in the sun or shade. In animals, plasticity primarily results from learning, and implies a memory to contain what is learned, and a neural circuitry to act upon what is learned. So an exceedingly general topic is to understand how the genotype and environment interact to produce the phenotype and, in turn, how this interaction affects evolution. One may ask, for example, what promotes the degree of learning versus genetic determination involved in the development of a trait, especially *vis-à-vis* environmental variability. Also one may ask what physiological and logical hardware is needed to implement biological learning, memory, and perception. Concerning the feedback of learning on evolution, the idea that learning affects the population-genetic selection coefficients on a trait, thereby altering its evolution in the spirit of Lamarckianism, is of great interest biologically (Hinton and Nowlan, Chapter 25). The idea from genetic programming in which a sophisticated trait is cobbled together from elementary subroutines by a selection process offers an illuminating metaphor for how intricate adaptation at the phenotypic level may result from inelegant and arbitrary mechanisms at the genetic and biochemical level (Koza, 1992). Thus, the potential seems limitless for adaptive computation to contribute pure idea models about the gene-environment interaction and its feedback on evolution.

2.2 TOPICS FOR MINIMAL SYSTEM MODELS

When we turn to particular types of ecological systems, many more issues arise where adaptive computation is relevant, which we now list in order of complexity. First, it is important to know if learning algorithms can lead to optimal strategies. Ecologists investigate many traits from the standpoint that they represent optimal solutions to some problem. Examples include the theory of optimal foraging and habitat choice. The issue then is whether learning algorithms eventually lead the animal to behave in an optimal way as predicted by optimal foraging theory. Second, much behavioral ecology concerns behavioral interactions, such as territoriality and courtship. Issues of learning and adaptation abound concerning interactions.

How does an animal learn where to put its territory? How does an animal learn how to court another? Third, interactive learning immediately leads to the possibility of signaling wherein one animal influences the learning of another. Interactive signaling in turn raises the issue of truth in signaling, and its converse, deceit. Do animals manipulate one another with signals, and if so, can learning algorithms be devised to ensure a cooperative (Pareto) equilibrium, or are animal social systems destined to implement a noncooperative (Nash) equilibrium? Fourth, some of the most ecologically influential species in the world (ants, termites, and humans) form complex societies, and it is of great interest to see how castes and other aspects of the internal organization of societies emerge from the presumably simple rules that guide the interaction of each individual with another. A more elementary but poorly understood issue is simply what determines the size of a social colony. And a more advanced issue is whether a colony can be said to innovate as a result of the interaction of its members.

3. ISSUES IN EVOLUTIONARY BIOLOGY

In evolutionary biology, models tend primarily to be idea models rather than minimal system models because evolutionary hypotheses are usually much harder to test than ecological hypotheses. Issues that merit theoretical study include, first, the processes that can account for phylogenetic relationships among species, and populations within species. Molecular genetic techniques have made phylogenetic trees easier to measure, and theory is needed to interpret the increasing database of trees that are being discovered for all kinds of organisms from viruses to primates. There is also medical interest in trees that can describe the epidemiological state of a virus—whether it is expanding and whether it is evolving drug resistance. Second, the process of speciation (one species splitting into two species) is poorly understood theoretically, although purely verbal accounts of how speciation occurs have been accepted as gospel for over fifty years. Simulations of speciation would be of great interest. Third, species selection and other forms of selection above the level of the individual (traditional Darwinian selection), are poorly understood, especially in relation to explaining the fossil record. The ideas of punctuation and the role of catastrophes would be interesting to explore with simulation. Fourth, the human genome project is producing data on genetic fragments that need to be spliced together somehow to yield the total sequence. Various adaptive searching algorithms might find application here.

REFERENCES

Koza, J. R. *Genetic Programming: On the Programming of Computers by Means of Natural Selection.* Cambridge, MA: MIT Press, 1992.

Ray, T. S. "An Approach to the Synthesis of Life." In *Artificial Life II*, edited by C. G. Langton, C. Taylor, J. D. Farmer, and S. Rasmussen. Santa Fe Institute Studies in the Sciences of Complexity, Proc. Vol. X, 371–408. Redwood City, CA: Addison-Wesley, 1992.

Reprinted Classics

Chapter 3:
The Classics in Their Context, and in Ours

Controversy and confusion about adaptive individuals in evolving populations has prevailed since the time of Lamarck. This is as long as the controversy *could rage*, since Lamarck's was the first theory of evolution, and it gave plastic individuals a crucial causal role. In this preface I will try to shine a modern light on the confusions of the past, with an optimistic look toward the controversies of the future. It is possible that, for the first time in almost two centuries, the scientific and philosophical climate can accommodate ideas that have long had scientific validity, but have been denied the attention they deserved. In the first part of this preface I will try to characterize the climate which has in the past been so inhospitable to these ideas. In the second part, I review some of the ideas themselves.

THE CLIMATE

Plastic individuals are (often and apparently) purposive, goal-driven, and intelligent. If plastic individuals play a causal role in evolution, then the question arises whether the evolutionary process itself is purposive, goal-driven, and/or intelligent. This is an embarrassing question. In the absence of a mechanistic science of purposive goal-driven intelligent systems, such an implication could hardly be acknowledged by the "scientific biology" of the twentieth century. Indeed, even now the notion that biological design might be the product of purposive processes has unscientific and almost religious connotations.

This "almost religious" implication was explicitly recognized and addressed in Lamarck's *Zoological Philosophy*, written at a time when the boundary between science and religion was relatively diffuse. It is to Lamarck's undying (but rarely acknowledged) credit that he attempted not only to devise a theory of evolution but also an associationist-inspired theory of neurophysiology, a nondualistic theory of psychology, and a philosophy of science which could appropriately nurture a new discipline (for which he coined the term "biology"). But it is to Lamarck's undying misfortune that his books were poorly written, and of uneven quality. Over the decades, scientific progress drove an increasingly firm wedge between traditional religion and modern science, and it became increasingly difficult to discuss, let alone investigate, some of these questions. Some important insights were neglected in the process.

Here are three revolutionary insights which Lamarck seems to have gotten right. (1) Adaptive plasticity in individual organisms can play a crucial role in biological evolution. (2) The processes by which individuals adapt to their environment may be very similar to the processes by which species adapt to *their* environments. (3) The study of purposiveness is in some fundamental way relevant to the study of evolution.

The problem is that Lamarck expressed these insights through a speculative scientific theory which turns out to have been wrong, and his philosophical challenges of Cartesian dualism fell on deaf ears. As is well known, Lamarck's theory presumed and required that acquired characteristics be biologically transmitted to offspring. That idea is now discredited and (to add insult to injury) is now known as "Lamarckian inheritance," even though it did not originate with him and was shared by Darwin, Spencer, and many others. This term is unfortunate, because it has resulted in the dismissal of valid theories which share Lamarck's insights, even when they do not presume or require "Lamarckian inheritance."

In fact, organic selection was propounded as a way of reconciling the Darwinian *mechanism* of evolution (natural selection) with a Lamarck-inspired *characterization* of evolution (as purposive and individual-driven). The early theorists were aware that their theory had application beyond the scope of intelligent animals (but that was the most interesting case), and discussion of conscious, purposive intelligence had not yet acquired a nonscientific "odor."

But in the ensuing decades the scientific cachet of such notions and connotations changed. Darwin's theory demonstrated that purely mechanistic processes (mutation and differential reproduction) could explain evolutionary adaptation without (necessarily) invoking "purposiveness." The study of brain physiology and neuroanatomy (especially as interpreted by Darwin's pugnacious and materialistic "bulldog" T. H. Huxley) seemed to support the reductionist view that purely mechanistic processes (reflexes) could explain in a bottom-up fashion the seemingly purposive character of behavioral and psychological processes in individual animals. Furthermore, Weismann's analysis of germ theory provided a compelling refutation of Lamarckian inheritance. The temper of the times dictated that organismic behavior and psychological phenomena be seen as results rather than causes of evolutionary progress.

In a separate chapter, I will demonstrate that in Darwin's own time, William James recognized and addressed all of these issues, anticipated (and refuted) the behaviorist and materialistic tendencies of the time, and advocated a multi-leveled, nonreductionist, adaptationist selectionism which spanned biology, psychology, and sociology. But James was a generalist and a holist in a time of specialization and reductionism. Biologists were able to ignore his transdisciplinary writings while making genuine progress in the understanding of biological mechanism.

Thus materialistic and behavioristic interpretations of biology and evolution prevailed, and created a climate which was constitutionally antipathetic to the broader implications of organic selection, even though they were actually quite compatible with them. The simplest demonstration of this is to note that B. F. Skinner, of all people, independently invented the Baldwin effect in his 1969 paper, "The Ontogeny and Phylogeny of Behavior." Skinner's case is particularly interesting because he was a behaviorist but *not* a reductionist, and it may have been this which allowed him to see that learned behavior (learned through Skinnerian reinforcement, of course) could change selection pressures and thereby influence the course of biological evolution. But Skinner's writings on such matters were not even mainstream *psychology*; they received virtually no attention in *biology*.

We can now see why the last few decades have made such a difference. Behaviorism has been replaced by a psychology that acknowledges the existence and scientific respectability of higher-level phenomena. Computer simulation in general and the discipline of artificial life in particular has produced an unambiguously scientific paradigm for the study of emergent properties and levels of organization. The mechanistic determinism presumed by Descartes and Newton has given way to the mechanistic indeterminism acknowledged by quantum physics and nonlinear systems theory. And the idea of intelligent machines is more "forward thinking" than the idea of unthinking robots. Science may, for the very first time, be in a position to deal with the ideas of Baldwin, Morgan, James, Bateson, and Piaget. So to those ideas we now turn.

THE IDEAS

James Mark Baldwin, C. Lloyd Morgan, and other early theorists originally saw "organic selection" as a natural-selection-based mechanism by which the adaptive achievements of individual animals could influence the direction of evolution. Their idea was that ontogenetic adaptation by animals to their local circumstances could prolong the race long enough for natural selection to accumulate genetic variations that would support the behaviors in question. These ontogenetic adjustments would often be mediated by learning and behavioral change on the part of individual animals. Over time, however, the need for intelligence adjustment would be eliminated and the burden for developmental production of the adaptive traits would be shifted entirely to inherited factors: the originally learned behavior would become innate. Credit for the *origin* of adaptations could thus be given to creative individuals, not to the inheritable adaptations which would eventually replace the individually achieved inventions.

It is worth noting that the "Baldwin effect" was the term James Mark Baldwin managed to attach to a family of processes about which he and several of his contemporaries (Morgan and Poulton) theorized. Baldwin waged a successful strategic battle for credit (see Richards, 1989) but in the course of doing so he substantially narrowed the scope and obscured the generality of the concept. By the time he was done, he had articulated a particular scenario which was intrinsically self-limiting: in this scenario, the effect of intelligent plasticity was to usher in nonintelligent, nonplastic instincts, and evolution (in this domain) would cease.

It was not until 1963 that a more general formulation of the theory was advanced, and when it was, it was put forth by a holistically inclined systems theorist with an interesting background. Gregory Bateson had been named after Gregor Mendel by his father, William. William Bateson had popularized Mendel's works, coined the term "genetics," and was such a vigorous anti-Lamarckian that he probably drove a Lamarck-leaning scientific adversary to suicide (see Koestler, 1971). In contradistinction to his father, Gregory Bateson considered Lamarck a greater scientist than Darwin, and in contradistinction to the behaviorists who then dominated psychology, Gregory was openly interested in accommodating (rather than exorcising) mental phenomena in a science which spanned psychology and biology (Bateson, 1972, dubbed this science "the ecology of mind"). Bateson's (1963) account was the first to explicitly addresses the self-limitation issue, and may still be the best. It can be applied to the full range of phenomena and all levels of organization. And it provides a general framework within which we can appreciate the sub-varieties of "organic selection" described by other authors.

Bateson pointed out that an individual's ability to adjust to its local environment in its own lifetime (an ability called "ontogenetic adaptation") is valuable but limited: an organism's investments in adapting to one domain will tend to limit its capacity to adapt to other domains or in other directions. It follows that just by being prevalent in a population, ontogenetically generated adaptations (physiological, somatic, behavioral, or psychological) will create and maintain selection pressures for genes that allow allied or complementary adaptations to be achieved more rapidly, with greater probability, or at lower cost. Furthermore, since the time scale for ontogenetic adaptation is so much shorter than that of natural selection, adaptive strategies will often be "invented" first through individual plasticity rather than genetic mutation and population-genetic evolution. Bateson suggests that the usual role of population-genetic evolution is often to "ratify" rather than generate the adaptations which set the course of evolution.

One nice thing about Bateson's formulation is that it allows for the possibility that the net "effect" of this dynamic can be *either* a self-limiting "Baldwinian" one (in which plasticity is replaced by fixity), or a "progressivist" one in which adaptive ontogenetic plasticity plays a more dynamic and innovative evolutionary role which Braestrup describes as "modificatory steering" (Schull, 1990).

I have suggested that Baldwin's original term "organic selection" be used to cover this broad range of processes covered by Bateson's abstract and general formulation (Schull, 1990), but whatever term is applied, the formulation allows us to recognize Waddington's and Piaget's accounts as discipline-specific special cases.

Thus, Waddington's notions of "genetic assimilation" and "canalization" can be seen as the Baldwinian scenario transposed from the level of intelligent behavior to the level of physiological adaptation. Waddington demonstrated that plasticity (in morphological development) could influence selection in a way that would lead to the selection of genes which facilitate, elaborate, or "canalize" the particular "adaptabilities" involved in generating a given adaptive response. Over several generations adaptive response could come to be so reliably and robustly achieved that the very notion of plasticity could come to seem irrelevant, even though it played a crucial role at the outset.

As detailed elsewhere in this volume, Jean Piaget applied the term "phenocopy" to this general pattern of results (giving credit to both Waddington and Baldwin) and applied the word to the more general case in which genetic adaptedness comes to supplant or replace phenotypic adaptability, with little change in the observable phenotype. However, Piaget also recognized (as did Bateson, 1963, and Braestrup, 1971) that another consequence of the evolutionary dynamic under discussion is that plasticity would persist or be enhanced due to its effects upon natural selection. It thus led him to characterize behavior as the "motor of evolution." But the phenomenon is more general even than this, because it is not restricted to the domain of behavior.

Indeed, the broader concept of "organic selection" is really more general still. While adaptive plasticity by individuals may indeed influence the selection pressures acting on populations of individuals, the results of those selection pressures are indeterminate. The role of adaptive individuals in evolving populations may well be to influence or even guide the evolution of those populations, just as the role of neural plasticity in individual animals is to influence or even guide the learning and behavior of those animals. But the result and direction of evolutionary change in the population is not necessarily the same as the direction of ontogenetic change in the individuals. There may well be emergent properties in populations which differ from the properties of individuals in populations. And there are certainly differences between the aggregate environment of a population (the "niche") and the local environment to which an individual adapts. So phenocopy is "just" another special case of the generic phenomena captured by Bateson's formula.

CONCLUSIONS

The essential point is that prevalent plasticity (at any level) in a population can systematically influence the selection pressures acting on that population. Even if (especially if) we believe that populational selection pressure is the driving force of evolution, we must recognize that the achievements of plastic individuals may "steer" that driving force, and that the yoking of individual plasticity of populational selection may significantly influence the adaptive power and "psychological character" of adapting systems. In short, the classical texts excerpted here are more than museum pieces. They are intellectual mother lodes still waiting to be mined.

REFERENCES

Bateson, G. "The Role of Somatic Change in Biological Evolution." *Evolution* **17** (1963): 529–539.

Bateson, G. *Steps to an Ecology of Mind.* New York: Ballantine Books, 1972.

Braestrup, F. W. "The Evolutionary Significance of Learning." *Vidensk abelige Meddelelser Fra Dansk Naturhistorisk Forening i Kjebenhaun* **134** (1971): 89–102

Koestler, A. *Case of the Midwife Toad.* New York: Vintage Books, 1971.

Richards, R. *Darwin and the Emergence of Evolutionary Theories of Mind and Behavior.* University of Chicago Press, 1987.

Schull, J. "Are Species Intelligent?" *Behav. & Brain Sci.* **13(1)** (1990): 63–108.

Chapter 4:
Of the Influence of the Environment on the Activities and Habits of Animals, and the Influence of the Activities and Habits of These Living Bodies in Modifying Their Organization and Structure

This excerpt originally appeared in *Zoological Philosophy* by J. B. Lamarck (London: Macmillan, 1914), Chapt. 7, 106–127. Reprinted by permission.

We are not here concerned with an argument, but with the examination of a positive fact—a fact which is of more general application than is supposed, and which has not received the attention that it deserves, no doubt because it is usually very difficult to recognise. This fact consists in the influence that is exerted by the environment on the various living bodies exposed to it.

It is indeed long since the influence of the various states of our organisation on our character, inclinations, activities and even ideas has been recognised; but I do not think that anyone has yet drawn attention to the influence of our activities and habits even on our organisation. Now since these activities and habits depend entirely on the environment in which we are habitually placed, I shall endeavour to show how great is the influence exerted by that environment on the general shape, state of the parts and even organisation of living bodies. It is, then, with this very positive fact that we have to do in the present chapter.

If we had not had many opportunities of clearly recognising the result of this influence on certain living bodies that we have transported into an environment altogether new and very different from that in which they were previously placed, and if we had not seen the resulting effects and alterations take place almost under our very eyes, the important fact in question would have remained for ever unknown to us.

The influence of the environment as a matter of fact is in all times and places operative on living bodies; but what makes this influence difficult to perceive is that its effects only become perceptible or recognisable (especially in animals) after a long period of time.

Before setting forth to examine the proofs of this fact, which deserves our attention and is so important for zoological philosophy, let us sum up the thread of the discussions that we have already begun.

In the preceding chapter we saw that it is now an unquestionable fact that on passing along the animal scale in the opposite direction from that of nature, we discover the existence, in the groups composing this scale, of a continuous but irregular degradation in the organisation of animals, an increasing simplification in their organisation, and, lastly, a corresponding diminution in the number of their faculties.

This well-ascertained fact may throw the strongest light over the actual order followed by nature in the production of all the animals that she has brought into existence, but it does not show us why the increasing complexity of the organisation of animals from the most imperfect to the most perfect exhibits only an *irregular gradation*, in the course of which there occur numerous anomalies or deviations with a variety in which no order is apparent.

Now on seeking the reason of this strange irregularity in the increasing complexity of animal organisation, if we consider the influence that is exerted by the infinitely varied environments of all parts of the world on the general shape, structure and even organisation of these animals, all will then be clearly explained.

It will in fact become clear that the state in which we find any animal, is, on the one hand, the result of the increasing complexity of organisation tending to form a regular gradation; and, on the other hand, of the influence of a multitude of very various conditions ever tending to destroy the regularity in the gradation of the increasing complexity of organisation.

I must now explain what I mean by this statement: *the environment affects the shape and organisation of animals*, that is to say that when the environment becomes very different, it produces in course of time corresponding modifications in the shape and organisation of animals.

It is true if this statement were to be taken literally, I should be convicted of an error; for, whatever the environment may do, it does not work any direct modification whatever in the shape and organisation of animals.

But great alterations in the environment of animals lead to great alterations in their needs, and these alterations in their needs necessarily lead to others in their activities. Now if the new needs become permanent, the animals then adopt new habits which last as long as the needs that evoked them. This is easy to demonstrate, and indeed requires no amplification.

It is then obvious that a great and permanent alteration in the environment of any race of animals induces new habits in these animals.

Now, if a new environment, which has become permanent for some race of animals, induces new habits in these animals, that is to say, leads them to new activities which become habitual, the result will be the use of some one part in preference to some other part, and in some cases the total disuse of some part no longer necessary.

Nothing of all this can be considered as hypothesis or private opinion; on the contrary, they are truths which, in order to be made clear, only require attention and the observation of facts.

We shall shortly see by the citation of known facts in evidence, in the first place, that new needs which establish a necessity for some part really bring about the existence of that part, as a result of efforts; and that subsequently its continued use gradually strengthens, develops and finally greatly enlarges it; in the second place, we shall see that in some cases, when the new environment and the new needs have altogether destroyed the utility of some part, the total disuse of that part has resulted in its gradually ceasing to share in the development of the other parts of the animal; it shrinks and wastes little by little, and ultimately, when there has been total disuse for a long period, the part in question ends by disappearing. All this is positive; I propose to furnish the most convincing proofs of it.

In plants, where there are no activities and consequently no habits, properly so-called, great changes of environment none the less lead to great differences in the development of their parts; so that these differences cause the origin and development of some, and the shrinkage and disappearance of others. But all this is here brought about by the changes sustained in the nutrition of the plant, in its absorption and transpiration, in the quantity of caloric, light, air and moisture that it habitually receives; lastly, in the dominance that some of the various vital movements acquire over others.

Among individuals of the same species, some of which are continually well fed and in an environment favourable to their development, while others are in an opposite environment, there arises a difference in the state of the individuals which gradually becomes very remarkable. How many examples I might cite both in animals and plants which bear out the truth of this principle! Now if the environment remains constant, so that the condition of the ill fed, suffering or sickly individuals becomes permanent, their internal organisation is ultimately modified, and these acquired modifications are preserved by reproduction among the individuals in question, and finally give rise to a race quite distinct from that in which the individuals have been continuously in an environment favourable to their development.

A very dry spring causes the grasses of a meadow to grow very little, and remain lean and puny; so that they flower and fruit after accomplishing very little growth.

A spring intermingled with warm and rainy days causes a strong growth in this same grass, and the crop is then excellent.

But if anything causes a continuance of the unfavourable environment, a corresponding variation takes place in the plants: first in their general appearance and condition, and then in some of their special characters.

Suppose, for instance, that a seed of one of the meadow grasses in question is transported to an elevated place on a dry, barren and stony plot much exposed to the winds, and is there left to germinate; if the plant can live in such a place, it will always be badly nourished, and if the individuals reproduced from it continue to exist in this bad environment, there will result a race fundamentally different from that which lives in the meadows and from which it originated. The individuals of this new race will have small and meager parts; some of their organs will have developed more than others, and will then be of unusual proportions.

Those who have observed much and studied large collections, have acquired the conviction that according as changes occur in environment, situation, climate, food, habits of life, etc., corresponding changes in the animals likewise occur in size, shape, proportions of the parts, colour, consistency, swiftness and skill.

What nature does in the course of long periods we do every day when we suddenly change the environment in which some species of living plant is situated.

Every botanist knows that plants which are transported from their native places to gardens for purposes of cultivation, gradually undergo changes which ultimately make them unrecognisable. Many plants, by nature hairy, become glabrous or nearly so; a number of those which used to lie and creep on the ground, become erect; others lose their thorns or excrescences; others again whose stem was perennial and woody in their native hot climates, become herbaceous in our own climates and some of them become annuals; lastly, the size of their parts itself undergoes very considerable changes. These effects of alterations of environment are so widely recognised, that botanists do not like to describe garden plants unless they have been recently brought into cultivation.

Is it not the case that cultivated wheat (*Triticum sativum*) is a plant which man has brought to the state in which we now see it? I should like to know in what country such a plant lives in nature, otherwise than as the result of cultivation.

Where in nature do we find our cabbages, lettuces, etc., in the same state as in our kitchen gardens? And is not the case the same with regard to many animals which have been altered or greatly modified by domestication?

How many different races of our domestic fowls and pigeons have we obtained by rearing them in various environments and different countries; birds which we should now vainly seek in nature?

Those which have changed the least, doubtless because their domestication is of shorter standing and because they do not live in a foreign climate, none the less display great differences in some of their parts, as a result of the habits which we have made them contract. Thus our domestic ducks and geese are of the same type as wild ducks and geese; but ours have lost the power of rising into high regions of the air and flying across large tracts of country; moreover, a real change has come about in the state of their parts, as compared with those of the animals of the race from which they come.

Who does not know that if we rear some bird of our own climate in a cage and it lives there for five or six years, and if we then return it to nature by setting it at liberty, it is no longer able to fly like its fellows, which have always been free? The slight change of environment for this individual has indeed only diminished its power of flight, and doubtless has worked no change in its structure; but if a long succession of generations of individuals of the same race had been kept in captivity for a considerable period, there is no doubt that even the structure of these individuals would gradually have undergone notable changes. Still more, if instead of a mere continuous captivity, this environmental factor had been further accompanied by a change to a very different climate; and if these individuals had by degrees been habituated to other kinds of food and other activities for seizing it, these factors when combined together and become permanent would have unquestionably given rise imperceptibly to a new race with quite special characters.

Where in natural conditions do we find that multitude of races of dogs which now actually exist, owing to the domestication to which we have reduced them? Where do we find those bull-dogs, grey-hounds, water-spaniels, spaniels, lap-dogs, etc., etc.; races which show wider differences than those which we call specific when they occur among animals of one genus living in natural freedom?

No doubt a single, original race, closely resembling the wolf, if indeed it was not actually the wolf, was at some period reduced by man to domestication. That race, of which all the individuals were then alike, was gradually scattered with man into different countries and climates; and after they had been subjected for some time to the influences of their environment and of the various habits which had been forced upon them in each country, they underwent remarkable alterations and formed various special races. Now man travels about to very great distances, either for trade or any other purpose; and thus brings into thickly populated places, such as a great capital, various races of dogs formed in very distant countries. The crossing of these races by reproduction then gave rise in turn to all those that we now know.

The following fact proves in the case of plants how the change of some important factor leads to alteration in the parts of these living bodies.

So long as *Ranunculus aquatilis* is submerged in the water, all its leaves are finely divided into minute segments; but when the stem of this plant reaches the surface of the water, the leaves which develop in the air are large, round and simply lobed. If several feet of the same plant succeed in growing in a soil that is merely damp without any immersion, their stems are then short, and none of their leaves are broken up into minute divisions, so that we get *Ranunculus hederaceus*, which botanists regard as a separate species.

There is no doubt that in the case of animals, extensive alterations in their customary environment produce corresponding alterations in their parts; but here the transformations take place much more slowly than in the case of plants; and for us therefore they are less perceptible and their cause less readily identified.

As to the conditions which have so much power in modifying the organs of living bodies, the most potent doubtless consist in the diversity of the places where they live, but there are many others as well which exercise considerable influence in producing the effects in question.

It is known that localities differ as to their character and quality, by reason of their position, construction and climate: as is readily perceived on passing through various localities distinguished by special qualities; this is one cause of variation for animals and plants living in these various places. But what is not known so well and indeed what is not generally believed, is that every locality itself changes in time as to exposure, climate, character and quality, although with such extreme slowness, according to our notions, that we ascribe to it complete stability.

Now in both cases these altered localities involve a corresponding alteration in the environment of the living bodies that dwell there, and this again brings a new influence to bear on these same bodies.

Hence it follows that if there are extremes in these alterations, there are also finer differences: that is to say, intermediate stages which fill up the interval. Consequently there are also fine distinctions between what we call species.

It is obvious then that as regards the character and situation of the substances which occupy the various parts of the earth's surface, there exists a variety of environmental factors which induces a corresponding variety in the shapes and structure of animals, independent of that special variety which necessarily results from the progress of the complexity of organisation in each animal.

In every locality where animals can live, the conditions constituting any one order of things remain the same for long periods: indeed they alter so slowly that man cannot directly observe it. It is only by an inspection of ancient monuments that he becomes convinced that in each of these localities the order of things which he now finds has not always been existent; he may thence infer that it will go on changing.

Races of animals living in any of these localities must then retain their habits equally long: hence the apparent constancy of the races that we call species,—a constancy which has raised in us the belief that these races are as old as nature.

But in the various habitable parts of the earth's surface, the character and situation of places and climates constitute both for animals and plants environmental influences of extreme variability. The animals living in these various localities must therefore differ among themselves, not only by reason of the state of complexity of organisation attained in each race, but also by reason of the habits which each race is forced to acquire; thus when the observing naturalist travels over large portions of the earth's surface and sees conspicuous changes occurring in the environment, he invariably finds that the characters of species undergo a corresponding change.

Now the true principle to be noted in all this is as follows:

1. Every fairly considerable and permanent alteration in the environment of any race of animals works a real alteration in the needs of that race.
2. Every change in the needs of animals necessitates new activities on their part for the satisfaction of those needs, and hence new habits.

3. Every new need, necessitating new activities for its satisfaction, requires the animal, either to make more frequent use of some of its parts which it previously used less, and thus greatly to develop and enlarge them; or else to make use of entirely new parts, to which the needs have imperceptibly given birth by efforts of its inner feeling; this I shall shortly prove by means of known facts.

Thus to obtain a knowledge of the true causes of that great diversity of shapes and habits found in the various known animals, we must reflect that the infinitely diversified but slowly changing environment in which the animals of each race have successively been placed, has involved each of them in new needs and corresponding alterations in their habits. This is a truth which, once recognised, cannot be disputed. Now we shall easily discern how the new needs may have been satisfied, and the new habits acquired, if we pay attention to the two following laws of nature, which are always verified by observation.

FIRST LAW. *In every animal which has not passed the limit of its development, a more frequent and continuous use of any organ gradually strengthens, develops and enlarges that organ, and gives it a power proportional to the length of time it has been so used; while the permanent disuse of any organ imperceptibly weakens and deteriorates it, and progressively diminishes its functional capacity, until it finally disappears.*

SECOND LAW. *All the acquisitions or losses wrought by nature on individuals, through the influence of the environment in which their race has long been placed, and hence through the influence of the predominant use or permanent disuse of any organ; all these are preserved by reproduction to the new individuals which arise, provided that the acquired modifications are common to both sexes, or at least to the individuals which produce the young.*

Here we have two permanent truths, which can only be doubted by those who have never observed or followed the operations of nature, or by those who have allowed themselves to be drawn into the error which I shall now proceed to combat.

Naturalists have remarked that the structure of animals is always in perfect adaptation to their functions, and have inferred that the shape and condition of their parts have determined the use of them. Now this is a mistake: for it may be easily proved by observation that it is on the contrary the needs and uses of the parts which have caused the development of these same parts, which have even given birth to them when they did not exist, and which consequently have given rise to the condition that we find in each animal.

If this were not so, nature would have had to create as many different kinds of structure in animals, as there are different kinds of environment in which they have to live; and neither structure nor environment would ever have varied.

This is indeed far from the true order of things. If things were really so, we should not have race-horses shaped like those in England; we should not have big draught-horses so heavy and so different from the former, for none such are produced in nature; in the same way we should not have basset-hounds with crooked legs, nor grey-hounds so fleet of foot, nor water-spaniels, etc.; we should not have fowls without tails, fantail pigeons, etc.; finally, we should be able to cultivate wild plants as long as we liked in the rich and fertile soil of our gardens, without the fear of seeing them change under long cultivation.

A feeling of the truth in this respect has long existed; since the following maxim has passed into a proverb and is known by all, *Habits form a second nature.*

Assuredly if the habits and nature of each animal could never vary, the proverb would have been false and would not have come into existence, nor been preserved in the event of any one suggesting it.

If we seriously reflect upon all that I have just set forth, it will be seen that I was entirely justified when in my work entitled *Recherches sur les corps vivants* (p. 50), I established the following proposition:

> *It is not the organs, that is to say, the nature and shape of the parts of an animal's body, that have given rise to its special habits and faculties; but it is, on the contrary, its habits, mode of life and environment that have in course of time controlled the shape of its body, the number and state of its organs and, lastly, the faculties which it possesses.*

If this proposition is carefully weighed and compared with all the observations that nature and circumstances are incessantly throwing in our way, we shall see that its importance and accuracy are substantiated in the highest degree.

Time and a favourable environment are as I have already said nature's two chief methods of bringing all her productions into existence: for her, time has no limits and can be drawn upon to any extent.

As to the various factors which she has required and still constantly uses for introducing variations in everything that she produces, they may be described as practically inexhaustible.

The principal factors consist in the influence of climate, of the varying temperatures of the atmosphere and the whole environment, of the variety of localities and their situation, of habits, the commonest movements, the most frequent activities, and, lastly, of the means of self-preservation, the mode of life and the methods of defence and multiplication.

Now as a result of these various influences, the faculties become extended and strengthened by use, and diversified by new habits that are long kept up. The conformation, consistency and, in short, the character and state of the parts, as well as of the organs, are imperceptibly affected by these influences and are preserved and propagated by reproduction.

These truths, which are merely effects of the two natural laws stated above, receive in every instance striking confirmation from facts; for the facts afford a clear indication of nature's procedure in the diversity of her productions.

But instead of being contented with generalities which might be considered hypothetical, let us investigate the facts directly, and consider the effects in animals of the use or disuse of their organs on these same organs, in accordance with the habits that each race has been forced to contract.

Now I am going to prove that the permanent disuse of any organ first decreases its functional capacity, and then gradually reduces the organ and causes it to disappear or even become extinct, if this disuse lasts for a very long period throughout successive generations of animals of the same race.

I shall then show that the habit of using any organ, on the contrary, in any animal which has not reached the limit of the decline of its functions, not only perfects and increases the functions of that organ, but causes it in addition to take on a size and development which imperceptibly alter it; so that in course of time it becomes very different from the same organ in some other animal which uses it far less.

The permanent disuse of an organ, arising from a change of habits, causes a gradual shrinkage and ultimately the disappearance and even extinction of that organ.

Since such a proposition could only be accepted on proof, and not on mere authority, let us endeavour to make it clear by citing the chief known facts which substantiate it.

The vertebrates, whose plan of organisation is almost the same throughout, though with much variety in their parts, have their jaws armed with teeth; some of them, however, whose environment has induced the habit of swallowing the objects they feed on without any preliminary mastication, are so affected that their teeth do not develop. The teeth then remain hidden in the bony framework of the jaws, without being able to appear outside; or indeed they actually become extinct down to their last rudiments.

In the right-whale, which was supposed to be completely destitute of teeth, M. Geoffroy has nevertheless discovered teeth concealed in the jaws of the foetus of this animal. The professor has moreover discovered in birds the groove in which the teeth should be placed, though they are no longer to be found there.

Even in the class of mammals, comprising the most perfect animals, where the vertebrate plan of organisation is carried to its highest completion, not only is the right-whale devoid of teeth, but the ant-eater (*Myrmecophaga*) is also found to be in the same condition, since it has acquired a habit of carrying out no mastication, and has long preserved this habit in its race.

Eyes in the head are characteristic of a great number of different animals, and essentially constitute a part of the plan of organisation of the vertebrates.

Yet the mole, whose habits require a very small use of sight, has only minute and hardly visible eyes, because it uses that organ so little.

Olivier's *Spalax* (*Voyage en Égypte et en Perse*), which lives underground like the mole, and is apparently exposed to daylight even less than the mole, has altogether lost the use of sight: so that it shows nothing more than vestiges of this organ. Even these vestiges are entirely hidden under the skin and other parts, which cover them up and do not leave the slightest access to light.

The *Proteus*, an aquatic reptile allied to the salamanders, and living in deep dark caves under the water, has, like the *Spalax*, only vestiges of the organ of sight, vestiges which are covered up and hidden in the same way.

The following consideration is decisive on the question which I am now discussing,

Light does not penetrate everywhere; consequently animals which habitually live in places where it does not penetrate, have no opportunity of exercising their organ of sight, if nature has endowed them with one. Now animals belonging to a plan of organisation of which eyes were a necessary part, must have originally had them. Since, however, there are found among them some which have lost the use of this organ and which show nothing more than hidden and covered up vestiges of them, it becomes clear that the shrinkage and even disappearance of the organ in question are the results of a permanent disuse of that organ.

This is proved by the fact that the organ of hearing is never in this condition, but is always found in animals whose organisation is of the kind that includes it: and for the following reason.

The substance of sound,[1] that namely which, when set in motion by the shock or the vibration of bodies, transmits to the organ of hearing the impression received,

---

[1] Physicists believe and even affirm that the atmospheric air is the actual substance of sound, that is to say, that it is the substance which, when set in motion by the shocks or vibrations of bodies, transmits to the organ of hearing the impression of the concussions received.

That this is an error is attested by many known facts, showing that it is impossible that the air should penetrate to all places to which the substance producing sound actually does penetrate.

See my memoir *On the Substance of Sound*, printed at the end of my *Hydrogéologie*, p. 225, in which I furnished the proofs of this mistake.

Since the publication of my memoir, which by the way is seldom cited, great efforts have been made to make the known velocity of the propagation of sound in air tally with the elasticity of the air, which would cause the propagation of its oscillations to be too slow for the theory. Now, since the air during oscillation necessarily undergoes alternate compressions and dilatations in its parts, recourse has been had to the effects of the caloric squeezed out during the sudden compressions of the air and of the caloric absorbed during the rarefactions of that fluid. By means of these effects, quantitatively determined by convenient hypotheses, geometricians now account for the velocity with which sound is propagated through air. But this is no answer to the fact that sound is also propagated through bodies which air can neither traverse nor set in motion.

These physicists assume forsooth a vibration in the smallest particles of solid bodies; a vibration of very dubious existence, since it can only be propagated through homogeneous bodies of equal density, and cannot spread from a dense body to a rarefied one or *vice versa*. Such a hypothesis offers no explanation of the well-known fact that sound is propagated through heterogeneous bodies of very different densities and kinds.

penetrates everywhere and passes through any medium, including even the densest bodies: it follows that every animal, belonging to a plan of organisation of which hearing is an essential part, always have some opportunity for the exercise of this organ wherever it may live. Hence among the vertebrates we do not find any that are destitute of the organ of hearing; and after them, when this same organ has come to an end, it does not subsequently recur in any animal of the posterior classes.

It is not so with the organ of sight; for this organ is found to disappear, re-appear and disappear again according to the use that the animal makes of it.

In the acephalic molluscs, the great development of the mantle would make their eyes and even their head altogether useless. The permanent disuse of these organs has thus brought about their disappearance and extinction, although molluscs belong to a plan of organisation which should comprise them.

Lastly, it was part of the plan of organisation of the reptiles, as of other vertebrates, to have four legs in dependence on their skeleton. Snakes ought consequently to have four legs, especially since they are by no means the last order of the reptiles and are farther from the fishes than are the batrachians (frogs, salamanders, etc.).

Snakes, however, have adopted the habit of crawling on the ground and hiding in the grass; so that their body, as a result of continually repeated efforts at elongation for the purpose of passing through narrow spaces, has acquired a considerable length, quite out of proportion to its size. Now, legs would have been quite useless to these animals and consequently unused. Long legs would have interfered with their need of crawling, and very short legs would have been incapable of moving their body, since they could only have had four. The disuse of these parts thus became permanent in the various races of these animals, and resulted in the complete disappearance of these same parts, although legs really belong to the plan of organisation of the animals of this class.

Many insects, which should have wings according to the natural characteristics of their order and even of their genus, are more or less completely devoid of them through disuse. Instances are furnished by many Coleoptera, Orthoptera, Hymenoptera and Hemiptera, etc., where the habits of these animals never involve them in the necessity of using their wings.

But it is not enough to give an explanation of the cause which has brought about the present condition of the organs of the various animals,—a condition that is always found to be the same in animals of the same species; we have in addition to cite instances of changes wrought in the organs of a single individual during its life, as the exclusive result of a great mutation in the habits of the individuals of its species. The following very remarkable fact will complete the proof of the influence of habits on the condition of the organs, and of the way in which permanent changes in the habits of an individual lead to others in the condition of the organs, which come into action during the exercise of these habits.

M. Tenon, a member of the Institute, has notified to the class of sciences, that he had examined the intestinal canal of several men who had been great drinkers for a large part of their lives, and in every case he had found it shortened to an extraordinary degree, as compared with the same organ in all those who had not adopted the like habit.

It is known that great drinkers, or those who are addicted to drunkenness, take very little solid food, and eat hardly anything; since the drink which they consume so copiously and frequently is sufficient to feed them.

Now since fluid foods, especially spirits, do not long remain either in the stomach or intestine, the stomach and the rest of the intestinal canal lose among drinkers the habit of being distended, just as among sedentary persons, who are continually engaged on mental work and are accustomed to take very little food; for in their case also the stomach slowly shrinks and the intestine shortens.

This has nothing to do with any shrinkage or shortening due to a binding of the parts which would permit of the ordinary extension, if instead of remaining empty these viscera were again filled; we have to do with a real shrinkage and shortening of considerable extent, and such that these organs would burst rather than yield at once to any demand for the ordinary extension.

Compare two men of equal ages, one of whom has contracted the habit of eating very little, since his habitual studies and mental work have made digestion difficult, while the other habitually takes much exercise, is often out-of-doors, and eats well; the stomach of the first will have very little capacity left and will be filled up by a very small quantity of food, while that of the second will have preserved and even increased its capacity.

Here then is an organ which undergoes profound modification in size and capacity, purely on account of a change of habits during the life of the individual.

The frequent use of any organ, when confirmed by habit, increases the functions of that organ, leads to its development and endows it with a size and power that it does not possess in animals which exercise it less.

We have seen that the disuse of any organ modifies, reduces and finally extinguishes it. I shall now prove that the constant use of any organ, accompanied by efforts to get the most out of it, strengthens and enlarges that organ, or creates new ones to carry on functions that have become necessary.

The bird which is drawn to the water by its need of finding there the prey on which it lives, separates the digits of its feet in trying to strike the water and move about on the surface. The skin which unites these digits at their base acquires the habit of being stretched by these continually repeated separations of the digits; thus in course of time there are formed large webs which unite the digits of ducks, geese, etc., as we actually find them. In the same way efforts to swim, that is to push against the water so as to move about in it, have stretched the membranes between the digits of frogs, sea-tortoises, the otter, beaver, etc.

On the other hand, a bird which is accustomed to perch on trees and which springs from individuals all of whom had acquired this habit, necessarily has longer digits on its feet and differently shaped from those of the aquatic animals that I have just named. Its claws in time become lengthened, sharpened and curved into hooks, to clasp the branches on which the animal so often rests.

We find in the same way that the bird of the water-side which does not like swimming and yet is in need of going to the water's edge to secure its prey, is continually liable to sink in the mud. Now this bird tries to act in such a way that its body should not be immersed in the liquid, and hence makes its best efforts to stretch and lengthen its legs. The long-established habit acquired by this bird and all its race of continually stretching and lengthening its legs, results in the individuals of this race becoming raised as though on stilts, and gradually obtaining long, bare legs, denuded of feathers up to the thighs and often higher still. (*Système des Animaux sans vertèbres*, p. 14.)

We note again that this same bird wants to fish without wetting its body, and is thus obliged to make continual efforts to lengthen its neck. Now these habitual efforts in this individual and its race must have resulted in course of time in a remarkable lengthening, as indeed we actually find in the long necks of all water-side birds.

If some swimming birds like the swan and goose have short legs and yet a very long neck, the reason is that these birds while moving about on the water acquire the habit of plunging their head as deeply as they can into it in order to get the aquatic larvae and various animals on which they feed; whereas they make no effort to lengthen their legs.

If an animal, for the satisfaction of its needs, makes repeated efforts to lengthen its tongue, it will acquire a considerable length (ant-eater, green-woodpecker); if it requires to seize anything with this same organ, its tongue will then divide and become forked. Proofs of my statement are found in the humming-birds which use their tongues for grasping things, and in lizards and snakes which use theirs to palpate and identify objects in front of them.

Needs which are always brought about by the environment, and the subsequent continued efforts to satisfy them, are not limited in their results to a mere modification, that is to say, an increase or decrease of the size and capacity of organs; but they may even go so far as to extinguish organs, when any of these needs make such a course necessary.

Fishes, which habitually swim in large masses of water, have need of lateral vision; and, as a matter of fact, their eyes are placed on the sides of their head. Their body, which is more or less flattened according to the species, has its edges perpendicular to the plane of the water; and their eyes are placed so that there is one on each flattened side. But such fishes as are forced by their habits to be constantly approaching the shore, and especially slightly inclined or gently sloping beaches, have been compelled to swim on their flattened surfaces in order to make a close approach to the water's edge. In this position, they receive more light from above than below and stand in special need of paying constant attention to what is passing above them; this requirement has forced one of their eyes to undergo a sort of displacement, and to assume the very remarkable position found in the soles, turbots, dabs, etc. (*Pleuronectes* and *Achirus*). The position of these eyes is not symmetrical, because it results from an incomplete mutation. Now this mutation is entirely completed in the skates, in which the transverse flattening of the body is altogether horizontal, like the head. Accordingly the eyes of skates are both situated on the upper surface and have become symmetrical.

Snakes, which crawl on the surface of the earth, chiefly need to see objects that are raised or above them. This need must have had its effect on the position of the organ of sight in these animals, and accordingly their eyes are situated in the lateral and upper parts of their head, so as easily to perceive what is above them or at their sides; but they scarcely see at all at a very short distance in front of them. They are, however, compelled to make good the deficiency of sight as regards objects in front of them which might injure them as they move forward. For this purpose they can only use their tongue, which they are obliged to thrust out with all their might. This habit has not only contributed to making their tongue slender and very long and contractile, but it has even forced it to undergo division in the greater number of species, so as to feel several objects at the same time; it has even permitted of the formation of an aperture at the extremity of their snout, to allow the tongue to pass without having to separate the jaws.

Nothing is more remarkable than the effects of habit in herbivorous mammals.

A quadruped, whose environment and consequent needs have for long past inculcated the habit of browsing on grass, does nothing but walk about on the ground; and for the greater part of its life is obliged to stand on its four feet, generally making only few or moderate movements. The large portion of each day that this kind of animal has to pass in filling itself with the only kind of food that it cares for, has the result that it moves but little and only uses its feet for support in walking or running on the ground, and never for holding on, or climbing trees.

From this habit of continually consuming large quantities of food-material, which distend the organs receiving it, and from the habit of making only moderate movements, it has come about that the body of these animals has greatly thickened, become heavy and massive and acquired a very great size: as is seen in elephants, rhinoceroses, oxen, buffaloes, horses, etc.

The habit of standing on their four feet during the greater part of the day, for the purpose of browsing, has brought into existence a thick horn which invests the extremity of their digits; and since these digits have no exercise and are never moved and serve no other purpose than that of support like the rest of the foot, most of them have become shortened, dwindled and, finally, even disappeared.

Thus in the pachyderms, some have five digits on their feet invested in horn, and their hoof is consequently divided into five parts; others have only four, and others again not more than three; but in the ruminants, which are apparently the oldest of the mammals that are permanently confined to the ground, there are not more than two digits on the feet and indeed, in the solipeds, there is only one (horse, donkey).

Nevertheless some of these herbivorous animals, especially the ruminants, are incessantly exposed to the attacks of carnivorous animals in the desert countries that they inhabit, and they can only find safety in headlong flight. Necessity has in these cases forced them to exert themselves in swift running, and from this habit their body has become more slender and their legs much finer; instances are furnished by the antelopes, gazelles, etc.

In our own climates, there are other dangers, such as those constituted by man, with his continual pursuit of red deer, roe deer and fallow deer; this has reduced them to the same necessity, has impelled them into similar habits, and had corresponding effects.

Since ruminants can only use their feet for support, and have little strength in their jaws, which only obtain exercise by cutting and browsing on the grass, they can only fight by blows with their heads, attacking one another with their crowns.

In the frequent fits of anger to which the males especially are subject, the efforts of their inner feeling cause the fluids to flow more strongly towards that part of their head; in some there is hence deposited a secretion of horny matter, and in others of bony matter mixed with horny matter, which gives rise to solid protuberances: thus we have the origin of horns and antlers, with which the head of most of these animals is armed.

It is interesting to observe the result of habit in the peculiar shape and size of the giraffe (*Camelo-pardalis*): this animal, the largest of the mammals, is known to live in the interior of Africa in places where the soil is nearly always arid and barren, so that it is obliged to browse on the leaves of trees and to make constant efforts to reach them. From this habit long maintained in all its race, it has resulted that the animal's fore-legs have become longer than its hind legs, and that its neck is lengthened to such a degree that the giraffe, without standing up on its hind legs, attains a height of six metres (nearly 20 feet).

Among birds, ostriches, which have no power of flight and are raised on very long legs, probably owe their singular shape to analogous circumstances.

The effect of habit is quite as remarkable in the carnivorous mammals as in the herbivores; but it exhibits results of a different kind.

Those carnivores, for instance, which have become accustomed to climbing, or to scratching the ground for digging holes, or to tearing their prey, have been under the necessity of using the digits of their feet: now this habit has promoted the separation of their digits, and given rise to the formation of the claws with which they are armed.

But some of the carnivores are obliged to have recourse to pursuit in order to catch their prey: now some of these animals were compelled by their needs to contract the habit of tearing with their claws, which they are constantly burying deep in the body of another animal in order to lay hold of it, and then make efforts to tear out the part seized. These repeated efforts must have resulted in its claws reaching a size and curvature which would have greatly impeded them in walking or running on stony ground: in such cases the animal has been compelled to make further efforts to draw back its claws, which are so projecting and hooked as to get in its way. From this there has gradually resulted the formation of those peculiar sheaths, into which cats, tigers, lions, etc. withdraw their claws when they are not using them.

Hence we see that efforts in a given direction, when they are long sustained or habitually made by certain parts of a living body, for the satisfaction of needs established by nature or environment, cause an enlargement of these parts and the acquisition of a size and shape that they would never have obtained, if these efforts had not become the normal activities of the animals exerting them. Instances are everywhere furnished by observations on all known animals.

Can there be any more striking instance than that which we find in the kangaroo? This animal, which carries its young in a pouch under the abdomen, has acquired the habit of standing upright, so as to rest only on its hind legs and tail; and of moving only by means of a succession of leaps, during which it maintains its erect attitude in order not to disturb its young. And the following is the result:

1. Its fore legs, which it uses very little and on which it only supports itself for a moment on abandoning its erect attitude, have never acquired a development proportional to that of the other parts, and have remained meagre, very short and with very little strength.
2. The hind legs, on the contrary, which are almost continually in action either for supporting the whole body or for making leaps, have acquired a great development and become very large and strong.
3. Lastly, the tail, which is in this case much used for supporting the animal and carrying out its chief movements, has acquired an extremely remarkable thickness and strength at its base.

These well-known facts are surely quite sufficient to establish the results of habitual use of an organ or any other part of animals. If on observing in an animal any organ particularly well-developed, strong, and powerful, it is alleged that its habitual use has nothing to do with it, that its continued disuse involves it in no loss, and finally, that this organ has always been the same since the creation of the species to which the animal belongs, then I ask, Why can our domestic ducks no longer fly like wild ducks? I can, in short, cite a multitude of instances among ourselves, which bear witness to the differences that accrue to us from the use or disuse of any of our organs, although these differences are not preserved in the new individuals which arise by reproduction: for if they were their effects would be far greater.

I shall show in Part II., that when the will guides an animal to any action, the organs which have to carry out that action are immediately stimulated to it by the influx of subtle fluids (the nervous fluid), which become the determining factor of the movements required. This fact is verified by many observations, and cannot now be called in question.

Hence it follows that numerous repetitions of these organised activities strengthen, stretch, develop and even create the organs necessary to them. We have only to watch attentively what is happening all around us, to be convinced that this is the true cause of organic development and changes.

Now every change that is wrought in an organ through a habit of frequently using it, is subsequently preserved by reproduction, if it is common to the individuals who unite together in fertilisation for the propagation of their species. Such a change is thus handed on to all succeeding individuals in the same environment, without their having to acquire it in the same way that it was actually created.

Furthermore, in reproductive unions, the crossing of individuals who have different qualities or structures is necessarily opposed to the permanent propagation of these qualities and structures. Hence it is that in man, who is exposed to so great a diversity of environment, the accidental qualities or defects which he acquires are not preserved and propagated by reproduction. If, when certain peculiarities of shape or certain defects have been acquired, two individuals who are both affected were always to unite together, they would hand on the same peculiarities; and if successive generations were limited to such unions, a special and distinct race would then be formed. But perpetual crossings between individuals, who have not the same peculiarities of shape, cause the disappearance of all peculiarities acquired by special action of the environment. Hence, we may be sure that if men were not kept apart by the distances of their habitations, the crossing in reproduction would soon bring about the disappearance of the general characteristics distinguishing different nations.

If I intended here to pass in review all the classes, orders, genera and species of existing animals, I should be able to show that the conformation and structure of individuals, their organs, faculties, etc., etc., are everywhere a pure result of the environment to which each species is exposed by its nature, and by the habits that the individuals composing it have been compelled to acquire; I should be able to show that they are not the result of a shape which existed from the beginning, and has driven animals into the habits they are known to possess.

It is known that the animal called the *ai* or sloth (*Bradypus vridactylus*) is permanently in a state of such extreme weakness that it only executes very slow and limited movements, and walks on the ground with difficulty. So slow are its movements that it is alleged that it can only take fifty steps in a day. It is known, moreover, that the organisation of this animal is entirely in harmony with its state of feebleness and incapacity for walking; and that if it wished to make other movements than those which it actually does make it could not do so.

Hence on the supposition that this animal had received its organisation from nature, it has been asserted that this organisation forced it into the habits and miserable state in which it exists.

This is very far from being my opinion; for I am convinced that the habits which the ai was originally forced to contract must necessarily have brought its organisation to its present condition.

If continual dangers in former times have led the individuals of this species to take refuge in trees, to live there habitually and feed on their leaves, it is clear that they must have given up a great number of movements which animals living on the ground are in a position to perform. All the needs of the ai will then be reduced to clinging to branches and crawling and dragging themselves among them, in order to reach the leaves, and then to remaining on the tree in a state of inactivity in order to avoid falling off. This kind of inactivity, moreover, must have been continually induced by the heat of the climate; for among warm-blooded animals, heat is more conducive to rest than to movement.

Now the individuals of the race of the *ai* have long maintained this habit of remaining in the trees, and of performing only those slow and little varied movements which suffice for their needs. Hence their organisation will gradually have come into accordance with their new habits; and from this it must follow:

1. That the arms of these animals, which are making continual efforts to clasp the branches of trees, will be lengthened;
2. That the claws of their digits will have acquired a great length and a hooked shape, through the continued efforts of the animal to hold on;
3. That their digits, which are never used in making independent movements, will have entirely lost their mobility, become united and have preserved only the faculty of flexion or extension all together;
4. That their thighs, which are continually clasping either the trunk or large branches of trees, will have contracted a habit of always being separated, so as to lead to an enlargement of the pelvis and a backward direction of the cotyloid cavities;
5. Lastly, that a great many of their bones will be welded together, and that parts of their skeleton will consequently have assumed an arrangement and form adapted to the habits of these animals, and different from those which they would require for other habits.

This is a fact that can never be disputed; since nature shows us in innumerable other instances the power of environment over habit and that of habit over the shape, arrangement and proportions of the parts of animals.

Since there is no necessity to cite any further examples, we may now turn to the main point elaborated in this discussion.

It is a fact that all animals have special habits corresponding to their genus and species, and always possess an organisation that is completely in harmony with those habits.

It seems from the study of this fact that we may adopt one or other of the two following conclusions, and that neither of them can be verified.

Conclusion adopted hitherto: Nature (or her Author) in creating animals, foresaw all the possible kinds of environment in which they would have to live, and endowed each species with a fixed organisation and with a definite and invariable shape, which compel each species to live in the places and climates where we actually find them, and there to maintain the habits which we know in them.

My individual conclusion: Nature has produced all the species of animals in succession, beginning with the most imperfect or simplest, and ending her work with the most perfect, so as to create a gradually increasing complexity in their organisation; these animals have spread at large throughout all the habitable regions of the globe, and every species has derived from its environment the habits that we find in it and the structural modifications which observation shows us.

The former of these two conclusions is that which has been drawn hitherto, at least by nearly everyone: it attributes to every animal a fixed organisation and structure which never have varied and never do vary; it assumes, moreover, that none of the localities inhabited by animals ever vary; for if they were to vary, the same animals could no longer survive, and the possibility of finding other localities and transporting themselves thither would not be open to them.

The second conclusion is my own: it assumes that by the influence of environment on habit, and thereafter by that of habit on the state of the parts and even on organisation, the structure and organisation of any animal may undergo modifications, possibly very great, and capable of accounting for the actual condition in which all animals are found.

In order to show that this second conclusion is baseless, it must first be proved that no point on the surface of the earth ever undergoes variation as to its nature, exposure, high or low situation, climate, etc., etc.; it must then be proved that no part of animals undergoes even after long periods of time any modification due to a change of environment or to the necessity which forces them into a different kind of life and activity from what has been customary to them.

Now if a single case is sufficient to prove that an animal which has long been in domestication differs from the wild species whence it sprang, and if in any such domesticated species, great differences of conformation are found between the individuals exposed to such a habit and those which are forced into different habits, it will then be certain that the first conclusion is not consistent with the laws of nature, while the second, on the contrary, is entirely in accordance with them.

Everything then combines to prove my statement, namely: that it is not the shape either of the body or its parts which gives rise to the habits of animals and their mode of life; but that it is, on the contrary, the habits, mode of life and all the other influences of the environment which have in course of time built up the shape of the body and of the parts of animals. With new shapes, new faculties have been acquired, and little by little nature has succeeded in fashioning animals such as we actually see them.

Can there be any more important conclusion in the range of natural history, or any to which more attention should be paid than that which I have just set forth?

Chapter 5:
A New Factor in Evolution

The original paper appeared in *The American Naturalist* **30** (June 1896): 441–451, 536–553. Reprinted by permission.

In several recent publications I have developed, from different points of view, some considerations which tend to bring out a certain influence at work in organic evolution which I venture to call "a new factor." I give below a list of references to these publications and shall refer to them by number as this paper proceeds.[1] The object of the present paper is to gather into one sketch an outline of the view of the process of development which these different publications have hinged upon.

The problems involved in a theory of organic development may be gathered up under three great heads: Ontogeny, Phylogeny, Heredity. The general consideration, the "factor" which I propose to bring out, is operative in the first instance, in the field of *Ontogeny*; I shall consequently speak first of the problem of Ontogeny, then of that of Phylogeny, in so far as the topic dealt with makes it necessary, then of that of Heredity, under the same limitation, and finally, give some definitions and conclusions.

[1] Editor's note: these references have been moved to the end of the paper, and called out in the same manner as other references in this book, e.g., author and year.

SECTION I.
ONTOGENY: ORGANIC SELECTION (SEE BALDWIN, 1895A, CHAP VII.)

The series of facts which investigation in this field has to deal with are those of the individual creature's development; and two sorts of facts may be distinguished from the point of view of the *functions which an organism performs in the course of his life history.* There is, in the first place, the development of his heredity impulse, the unfolding of his heredity in the forms and functions which characterize his kind, together with the congenital variations which characterize the particular individual—the phylogenetic variations, which are constitutional to him; and there is, in the second place, the series of functions, acts, etc., *which he learns to do himself in the course of his life.* All of these latter, the *special modifications which an organism undergoes during its ontogeny,* thrown together, have been called "acquired characters," and we may use that expression or adopt one recently suggested by Osborn,[2] "ontogenic variations" (except that I should prefer the form "ontogenetic variations"), if the word variations seems appropriate at all.

Assuming that there are such new or modified functions, in the first instance, and such "acquired characters," arising by the law of "use and disuse" from these new functions, our farther question is about them. And the question is this: How does an organism come to be modified during its life history?

In answer to this question we find that there are three different sorts of ontogenic agencies which should be distinguished—each of which works to produce ontogenetic modifications, adaptations, or variations. These are: first, the physical agencies and influences in the environment which work upon the organism to produce modifications of its form and functions. They include all chemical agents, strains, contacts, hindrances to growth, temperature changes, etc. As far as these forces work changes in the organism, the changes may be considered largely "fortuitous" or accidental. Considering the forces which produce them I propose to call them "physico-genetic." Spencer's theory of ontogenetic development rests largely upon the occurrence of lucky movements brought out by such accidental influences. Second, there is a class of modifications which arise from the spontaneous activities of the organism itself in the carrying out of its normal congenital functions. These variations and adaptations are seen in a remarkable way in plants, in unicellular creatures, in very young children. There seems to be a readiness and capacity on the part of the organism to "rise to the occasion," as it were, and make gain out of the circumstances of its life. The facts have been put in evidence (for plants) by

---

[2]Reported in *Science*, April 3rd.; also used by him before N. Y. Acad. of Sci., April 13th. There is some confusion between the two terminations "genic" and "genetic." I think the proper distinction is that which reserves the former, "genic," for application in cases in which the word to which it is affixed qualifies a term used *actively*, while the other, "genetic" conveys similarly a *passive* signification; thus agencies, causes, influences, etc., [are] "ontogenic, phylogenic, etc.," while effects, consequences, etc., [are] "ontogenetic, phylogenetic, etc."

Henslow, Pfeffer, Sachs; (for micro-organisms) by Binet, Bunge; (in human pathology) by Bernheim, Janet; (in children) by Baldwin (1895a, chap. vi.) (See citations in Baldwin, 1895a, chap. ix, and in Orr, *Theory of Development*, chap. iv). These changes I propose to call "neuro-genetic," laying emphasis on what is called by Romanes, Morgan and others, the "selective property" of the nervous system, and of life generally. Third, there is the great series of adaptations secured by conscious agency, which we may throw together as "psycho-genetic." The processes involved here are all classed broadly under the term "intelligent," i.e., imitation, gregarious influences, maternal instruction, the lessons of pleasure and pain, and of experience generally, and reasoning from means to ends, etc.

We reach, therefore, the following scheme:

Ontogenic Modifications	*Ontogenic Agencies*
1. Physico-genetic	1. Mechanical
2. Neuro-genetic	2. Nervous
3. Psycho-genetic	3. Intelligent
	Imitation
	Pleasure and Pain
	Reasoning

Now it is evident that there are two very distinct questions which come up as soon as we admit modifications of function and of structure in ontogenic development: first, there is the question as to how these modifications can come to be adaptive in the life of the individual creature. Or in other words: What is the method of the individual's growth and adaptation as shown in the well known law of "use and disuse?" Looked at functionally, we see that the organism manages somehow to accommodate itself to conditions which are favorable, to repeat movements which are adaptive, and so to grow by the principle of use. This involves some sort of selection, from the actual ontogenetic variations, of certain ones—certain functions, etc. Certain other possible and actual functions and structures decay from disuse. Whatever the method of doing this may be, we may simply, at this point, claim the law of use and disuse, as applicable in ontogenetic development, and apply the phrase, "Organic Selection," to the organism's behavior in acquiring new modes of modifications of adaptive function with its influence of structure. The question of the method of "Organic Selection" is taken up below (IV); here, I may repeat, we simply assume what every one admits in some form, that such adaptations of function—"accommodations" the psychologist calls them, the processes of learning new movements, etc.—*do occur*. We then reach another question, second; what place these adaptations have in the general theory of development.

EFFECTS OF ORGANIC SELECTION

First, we may note the results of this principle in the creature's own private life.

1. *By securing adaptations, accommodations, in special circumstances the creature is kept alive* (Baldwin, 1895a, 1st ed., pp. 172 ff.). This is true in all the three spheres of ontogenetic variation distinguished in the table above. The creatures which can stand the "storm and stress" of the physical influences of the environment, and of the changes which occur in the environment, *by undergoing modifications of their congenital functions or of the structures which they get congenitally—these creatures will live; while those which cannot, will not*. In the sphere of neurogenetic variations we find a superb series of adaptations by lower as well as higher organisms during the course of ontogenetic development (Baldwin, 1895a, chap. ix). And in the highest sphere, that of intelligence (including the phenomena of consciousness of all kinds, experience of pleasure and pain, imitation, etc.), we find individual accommodations on the tremendous scale which culminates in the skilful [sic] performances of human volition, invention, etc. The progress of the child in all the learning processes which lead him on to be a man, just illustrates this higher form of ontogenetic adaptation (Baldwin, 1895a, chap. x–xiii).
   All these instances are associated in the higher organisms, and all of them unite to *keep the creature alive*.

2. By this means *those congenital or phylogenetic variations are kept in existence, which lend themselves to intelligent, imitative, adaptive, and mechanical modification during the lifetime of the creatures which have them*. Other congenital variations are not thus kept in existence. So there arises a more or less widespread series of *determinate variations in each generation's ontogenesis* (Baldwin, 1895b, 1896a, 1896b, 1896c, 1896d).[3]

The further applications of the principle lead us over into the field of our second question, i.e., phylogeny.

---

[3] "It is necessary to consider further how certain reactions of one single organism can be selected so as to adapt the organism better and give it a life history. Let us at the outset call this process "Organic Selection" in contrast with the Natural Selection of whole organisms. . . . If this (natural selection) worked alone, every change in the environment would weed out all life except those organisms, which by accidental variation reacted already in the way demanded by the changed conditions—in every case new organisms showing variations, not, in any case, new elements of life-history in the old organisms. In order to the latter we would have to conceive. . .some modification of the old reactions in an organism through the influence of new conditions. . . . We are, accordingly, left to the view that the new stimulations brought by changes in the environment themselves modify the reactions of an organism. . . . The facts show that individual organisms do acquire new adaptations in their lifetime, and that is our first problem. If in solving it we find a principle which may also serve as a principle of race-development, then we may possibly use it against the 'all sufficiency of natural selection' or in its support" (Baldwin, 1895a, 1st. ed., pp. 175–6.)

SECTION II.

PHYLOGENY: PHYSICAL HEREDITY

The question of phylogenetic development considered apart, in so far as may be, from that of heredity, is the question as to what the factors really are which show themselves in evolutionary progress from generation to generation. The most important series of facts recently brought to light are those which show what is called "determinate variation" from one generation to another. This has been insisted on by the paleontologists. Of the two current theories of heredity, only one, Neo-Lamarkism [sic]—by means of its principle of the inheritance of acquired characters—has been able to account for this fact of determinate phylogenetic change. Weismann admits the inadequacy of the principle of natural selection, as operative on rival organisms, to explain variations when they are wanted or, as he puts it, "the right variations in the right place" (*Monist*, Jan., 1896).

I have argued, however, in detail that the assumption of determinate variations of function in ontogenesis, under the principle of neuro-genetic and psycho-genetic adaptation, does away with the need of appealing to the Lamarkian [sic] factor. In the case i.e., of instincts, "if we do not assume consciousness, then natural selection is inadequate; but if we do assume consciousness, then the inheritance of acquired characters is unnecessary" (Baldwin, 1896d).

"The intelligence which is appealed to, to take the place of instinct and to give rise to it, uses just these partial variations which tend in the direction of the instinct; so the intelligence *supplements* such partial co-ordinations, makes them functional, and *so keeps the creature alive*. In the phrase of Prof. Lloyd Morgan, this prevents the 'incidence of natural selection.' So the supposition that intelligence is operative turns out to be just the supposition which makes use-inheritance unnecessary. Thus kept alive, the species has all the time necessary to perfect the variations required by a complete instinct. And when we bear in mind that the variation required is not on the muscular side to any great extent, but in the central brain connections, and is a slight variation for functional purposes at the best, the hypothesis of use-inheritance becomes not only unnecessary, but to my mind quite superfluous" (Baldwin, 1896b, p. 439). And for adaptations generally, "the most plastic individuals will be preserved to do the advantageous things for which their variations show them to be the most fit, and the next generation will show an emphasis of just this direction in its variations" (Baldwin, 1895b, p. 221).

We get, therefore, from Organic Selection, certain results in the sphere of phylogeny:

1. *This principle secures by survival certain lines of determinate phylogenetic variation in the directions of the determinate ontogenetic adaptations of the earlier generation.* The variations which were utilized for ontogenetic adaptation in the earlier generation, being thus kept in existence, are utilized more widely in the subsequent generation (Baldwin, 1895b, 1896a, 1896b, 1896c). "Congenital

variations, on the one hand, are kept alive and made effective by their use for adaptations in the life of the individual; and, on the other hand, adaptations become congenital by further progress and refinement of variation in the same lines of function as those which their acquisition by the individual called into play. But there is no need in either case to assume the Lamarkian [sic] factor" (Baldwin, 1895b). And in cases of conscious adaptation: "We reach a point of view which gives to organic evolution a sort of intelligent direction after all; for of all the variations tending in the direction of an adaptation, but inadequate to its complete performance, *only those will be supplemented and kept alive which the intelligence ratifies and uses.* The principle of 'selective value' applies to the others or to some of them. So natural selection kills off the others; and the *future development at each stage of a species' development must be in the directions thus ratified by intelligence.* So also with imitation. Only those imitative actions of a creature which are useful to him will survive in the species, for in so far as he imitates actions which are injurious he will aid natural selection in killing himself off. So intelligence, and the imitation which copies it, will set the direction of the development of the complex instincts even on the Neo-Darwinian theory; and in this sense we may say that consciousness is a 'factor'" (Baldwin, 1896b).

2. *The mean of phylogenetic variation being thus made more determinate, further phylogenetic variations follow about this mean, and these variations are again utilized by Organic Selection for ontogenetic adaptation.* So there is continual phylogenetic progress in the directions set by ontogenetic adaptation (Baldwin, 1895b, 1896b, 1896d). "The intelligence supplements slight co-adaptations and so gives them selective value; but it does not keep them from getting farther selective value as instincts, reflexes, etc., by farther variation" (Baldwin, 1896d). "The imitative function, by using muscular co-ordinations, supplements them, secures adaptations, keeps the creature alive, prevents the 'incidence of natural selection,' and so gives the species all the time necessary to get the variations required for the full instinctive performance of the function" (Baldwin, 1896b). But, "Conscious imitation, while it prevents the incidence of natural selection, as has been seen, and so keeps alive the creatures which have no instincts for the performance of the actions required, nevertheless does not subserve the utilities which the special instincts do, nor prevent them from having the selective value of which Romanes speaks. Accordingly, on the more general definition of intelligence, which includes in it all conscious imitation, use of maternal instruction, and that sort of thing—no less than on the more special definition—we still find the principal of natural selection operative" (Baldwin, 1896d).

3. *This completely disposes of the Lamarkian [sic] factor as far as two lines of evidence for it are concerned.* First, the evidence drawn from function, "use and disuse," is discredited; since "organic selection," the reappearance, in subsequent generations, of the variations first secured in ontogenesis is accounted for without the inheritance of acquired characters. So also the evidence drawn

from paleontology which cites progressive variations resting on functional use and disuse. Second, the evidence drawn from the facts of "determinate variations;" since by this principle we have the preservation of such variations in phylogeny without the inheritance of acquired characters.

4. *But this is not Preformism in the old sense; since the adaptations made in ontogenetic development which "set" the direction of evolution are novelties of function in whole or part* (although they utilize congenital variations of structure). And it is only by the exercise of these novel functions that the creatures are kept alive to propagate and thus produce further variations of structure which may in time make the whole function, with its adequate structure, congenital. Romanes' argument from "partial co-adaptations" and "selective value," seem[s] to hold in the case of reflex and instinctive functions (Baldwin, 1896b, 1896d), as against the old preformist or Weismannist view, although the operation of Organic Selection, as now explained, renders them ineffective when urged in support of Lamarkism [sic]. "We may imagine creatures, whose hands were used for holding only with the thumb and fingers on the same side of the object held, to have first discovered, under stress of circumstances and with variations which permitted the further adaptation, how to make use of the thumb for grasping opposite to the fingers, as we now do. Then let us suppose that this proved of such utility that all the young that did not do it were killed off; the next generation following would be plastic, intelligent, or imitative, enough to do it also. They would use the same co-ordinations and prevent natural selection getting its operation on them; and so instinctive 'thumb-grasping' might be waited for indefinitely by the species and then be got as an instinct altogether apart from use-inheritance" (Baldwin, 1896b). "I have cited 'thumb-grasping' because we can see in the child the anticipation, by intelligence and imitation, of the use of the thumb for the adaptation which the Simian probably gets entirely by instinct, and which I think an isolated and weak-minded child, say, would also come to do by instinct" (Baldwin, 1896b).

5. It seems to me also—though I hardly dare venture into a field belonging so strictly to the technical biologist—that *this principle might not only explain many cases of widespread "determinate variations" appearing suddenly, let us say, in fossil deposits, but the fact that variations seem often to be "discontinuous."* Suppose, for example, certain animals, varying, in respect to a certain quality, from a to n about a mean x. The mean x would be the case most likely to be preserved in fossil form (seeing that there are vastly more of them). Now suppose a sweeping change in the environment, in such a way that only the variations lying near the extreme n can accommodate to it and live to reproduce. The next generation would then show variations about the mean n. And the chances of fossils from this generation, and the subsequent ones, would be of creatures approximating n. Here would be a great discontinuity in the chain and also a widespread prevalence of these variations in a set direction. This seeing especially evident when we consider that the paleontologist does not deal with successive generations, but with widely remote periods, and

the smallest lapse of time which he can take cognizance of is long enough to give the new mean of variation, n, a lot of generations in which to multiply and deposit its representative fossils. Of course, this would be only the action of natural selection upon "preformed" variations in those cases which did not involve positive challenges, in structure and function, *acquired in ontogenesis*; but in so far as such ontogenetic adaptations were actually there, the extent of difference of the n mean from the x mean would be greater, and hence the resources of explanation, both of the sudden prevalence of the new type and of its discontinuity from the earlier, would be much increased. This additional resource, then, is due to the "Organic Selection" factor.

We seem to be able also to utilize all the evidence usually cited for the functional origin of specific characters and groupings of characters. So far as the Lamarkians [sic] have a strong case here, it remains as strong if Organic Selection be substituted for the "inheritance of acquired characters." This is especially true where intelligent and imitative adaptations are involved, as in the case of instinct. This "may give the reason, e.g., that instincts are so often coterminous with the limits of species. Similar structures find the similar uses for their intelligence, and they also find the same imitative actions to be to their advantage. So the interaction of these conscious factors with natural selection brings it about that the structural definition which represents species, and the functional definition which represents instinct, largely keep to the same lines" (Baldwin, 1896d).

6. It seems proper, therefore, to call the influence of Organic Selection "a new factor"; for it gives a method of deriving the determinate gains of phylogeny from the adaptations of ontogeny without holding to either of the two current theories. *The ontogenetic adaptations are really new, not [preformed]; and they are really reproduced in succeeding generations, although not physically inherited.*

SECTION III.
SOCIAL HEREDITY

There follows also another resource in the matter of development. In all the higher reaches of development we find certain co-operative or "social" processes which directly supplement or add to the individual's private adaptations. In the lower forms it is called gregariousnes [sic], in man sociality, and in the lowest creatures (except plants) there are suggestions of a sort of imitative and responsive action between creatures of the same species and in the same habitat. In all these cases it is evident that other living creatures constitute part of the environment of each, and many neuro-genetic and psycho-genetic accommodations have reference to or involve these other creatures. It is here that the principle of imitation gets tremendous significance; intelligence and volition, also, later on; and in human affairs it becomes social co-operation. Now it is evident that when young creatures have these imitative, intelligent, or quasi-social tendencies to any extent, they are able to pick up *for themselves*, by imitation, instruction, experience generally, the functions which their parents and other creatures perform in their presence. This then is a form of ontogenetic adaptation; it keeps these creatures alive, and so produces determinate variations in the way explained above. It is, therefore, a special, and from its wide range, an extremely important instance of the general principle of Organic Selection.

But it has a farther value. *It keeps alive a series of functions which either are not yet, or never do become, congenital at all.* It is a means of extra-organic transmission from generation to generation. It is really a form of heredity because (1) *it is a handing down of physical functions*; while it is not physical heredity. It is entitled to be called heredity for the further reason (2) that *it directly influences physical heredity in the way mentioned*, i.e., it keeps alive variations, thus sets the direction of ontogenetic adaptation, thereby influences the direction of the available congenital variations of the next generation, and so determines phylogenetic development. I have accordingly called it "Social Heredity" (Baldwin, 1895a, chap. xii; Baldwin, 1895b). In "Social Heredity," therefore, we have a more or less conservative, progressive, ontogenic atmosphere of which we may make certain remarks as follows:—

1. *It secures adaptations of individuals all through the animal world.* "Instead of limiting this influence to human life, we have to extend it to all the gregarious animals, to all the creatures that have any ability to imitate, and finally to all animals who have consciousness sufficient to enable them to make adaptations of their own; for such creatures will have children that can do the same, and it is unnecessary that the children must inherit what their fathers did by intelligence, when they can do the same things by intelligence" (Baldwin, 1896e).

2. *It tends to set the direction of phylogenetic progress* by Organic Selection, Sexual Selection, etc., i.e., it tends not only to give the young the adaptations which the adults already have, but also *to produce adaptations which depend upon social coöperation; thus variations in the direction of sociality are selected and made determinate.* "When we remember that the permanence of a habit learned by one individual is largely conditioned by the learning of the same habits by others (notably of the opposite sex) in the same environment, we see that an enormous premium must have been put on variations of a social kind—those which brought different individuals into some kind of joint action or coöperation. Wherever this appeared, not only would habits be maintained, but new variations, having all the force of double hereditary tendency, might also be expected" (Baldwin, 1896b). Why is it, for example, that a race of Mulattoes does not arise faster, and possess our Southern States? Is it not just the social repugnance to black-white marriages? Remove or reverse *this*

influence of education, imitation, etc., and the result *on phylogeny* would show in our faces, and even appear in our fossils when they are dug up long hence by the paleontologist of the succeeding aeons!

3. *In man it becomes the law of social evolution.* "Weismann and others have shown that the influence of animal intercourse, seen in maternal instruction, imitation, gregarious cooperation, etc., is very important. Wallace dwells upon the actual facts which illustrate the 'imitative factor,' as we may call it, in the personal development of young animals. I have recently argued that Spencer and others are in error in holding that social progress demands use-inheritance; since the socially-acquired actions of a species, notably man, are socially handed down, giving a sort of 'social heredity' which supplements natural heredity" (Baldwin, 1896b). The social "sport," the genius, is very often the controlling factor in social evolution. He not only sets the direction of future progress, but he may actually lift society at a bound up to a new standard of attainment (Baldwin, 1896e). "So strong does the case seem for the Social Heredity view in this matter of intellectual and moral progress that I may suggest an hypothesis which may not stand in court, but which I find interesting. May not the rise of social life be justified from the point of view of a second utility in addition to that of its utility in the struggle for existence as ordinarily understood, the second utility, i.e., of giving to each generation the attainments of the past which natural inheritance is inadequate to transmit. When social life begins, we find the beginning of the artificial selection of the unfit; and this negative principle begins to work directly in the teeth of progress, as many writers on social themes have recently made clear. This being the case, some other resource is necessary besides natural inheritance. On my hypothesis it is found in the common or social standards of attainment which the individual is fitted to grow up to and to which he is compelled to submit. This secures progress in two ways: First, by making the individual learn what the race has learned, thus preventing social retrogression, in any case; and second, by putting a direct premium on variations which are socially available" (Baldwin, 1895b).

4. The two ways of securing development in determinate directions—the purely extra-organic way of Social Heredity, and the way by which Organic Selection in general (both by social and by other ontogenetic adaptations) secures the fixing of phylogenetic variations, as described above—seem to run parallel. Their conjoint influence is seen most interestingly in the complex instincts (Baldwin, 1896b, 1896d). We find in some instincts completely reflex or congenital functions which are accounted for by Organic Selection. In other instincts we find only partial coördinations ready given by heredity, and the creature actually depending upon some conscious resource (imitation, instruction, etc.) to bring the instinct into actual operation. But as we come up in the line of phylogenetic development, both processes may be present *for the same function;* the intelligence of the creature may lead him to do consciously what he also does instinctively. In these cases the additional utility gained by the double performance accounts for the duplication. It has arisen either (1) by the accumulation

of congenital variations in creatures which already performed the action (by ontogenetic adaptation and handed it down socially), or (2) the reverse. In the animals, the social transmission seems to be mainly useful as enabling a species to get instincts slowly in determinate directions, by keeping off the operation of natural selection. Social Heredity is then the lesser factor; it serves Biological Heredity. But in man, the reverse. Social transmission is the important factor, and the congenital equipment of instincts is actually broken up in order to allow the plasticity which the human being's social learning requires him to have. So in all cases both factors are present, but in a sort of inverse ratio to each other. In the words of Preyer, "the more kinds of co-ordinated movement an animal brings into the world, the fewer is he able to learn afterwards." The child is the animal which inherits the smallest number of congenital co-ordinations, but he is the one that learns the greatest number (Baldwin, 1895a, p. 297).

"It is very probable, as far as the early life of the child may be taken as indicating the factors of evolution, that the main function of consciousness is to enable him to learn things which natural heredity fails to transmit; and with the child the fact that consciousness is the essential means of all his learning is correlated with the other fact that the child is the very creature for which natural heredity gives few independent functions. It is in this field only that I venture to speak with assurance; but the same point of view has been reached by Weismann and others on the purely biological side. The instinctive equipment of the lower animals is replaced by the plasticity for learning by consciousness. So it seems to me that the evidence points to some inverse ratio between the importance of consciousness as factor in development and the need of inheritance of acquired characters, as factor in development" (Baldwin, 1896f).

"Under this general conception we may bring the biological phenomena of infancy, with all their evolutionary significance: the great plasticity of the mammal infant as opposed to the highly developed instinctive equipment of other young; the maternal care, instruction and example during the period of dependence, and the very gradual attainment of the activities of self-maintenance in conditions in which social activities are absolutely essential. All this stock of the development theory is available to confirm this view" (Baldwin, 1896b).

But these two influences furnish a double resort against Neo-Lamarkism [sic]. And I do not see anything in the way of considering the fact of Organic Selection, from which both these resources spring, as being a sufficient supplement to the principle of natural selection. The relation which it bears to natural selection, however, is a matter of further remark below (V).

"We may say, therefore, that there are two great kinds of influence, each in a sense hereditary; there is *natural heredity* by which variations are congenitally transmitted with original endowment, and there is *'social heredity'* by which functions socially acquired (i.e., imitatively, covering all the conscious acquisitions made through intercourse with other animals) are also socially transmitted. The one is phylogenetic; the other ontogenetic. But these two lines of hereditary influence are not separate nor uninfluential on each other.

Congenital variations, on the one hand, are kept alive and made effective by their conscious use for intelligent and imitative adaptations in the life of the individual; and, on the other hand, intelligent and imitative adaptations become congenital by further progress and refinement of variation in the same lines of function as those which their acquisition by the individual called into play. But there is no need in either case to assume the Lamarkian [sic] factor" (Baldwin, 1896b).

5. "The only hindrance that I see to the child's learning everything that his life in society requires would be just the thing that the advocates of Lamarkism [sic] argue for—the inheritance of acquired characters. For such inheritance would tend so to bind up the child's nervous substance in fixed forms that he would have less or possibly no unstable substance left to learn anything with. So, in fact, it is with the animals in which instinct is largely developed; they have no power to learn anything new, just because their nervous systems are not in the mobile condition represented by high consciousness. They have instinct and little else" (Baldwin, 1895b).

SECTION IV.
THE PROCESS OF ORGANIC SELECTION

So far we have been dealing exclusively with facts. By recognizing certain facts we have reached a view which considers ontogenetic selection an important factor in development. Without prejudicing the statement of fact at all we may enquire into the actual working of the organism [in] making its organic selections or adaptations. The question is simply this: how does the organism secure, from the multitude of possible ontogenetic changes which it might and does undergo, those which are adaptive? As a matter of fact, all personal growth, all motor acquisitions made by the individual, show that it succeeds in doing this; the further question is, how? Before taking this up, I must repeat with emphasis that the position taken in the foregoing pages, which simply makes the fact of ontogenetic adaptation a factor in development, is not involved in the solution of the further question as to how the adaptations are secured. But from the answer to this latter question we may get further light of the interpretation of the facts themselves. So we come to ask how Organic Selection actually operates in the case of a particular adaptation of a particular creature (Baldwin, 1894; Baldwin, 1895a, chap. vii, iii; Baldwin, 1896e, 1896f).

I hold that the organism has a way of doing this which is peculiarly its own. The point is elaborated at such great length in the book referred to (Baldwin, 1895a) that I need not repeat details here. The summary in this journal (Baldwin, 1896e) may have been seen by its readers. There is a fact of physiology which, taken together with the facts of psychology, serves to indicate the method of the adaptations or accommodations of the individual organism. The general fact is that the organism concentrates it energies upon the locality stimulated, for the continuation of the conditions, movements, stimulations which are vitally beneficial, and for the cessation of the conditions, movements, stimulations, which are vitally depressing and harmful. In the case of beneficial conditions we find a general *increase of movement, an excess discharge of the energies of movement in the channels already open and habitual; and with this, on the psychological side, pleasurable consciousness and attention.* Attention to a member is accompanied by increased vasomotor activity, with higher muscular power, and a *general dynamogenic heightening in that member.* "The thought of a movement tends to discharge motor energy into the channels as near as may be to those necessary for that movement" (Baldwin, 1895b). By this organic concentration and excess of movement many combinations and variations are rendered possible, from which the advantageous and adaptive movements may be selected for their utility. These then give renewed pleasure, excite pleasurable associations, and again stimulate the attention, and *by these influences the adaptive movements thus struck are selected and held as permanent acquisitions.* This form of concentration of energy upon stimulated localities, with the resulting renewal by movements of conditions that are pleasure-giving and beneficial, and the subsequent repetitions of the movements, is called the "circular reaction"[4] (Baldwin, 1894, 1895a). It is the selective property which Romanes pointed out as characterizing and differentiating life. It characterizes the responses of the organism, however low in the scale, to all stimulations—even those of a mechanical and chemical (physico-genic) nature. Pfeffer has shown such a determination of energy toward the parts stimulated even in plants. And in the higher animals it finds itself exactly reproduced in the nervous reaction seen in imitation and—through processes of association, substitution, etc.—in all the higher mental acts of intelligence and volition. These are developed phylogenetically as variations whose direction is constantly determined, by this form of adaptation in ontogenesis. If this be true—and the biological facts seem fully to confirm it—this is the adaptive process in all life, and this process is that with which the development of mental life has been associated.

It follows, accordingly, that the three forms of ontogenetic adaptation distinguished above—physico-genetic, neuro-genetic, psycho-genetic—all involve the sort of response on the part of the organism seen in this circular reaction with excess discharge; and we reach one general law of ontogenetic adaptation and of Organic Selection. "The accommodation of an organism to a new stimulation is secured—not by the selection of this stimulation beforehand (nor of the necessary movements)—but by the reinstatement of it by a discharge of the energies of the organism, concentrated as far as may be for the excessive stimulation of the organs (muscles, etc.) most nearly fitted by former habit to get this stimulation again (in which the "stimulation" stands for the condition favorable to adaptation). After

---

[4]With the opposite (withdrawing, depressive affects) in injurious and painful conditions.

several trials the child (for example) gets the adaptation aimed at more and more perfectly, and the accompanying excessive and useless movements fall away. This is the kind of selection that intelligence does in its acquisition of new movements" (Baldwin, 1895a, p. 179; Baldwin, 1896e).

Accordingly, *all ontogenetic adaptations are neurogenetic.*[5] The general law of "motor excess" is one of *overproduction*; from movements thus overproduced, adaptations survive; these adaptations set the determinate direction of ontogenesis; and by their survival the same determination of direction is set in phylogenesis also.

The following quotation from an earlier paper (Baldwin, 1896f) will show some of the bearings of this position:

"That there is some general principle running through all the adaptations of movement which the individual creature makes is indicated by the very unity of the organism itself. The principle of Habit must be recognized in some general way which will allow the organism to do new things without utterly undoing what it has already acquired. This means that old habits must be substantially preserved *in the new functions*; that all new functions must be reached by gradual modifications. And we will all go further and say, I think, that the only way that these modifications can be got at all is through some sort of interaction of the organism with its environment. Now, as soon as we ask how the stimulations of the environment can produce new adaptive movements, we have the answer of Spencer and Bain—an answer directly confirmed, I think, without question, by the study both of the child and of the adult—i.e., by the selection of fit movements from excessively produced movements, that is, from *movement variations.* So granting this, we now have the further question: How do these movement variations come to be produced *when and where they are needed?*[6] And with it, the question: How does the organism *keep those movements going* which are thus selected, and suppress those which are not selected?

"Now these two questions are the ones which the biologists fail to answer. But the force of the facts leads to the hypotheses of 'conscious force,' 'self-development' of Henslow and 'directive tendency' of the American school—all aspects of the new Vitalism which just these questions and the facts which they rest upon are now forcing to the front. Have we anything definite, drawn from the study of the individual on the psychological side, to substitute for these confessedly vague biological

[5]Barring, of course, those violent compelling physical influences under the action of which the organism is quite helpless.

[6]This is just the question that Weismann seeks to answer (in respect to the supply of variations in forms which the paleontologists require), with his doctrine of 'Germinal Selection' (*Monist*, Jan., 1896). Why are not such applications of the principle of natural selection to variations *in the parts and functions of the single organism* just as reasonable and legitimate as it is to variations in separate organisms? As against "germinal selection," however, I may say, that in the cases in which ontogenetic adaptation sets the direction of survival of phylogenetic variations (as held in this paper) the hypothesis of germinal selection is in so far unnecessary. This view finds the operation of selection *on functions in ontogeny* the means of securing "variations when and where they are wanted;" while Weismann supposes competing germinal units.

phrases? Spencer gave an answer in a general way long ago to the *second* of these questions, by saying that in consciousness the function of pleasure and pain is just to keep some actions or movements going and to suppress others.

"But as soon as we enquire more closely into the actual working of pleasure and pain reactions, we find an answer suggested to the *first* question also, i.e., the question as to how the organism comes to make the kind and sort of movements which the environment calls for—the *movement variations when and where they are required.* The pleasure or pain produced by a stimulus—and by a movement also, for the utility of movement is always that it secures stimulation of this sort or that—does not lead to diffused, neutral, and characterless movements, as Spencer and Bain suppose; this is disputed no less by the infant's movements than by the actions of unicellular creatures. There are characteristic differences in vital movements wherever we find them. Even if Mr. Spencer's undifferentiated protoplasmic movements had existed, natural selection would very soon have put an end to it. There is a characteristic antithesis in vital movements always. Healthy, overflowing, outreaching, expansive, vital effects are associated with pleasure; and the contrary, the withdrawing, depressive, contractive, decreasing vital effects are associated with pain. This is exactly the state of things which the theory of selection of movements from overproduced movements requires, i.e., that increased vitality, represented by pleasure, should give the excess movements, from which new adaptations are selected; and that decreased vitality represented by pain should do the reverse, i.e., draw off energy and suppress movement.[7]

"If, therefore, we say that here is a type of reaction which all vitality shows, we may give it a general descriptive name, i.e., the "Circular Reaction," in that its significance for evolution is that it is not a random response in movement to all stimulations alike, but that it distinguishes in its very form and amount between stimulations which are vitally good and those which are vitally bad, tending to retain the good stimulations and to draw away from and so suppress the bad. The term 'circular' is used to emphasize the way such a reaction tends to keep itself going, over and over, by reproducing the conditions of its own stimulation. It represents habit, since it tends to keep up old movements; but it secures new adaptations, since it provides for the overproduction of movement variations for the

---

[7]It is probable that the origin of this antithesis is to be found in the waxing and waning of the nutritive processes. "We find that if by an organism we mean a thing merely of contractility or irritability, whose round of movements is kept up by some kind of nutritive process supplied by the environment—absorption, chemical action of atmospheric oxygen, etc.—and whose existence is threatened by dangers of contact and what not, the first thing to do is to secure a regular supply to the nutritive processes, and to avoid these contacts. But the organism can do nothing but move, as a whole or in some of its parts. So then if one of such creatures is to be fitter than another to survive, it must be the creature which by its movements secures more nutritive processes and avoids more dangerous contacts. But movements toward the source of stimulation keep hold on the stimulation, and movements away from contacts break the contacts, that is all. Nature selects these organisms; how could she do otherwise?... We only have to suppose, then, that the nutritive growth processes are by natural selection drained off in organic expansions, to get the division in movements which represents this earliest bifurcate adaptation" (Baldwin, 1895a, p. 201).

operation of selection. This kind of selection, since it requires the direct coöperation of the organism itself, I have called 'Organic Selection.'"

The advantages of this view seem to be somewhat as follows:

1. It gives a method of the individual's adaptations of function which is *one in principle with the law of overproduction and survival now so well established in the case of competing organisms.*

2. It reduces nervous and mental evolution to strictly parallel terms. The intelligent use of phylogenetic variations for functional purposes in the way indicated, puts a premium on variations which can be so used, and thus sets phylogenetic progress *in directions of constantly improved mental endowment.* The circular reaction which is the method of intelligent adaptations is liable to variation in a series of complex ways which represent phylogenetically the development of the mental functions known as memory, imagination, conception, thought, etc. We thus reach a phylogeny of mind which proceeds in the direction set by the ontogeny of mind,[8] just as on the organic side the phylogeny of the organism gets its determinate direction from the organism's ontogenetic adaptations. And since it is the one principle of Organic Selection working by *the same functions* to set the direction of both phylogenies, the physical and the mental, the two developments are not two, but one. Evolution is, therefore, not more biological than psychological (Baldwin, 1895a, chap. x, xi, and especially pp. 383–388).

3. It secures the relation of structure to function required by the principle of "use and disuse" in ontogeny.

4. The only alternative theory of the adaptations of the individual are those of "pure chance," on the one hand, and a "creative act" of consciousness, on the other hand. Pure chance is refuted by all the facts which show that the organism does not wait for chance, but goes right out and effects new adaptations to its environment. Furthermore, ontogenetic adaptations are determinate; they proceed in definite progressive lines. A short study of the child will disabuse any man, I think, of the "pure chance" theory. But the other theory which holds that consciousness makes adaptations and changes structures directly by its *fiat*, is contradicted by the psychology of voluntary movement (Baldwin, 1896b, 1896e, 1896f). Consciousness can bring about no movement without having first an adequate experience of that movement to serve on occasion as a stimulus to the innervation of the appropriate motor centers. "This point is no longer subject to dispute; for pathological cases show that unless some adequate idea of a former movement made by the same muscles, or by association some other idea which stands for it, can be brought up in mind the intelligence is helpless. Not only can it not make new movements; it can not even repeat old habitual movements. So we may say that intelligent adaptation does not create

---

[8]Prof. C. S. Minot suggests to me that the terms "ontopsychic" and "phylopsychic" might be convenient to mark this distinction.

coördinations; it only makes functional use of coördinations which were alternatively present already in the creature's equipment. Interpreting this in terms of congenital variations, we may say that the variations which the intelligence uses are alternative possibilities of muscular movement" (Baldwin, 1896b). So the only possible way that a really new movement can be made is *by making the movements already possible so excessively and with so many varieties of combination, etc., that new adaptations may occur.*

5. The problem seems to me to duplicate the conditions which led Darwin to the principle of natural selection. The alternatives before Darwin were "pure chance" or "special creation." The law of "overproduction with survival of the fittest" came as the solution. So in this case. Let us take an example. Every child has to learn how to write. If he depended upon chance movements of his hands he would never learn how to write. But on the other hand, he can not write simply by willing to do so; he might will forever without effecting a "special creation" of muscular movement. What he actually does is to *use his hand in a great many possible ways as near as he can to the way required*; and from these excessively produced movements, and after excessively varied and numerous trials, he gradually selects and fixes the slight successes made in the direction of correct writing. It is a long and most laborious accumulation of slight Organic Selections from overproduced movements (ref. for handwriting in detail, Baldwin, 1895a, chap. v; also Baldwin, 1895a, pp. 373, ff.).

6. The only resort left to the theory that consciousness is some sort of an *actus purus* is to hold that it *directs* brain energies or selects between possible alternatives of movement; but besides the objection that it is as hard to direct movement as it is to make it (for nothing short of a force could release or direct brain energies), we find nothing of the kind necessary. The attention is what determines the particular movement in developed organisms, and the attention is no longer considered an *actus purus* with no brain process accompanying it. The attention is a function of memories, movements, organic experiences. We do not attend to a thing because we have already selected it, or because the attention selects it; but *we select it because we—consciousness and organism—are attending to it.* "It is clear that this doctrine of selection as applied to muscular movement does away with all necessity for holding that consciousness even directs brain energy. The need of such direction seems to me to be as artificial as Darwin showed the need of special creation to be for the teleological adaptations of the different species. This need done away, in this case of supposed directive agency as in that, the question of the relation of consciousness to the brain becomes a metaphysical one, just as that of teleology in nature became a metaphysical one; and it is not to much profit that science meddles with it. And biological as well as psychological science should be glad that it is so, should it not?" (Baldwin, 1896e; and on the metaphysical question, Baldwin, 1896f).

SECTION V.

A word on the relation of this principle of Organic Selection to Natural Selection. Natural Selection is too often treated as a positive agency. It is not a positive agency; it is entirely negative. It is simply a statement of what occurs when an organism does not have the qualifications necessary to enable it to survive in given conditions of life; it does not in any way define positively the qualifications which do enable other organisms to survive. Assuming the principle of Natural Selection in any case, and saying that, according to it, if an organism do [sic] not have the necessary qualifications it will be killed off, it still remains in that instance to find what the qualifications are which this organism is to have if it is to be kept alive. So we may say that *the means of survival is always an additional question* to the negative statement of the operation of natural selection.

This latter question, of course, the theory of variations aims to answer. The positive qualifications which the organism has arise as congenital variations of a kind which enable the organism to cope with the conditions of life. This is the positive side of Darwinism, as the principle of Natural Selection is the negative side.

Now it is in relation to the theory of variations, and not in relation to that of natural selection, that Organic Selection has its main force. Organic Selection presents *a new qualification of a positive kind* which enables the organism to meet its environment and cope with it, while natural selection remains exactly what it was, the negative law that if the organism does not succeed in living, then it dies, and as such a qualification on the part of the organism, Organic Selection presents several interesting features.

1. If we hold, as has been argued above, that the method of Organic Selection is always the same (that is, that it has a natural method), being always accomplished by a certain typical sort of nervous process (i.e., being always neuro-genetic), then we may ask whether that form of nervous process—and the consciousness which goes with it—may not be a variation appearing early in the phylogenetic series. I have argued elsewhere (Baldwin, 1895a, pp. 200 ff. and 208 ff.) that this is the most probable view. Organisms that did not have some form of selective response to what was beneficial, as opposed to what was damaging in the environment, could not have developed very far; and as soon as such a variation did appear it would have immediate preëminence. So we have to say either that selective nervous property, with consciousness, is a variation, or that it is a fundamental endowment of life and part of its final mystery. "The intelligence holds a remarkable place. It is itself, as we have seen, a congenital variation; but it is also the great agent of the individual's personal adaptation both to the physical and to the social environment" (Baldwin, 1896b).
"The former (instinct) represents a tendency to brain variation in the direction of fixed connections between certain sense-centers and certain groups of coördinated muscles. This tendency is embodied in the white matter and the

lower brain centers. The other (intelligence) represents a tendency to variation in the direction of alternative possibilities of connection of the brain centers with the same or similar coördinated muscular groups. This tendency is embodied in the cortex of the hemispheres" (Baldwin, 1896b).

2. But however that may be, whether ontogenetic adaptation by selective reaction and consciousness be considered a variation or a final aspect of life, it is a *life-qualification of a very extraordinary kind.* It opens a new sphere for the application of the negative principle of natural selection upon organisms, i.e., with reference to *what they can do*, rather than to what they are; to the new use they make of their congenital functions, rather than to the mere possession of the functions (Baldwin, 1896b, pp. 202 f.). A premium is set on congenital plasticity and adaptability of function rather than on congenital fixity of function; and this adaptability reaches its highest in the intelligence.

3. It opens another field also for the operation of natural selection—still viewed as a negative principle—through the survival of particular overproduced and modified reactions of the organism, by which the determination of the organism's own growth and life-history is secured. If the young chick imitated the old duck instead of the old hen, it would perish; it can only learn those new things which its present equipment will permit—not swimming. So the chick's own possible actions and adaptations in ontogeny have to be selected. We have seen how it may be done by a certain competition of functions with survival of the fit. But this is an application of natural selection. I do not see how Henslow, for example, can get the so-called "self-adaptations"—apart from "special creation"—which justify an attack on natural selection. Even plants must grow in determinate or "select" directions in order to live.

4. So we may say, finally, that Organic Selection, while itself probably a congenital variation (or original endowment) works to secure new qualifications for the creature's survival; and its very working proceeds by securing a new application of the principle of natural selection to the possible modifications which the organism is capable of undergoing. Romanes says: "it is impossible that heredity can have provided in advance for innovations upon or alterations in its own machinery during the lifetime of a particular individual." To this we are obliged to reply in summing up—as I have done before (Baldwin, 1895a, p. 220)—we reach "just the state of things which Romanes declares impossible—heredity providing for the modification of its own machinery. Heredity not only leaves the future free for modifications, it also provides a method of life in the operation of which modifications are bound to come."

SECTION VI.
THE MATTER OF TERMINOLOGY

I anticipate criticism from the fact that several new terms have been used in this paper. Indeed one or two of these terms have already been criticised. I think, however, that novelty in terms is better than ambiguity in meanings. And in each case the new term is intended to mark off a real meaning which no current term seems to express. Taking these terms in turn and attempting to define them, as I have used them, it will be seen whether in each case the special term is justified; if not, I shall be only two [sic] glad to abandon it.

ORGANIC SELECTION

The process of ontogenetic adaptation considered as keeping single organisms alive and so securing determinate lines of variation in subsequent generations. Organic Selection is, therefore, a general principle of development which is a direct substitute for the Lamarkian [sic] factor in most if not in all instances. If it is really a new factor, then it deserves a new name, however contracted its sphere of application may finally turn out to be. The use of the word "Organic" in the phrase was suggested from the fact that the organism itself coöperates in the formation of the adaptations which are effected, and also from the fact that, in the results, the organism is itself selected; since those organisms which do not secure the adaptations fall by the principle of natural selection. And the word "Selection" used in the phrase is appropriate for just the same two reasons.

SOCIAL HEREDITY

The acquisition of functions from the social environment, also considered as a method of determining phylogenetic variations. It is a form of Organic Selection but it deserves a special name because of its special way of operation. It is really heredity, since it influences the direction of phylogenetic variation by keeping socially adaptive creatures alive while others which do not adapt themselves in this way are cut off. It is also heredity since it is a continuous influence from generation to generation. Animals may be kept alive let us say in a given environment by social cooperation only; these transmit this social type of variation to posterity; *thus social adaptation sets the direction of physical phylogeny and physical heredity is determined in part by this factor.* Furthermore the process is all the while, from generation to generation, aided by the continuous chain of extra-organic or purely social transmissions. Here are adequate reasons for marking off this influence with a name.
    The other terms I do not care so much about. "Physico-genetic," "neurogenetic," "psycho-genetic," and their correlatives in "genic," seem to me to be

convenient terms to mark distinctions which would involve long sentences without them, besides being self-explanatory. The phrase "circular reaction" has now been welcomed as appropriate by psychologists. "Accommodation" is also current among psychologists as meaning single functional adaptations, especially on the part of consciousness; the biological word "adaptation" refers more, perhaps, to racial or general functions. As between them, however, it does not much matter.[9]

[9]I have already noted in print (Baldwin, 1896b, 1896e) that Prof. Lloyd Morgan and Prof. H. F. Osborn have reached conclusions similar to my main one on Organic Selection. I do not know whether they approve of this name for the "factor"; but as I suggested it in the first edition of my book (April, 1895) and used it earlier, I venture to hope that it may be approved by the biologists.

REFERENCES

Baldwin, J. M. *Imitation: A Chapter in the Natural History of Consciousness, Mind.* London, Jan., 1894. Citations from earlier papers will be found in this article and in the next reference.

Baldwin, J. M. *Mental Development in the Child and the Race*, 1st. ed. April, 1895a; 2nd. ed., Oct., 1895. Macmillan & Co. The present paper expands an additional chapter (Chap. XVII) added in the German and French editions and to be incorporated in the third English edition.

Baldwin, J. M. "Consciousness and Evolution." *Science* (August 23, 1895b). Reprinted in the *American Naturalist* (April, 1896a).

Baldwin, J. M. "Heredity and Instinct (I)." *Science* (March 20, 1896b). Discussion before N. Y. Acad. of Sci., Jan. 31, 1896 (1896c).

Baldwin, J. M. "Heredity and Instinct (II)." *Science* (April 10, 1896d).

Baldwin, J. M. "Physical and Social Heredity." *American Naturalist* (May, 1896e).

Baldwin, J. M. "Consciousness and Evolution." *Psychol. Review* (May, 1896f). Discussion before Amer. Psychol. Association, Dec. 28, 1895.

Chapter 6:
On Modification and Variation[1]

The original paper appeared in *Science* **IV(99)** (November 20, 1896): 733–740. Reprinted by permission. Footnote in the title appeared in the original work.

Up to a date still comparatively recent, the transmission to offspring, in greater or less degree, of those modifications of habit or structure which the parents had acquired in the course of their individual lifetime, was generally accepted. Lamarck is regarded as the intellectual father of the transmissionists. In his 'Historic Naturelle' he said: "The development of organs and their power of action are continually determined by the use of these organs." This is known as his third law. In the fourth he insisted on the hereditary nature of the effects of such use. "All that has been acquired, begun or changed," he said, "in the course of their life is preserved in reproduction and transmitted to the new individuals which spring from those which have experienced the changes."

---

[1]Being a chapter from a forthcoming work on *Habit and Instinct* communicated at the request of Prof. Henry F. Osborn.

Darwin accepted such transmission as subordinate to natural selection, and attempted to account for it by his theory of pangenesis. According to that hypothesis all the component cells of an organism throw off minute gemmules, and these and their like, collecting in the reproductive cells, are the parental germs from which all the cells of the offspring of that organism are developed. This theory, here given in briefest outline, came in for its full share of criticism. The problems of heredity were recognized as being of supreme biological importance and were warmly discussed. Meanwhile a different view of the relation between the organism and its reproductive cells came into prominence. With it the names of Francis Galton, in England, and August Weismann, in Germany, are inseparably connected. Of late years it has gained the approval of many, though by no means all, of our foremost biologists. This view, again given in briefest possible outline, is as follows: The fertilized egg of any many-celled organism gives origin to all the cells of which that organism is composed. In some of these, the reproductive cells, germinal substance is set aside for the future continuance of the race; the rest give rise to all the other cells of the body, those which constitute or give rise to muscle, nerve, bone, gland and so forth. Thus we have a division into germ-substance and body-substance. Germ gives origin to germ plus body; but the body takes no share, according to Prof. Weismann, in giving origin to—though it ministers to, protects, and may exercise an influence on—the germinal substance of the reproductive cells.

The logical development of this theory led Prof. Weismann to doubt the inheritance of characters acquired by the bodily substance in the course of individual life, and to examine anew the supposed evidence in its favor. For if brain substance, for example, contributes nothing to the reproductive cells, any modification it acquires during individual life can only reach the germ through some indirect mode of influence. But does it—does any modification of the body substance—so affect the germ as to become hereditary? Prof. Weismann answers this question by asserting that the evidence for the direct transmission of acquired characters is wholly insufficient, and by contending that, until satisfactory evidence is forthcoming, we may not accept transmission as a factor in evolution.

How, then, is progress possible if none of the modifications which the body suffers is transmitted from parent to offspring? To this question we must reply that though modification is, on this view, excluded from taking any direct share in race-progress, yet there is still variation. By modifications I mean those changes which are in some way wrought in the body-structure, and by variations those differences which are of germinal origin. That variation of germinal origin is a fact in organic nature is admitted on all hands, and that some variations are adaptive is also unquestioned. Transmissionists contend that modification in a particular direction in one generation is, through the transmission of the change in some way from the bodily tissues to the germinal cells, a source of variation in the same direction in the next generation. Selectionists, on the other hand, exclude this source of variation, contending that the supposed evidence in its favor is insufficient or unsatisfactory. But their whole theory depends on the occurrence of variations, of which those that are in unfavorable directions are weeded out, while those that are useful and adaptive remain in possession of the field. How these variations originate in the germ we need not here discuss. Let us assume that variations of germinal origin in a great number of directions do as a matter of fact occur.

This, then, is how the matter stands. All acknowledge the existence of variations and admit that their proximate source is in the fertilized ovum. All admit that the individual is, through its plasticity, in greater or less degree capable of adaptive modification. Transmissionists contend that the effects of modification are somehow transferred to the germinal substance there to give origin to variations. Selectionists deny this transmission and contend that adaptive variations are independent of adaptive modifications.

Now, what is natural selection, at any rate as understood by the master— Darwin? It is a process whereby, in the struggle for existence, individuals possessed of favorable and adaptive variations survive and hand on their good seed, while individuals possessed of unfavorable variations succumb, are sooner or later eliminated, standing therefore a less chance of begetting offspring. This is the natural selection of Darwin. But it is clear that to make the difference between survival and elimination the favorableness of the variation must reach a certain amount— varying with the keenness of the struggle. This was termed by Romanes 'selection value.' And one of the difficulties which critics of natural selection have felt is that the little more or the little less of variation must often be too small in amount to be of selection value so as to determine survival. This difficulty is admitted by Prof. Weismann as a real one. "The Lamarckians were right," he says, "when they maintained that the factor for which hitherto the name of natural selection had been exclusively reserved, viz., personal selection [*i.e.*, the selection of individuals], was insufficient for the explanation of the phenomena."[2] And again:[3] "Something is still wanting to the selection of Darwin and Wallace, which it is obligatory on us to discover, if we possibly can."

The additional factor which Dr. Weismann suggests is what he terms germinal selection. This, briefly stated, is as follows: There is a competition for nutrient among those parts of the germ from which the several organs or groups of organs are developed. These he names determinants; in this competition the stronger determinants get the best of it, and are further developed at the expense of the weaker determinants, which are starved and tend to dwindle and eventually disappear. The suggestion is an interesting one, but one well-nigh impossible to put to the test of observation. It must at present be placed among the 'may-bes' of biology. If accepted as a factor, it would serve to account for the existence of determinate variations, that is to say, variations along special or particular lines of adaptation.

Such determinate variations are, however, explicable on the theory of natural selection—a term which, in my opinion, should be reserved for that process of individual survival and elimination to which it was applied by Darwin. Writing in

[2] "Germinal Selection," *Monist* Jan., 1896, p. 290.
[3] Op cit. p. 264.

1892 I put the matter thus:[4] "Take the case of an organism which has in some way reached harmony with its environment. Slight variations occur in many directions, but these are bred out by intercrossing. It is as if a hundred pendulums were swinging just a little in many directions, but were at once damped down. Now, place such an organism in changed conditions. The swing of one or two of the pendulums is found advantageous; the organisms in which these two pendulums are swinging are selected; they mate together and in their offspring, while these 2 pendulums are by congenital inheritance kept a-swinging, the other 98 pendulums are rapidly damped down as before.

"Let us suppose, then, that the variation in tooth structure, in a certain mechanically advantageous direction, be such a selected pendulum swing. That particular pendulum, swinging in that particular direction, will be the subject of selection. The other pendulums will still be damped down as before, and in that particular pendulum variations from the particular direction will be similarly damped down. It will wobble a little, but its wobbling will be as nothing compared with the swing that is fostered by selection. In this case, then, selection will choose between the little more complexity that is advantageous and the little less complexity that is disadvantageous. The little less complexity will be eliminated, the little more complexity will survive. The little less and the little more are, however, in the same line of developmental swing. Hence, the variations discoverable in fossil mammals in which tooth development along special lines is in progress, will, on the hypothesis of selection, be plus and minus along a given line; in other words, the variations will be determinate, and in the direction of special adaptation."

Prof. Weismann adopts a similar position in his recent paper on germinal selection.[5] "By the selection alone," he says, "of the plus or minus variations of a character is the constant modification of that character in the plus or minus direction determined. We may assert therefore, in general terms, that a definitely directed progressive variation of a given part is produced by continued selection in that definite direction. This is no hypothesis, but a direct inference from the facts and may also be expressed as follows: By selection of the kind referred to, the germ is progressively modified in a manner corresponding with the production of a definitely directed progressive variation of the part."

In his Romanes Lecture, Prof. Weismann makes another suggestion which is valuable and helpful and which, I think, may be further developed and extended. He is there dealing with what he terms 'intra-selection,' or that individual plasticity to which I have frequently made reference. One of the examples that he adduces is the structure of bone. "Herman Meyer," he says,[6] "seems to have been the first to call attention to the adaptiveness as regards minute structure in animal tissues, which is most strikingly exhibited in the structure of the spongy substance of the long bones in the higher vertebrates. This substance is arranged on a similar mechanical

---

[4] *Natural Sciences*, Vol. I., April, 1892, pp. 100–101.

[5] *Monist*, Jan., 1896, p. 208

[6] Romanes Lecture on *The Effect of External Influences on Development*, pp. 11, 19.

principle to that of arched structures in general; it is composed of numerous fine bony plates so arranged as to withstand the greatest amount of tension and pressure, and to give the utmost firmness with a minimum expenditure of material. But the direction, position and strength of these long bony plates are by no means congenital or determined in advance; they depend on circumstances. If the bone is broken and heals out of the straight, the plates of the spongy tissue become rearranged so as to be in the new direction of greatest tension and pressure; thus they can adapt themselves to changed circumstances."

Then, after referring to the explanation, by Wilhelm Roux, of the cause of these wonderfully fine adaptations by applying the principle of selection to the parts of the organism in which, it is assumed, there is a struggle for existence among each other, Prof. Weismann proceeds to show[7] that "it is not the particular adaptive structures themselves that are transmitted, but only the quality of the material from which intra-selection forms these structures anew in each individual life. It is not the particular spongy plates which are transmitted, but a cell mass, that from the germ onwards so reacts to tension and pressure that the spongy structure necessarily results." In other words it is not the more or less definite congenital adaptation that is handed on through heredity, but an innate plasticity which renders possible adaptive modification in the individual.

This individual plasticity is undoubtedly of great advantage in race progress. The adapted individual will escape elimination in the life-struggle, and it matters not whether the adaptation [is] reached through individual modification of the bodily tissues, or through racial variation of germinal origin. So long as the adaptation is there—no matter how it originated—that is sufficient to secure survival. Prof. Weismann applies this conception to one of those difficulties which have been urged by critics of natural selection. "Let us take," he says,[8] "the well-known instance of the gradual increase in development of the deers' antlers, in consequence of which the head, in the course of generations, has become more and more heavily loaded. The question has been asked as to how it is possible for the parts of the body which have to support and move this weight to vary simultaneously and harmoniously if there is no such thing as the transmission of the effects of use or disuse, and if the changes have resulted from processes of selection only. This is the question put by Herbert Spencer as to 'co-adaptation,' and the answer is to be found in connection with the process of intra-selection. It is by no means necessary that all the parts concerned—skull, muscles and ligaments of the neck, cervical vertebrae, bones of the fore-limbs, etc.—should simultaneously adapt themselves by variation of the germ to the increase of the size of the antlers, for in each separate individual the necessary adaptation will be temporarily accomplished by intra-selection," that is, by individual modification due to the innate plasticity of the parts concerned. "The improvement of the parts in question," Prof. Weismann urges, "when so acquired, will certainly not be transmitted, but yet the primary variation is not lost. Thus

[7]Romanes Lecture, p. 15

[8]Romanes Lecture, pp. 18, 19.

when an advantageous increase in the size of the antlers has taken place, it does not lead to the destruction of the animal in consequence of other parts being unable to suit themselves to it. All parts of the organism are in a certain degree variable [*i.e.,* modifiable] and capable of being determined by the strength and nature of the influences that affect them; and this capacity to respond conformably to functional stimulus must be regarded as the means which make possible the maintenance of a harmonious co-adaptation of parts in the course of the phyletic metamorphosis of a species. As the primary variations in the phyletic metamorphosis occurred little by little, the secondary adaptations would as a rule be able to keep pace with them."

So far Prof. Weismann. According to his conception, variations of germinal origin occur from time to time. By its innate plasticity the several parts of an organism implicated by their association with the varying part are modified in individual life in such a way that their modifications cooperate with the germinal variation in producing an adaption of double origin, partly congenital, partly acquired. The organism then waits, so to speak, for a further congenital variation, when a like process of adaptation again occurs; and thus race-progress is effected by a series of successive variational steps, assisted by a series of cooperating individual modifications.

If now it would be shown that, although on selectionist principles there is no transmission of modification due to individual plasticity, yet these modifications afford the conditions under which variations of like nature are afforded an opportunity of occurring and of making themselves felt in race-progress, a further step would be taken towards a reconciliation of opposing views. Such it appears to me, may well be the case.

To explain the connection which may exist between modifications of the bodily tissues due to innate plasticity (intra-selection) and variations of germinal origin in similar adaptive directions, we may revert to the pendulum analogy which was adduced a few pages back. Assuming that variations do tend to occur in a great number of divergent directions we may liken each to a pendulum which tends to swing; nay, which is swinging through a small arc. The organism, so far as variation is concerned, is a complex aggregate of such pendulums. Suppose then that it has reached congenital harmony with its environment. The pendulums are all swinging through the small arc implied by the slight variations which occur even among the offspring of the same parents. No pendulum can materially increase its swing; for since the organism has reached congenital harmony with its environment, any marked variation will be out of harmony and the individual in which it occurs will be eliminated. Natural selection, then, will ensure the damping down of the swing of all the pendulums within comparatively narrow limits.

But now suppose that the conditions of the environment somewhat rapidly change. Congenital variations will not be equal to the occasion. The swing of the pendulums concerned cannot be rapidly augmented. Here individual plasticity steps in to save some of the members of the race from extinction. They adapt themselves to the changed conditions through a modification of bodily tissues. If no members of the race have sufficient plasticity to effect this accommodation the race will become extinct, as has indeed occurred again and again in the course of geological history. The stereotyped races have succumbed; the plastic races have survived. Let us grant, then, that certain organisms accommodate themselves to the new conditions by plastic modification of the bodily tissues, say by the adaptive strengthening of some bony structure. What is the effect on congenital variations? Whereas the other pendulums are still damped down by natural selection as before, the oscillation of the pendulum, which represents a variation in this bony structure, is no longer checked. It is free to swing as much as it can. Congenital variations in the direction of adaptive modification will be so much to the good of the individual concerned. They will constitute a congenital predisposition to that strengthening of the part which is essential for survival. Variations in the opposite direction, tending to thwart the adaptive modification, will be disadvantageous and will be eliminated. Thus, if the conditions remain constant for many generations, congenital variation will gradually render hereditary the same strengthening of bone structure that was provisionally attained by plastic modification. The effects are precisely the same as they would be if the modification in question were directly transmitted in a slight but cumulatively increasing degree. They are reached, however, in a manner which involves no such transmission.

To take a particular case: Let us grant that, in the evolution of the horse tribe, it was of advantage to this line of vertebrate life that the middle digit of each foot should be largely developed and the lateral digits reduced in size; and let us grant that this took its rise in adaptive modification through the increased use of the middle digit and the relative disuse of the lateral digits. Variations in these digits are no longer suppressed and eliminated. Any congenital predisposition to increased development of the middle digit and decreased size in the lateral digits will tend to assist the adaptive modification and to supplement its deficiencies. Any congenital predisposition in the contrary direction will tend to thwart the adaptive modification and to render it less efficient. The former will let adaptive modification start at a higher level, so to speak, and thus enable it to be carried a step further. The latter will force it to start at a lower level, and will prevent its going so far. If natural selection take[s] place at all, we may well believe that it would do so under such circumstances.[9] And it would work along the lines laid down for it in adaptive modification. Modification would lead; variation [would] follow in its wake. It is not surprising that for long we believed that modification was transmitted as hereditary variation. Such an interpretation of the facts is the simpler and more obvious. But simple and obvious interpretations are not always correct. And if, on closer examination, in the light of fuller knowledge, they are found to present grave difficulties, a less simple and less obvious interpretation may claim our provisional acceptance.

---

[9]Prof. Weismann's 'Germinal Selection' if a *vera causa* would be a cooperating factor and assist in producing the requisite variations.

In his recent paper on Germinal Selection Prof. Weismann says:[10] "I am fain to relinquish myself to the hope that now, after another explanation has been found, a reconciliation and unification of the hostile views is not so very distant, and that then we can continue our work together on the newly laid foundations." As one to whom Prof. Weismann alludes as having expressed the opinion that the Lamarckian principle must be admitted as a working hypothesis, I am now ready to relinquish myself also to the same hope. Germinal Selection does not convince me, though I regard it as a suggestive hypothesis; and assuredly I am not convinced by the argument that because in certain cases, such as the changes in the chitinous parts of the skeleton of insects and crustacea, and in the teeth of mammals, use and disuse can have played no part, therefore in no other cases has use-inheritance prevailed. Even Homer sometimes nods, and Prof. Weismann's logical acumen seems to have deserted him here. But it appears to me that on the lines I have sketched out, it is open to us to accept the facts adduced by the transmissionists and at the same time interpret them on selectionist principles.

It may be well now briefly to summarize the line of argument in a series of numbered paragraphs.

1. In addition to what is congenitally definite in structure or mode of response, an organism inherits a certain amount of innate modifiability or plasticity[.]

2. Natural selection secures:
   a.   such congenital definiteness as is advantageous.
   b.   such innate plasticity as is advantageous.

3. Both a and b are commonly present; but uniformity of conditions tends to emphasize the former variable conditions of life, the latter.

4. The organism is subject to:
   a.   variation of germinal origin.
   b.   modification of environmental origin, affecting the soma or body tissues.

5. Transmissionists contend that somatic modification in a given direction in one generation is transmitted to the reproductive cells to constitute a source of germinal variation in the same direction in the next generation.

6. It is here suggested that persistent modification through many generations, though not transmitted to the germ, nevertheless affords the opportunity for the occurrence of germinal variation of like nature.

7. Under constant conditions of life, though variations in many directions are occurring in the organisms which have reached harmonious adjustment to these conditions, yet natural selection eliminates all those which are of such amount as to be disadvantageous, and thus acts as a check on all variations, repressing them to within narrow limits.

---

[10] *Monist, loc cit*, p. 290

8. Let us suppose, however, that a group of organisms belonging to a plastic species is placed under new conditions of environment.

9. Those whose innate somatic plasticity is equal to the occasion survive. They are modified. Those whose innate plasticity is not equal to the occasion are eliminated.

10. Such modification takes place generation after generation, but, as such, is not inherited. There is no transmission of the effects of modification to the germinal substance.

11. But variations in the same direction as the somatic modification are now no longer repressed and are allowed full scope.

12. Any congenital variations antagonistic in direction to those modifications will tend to thwart them and to render the organism in which they occur liable to elimination.

13. Any congenital variations similar in direction to these modifications will tend to support them and to favor the individuals in which they occur.

14. Thus will arise a congenital predisposition to the modifications in question.

15. The longer this process continues, the more marked will be the predisposition and the greater the tendency of the congenital variations to conform in all respects to the persistent plastic modifications; while

16. The plasticity continuing the operation, the modifications become yet further adaptive.

17. Thus plastic modification leads and germinal variation follows; the one paves the way for the other.

18. Natural selection will tend to foster variability in given advantageous lines when once initiated, for (a) the constant eliminations of variations leads to the survival of the relatively invariable; but (b) the perpetuation of variations in any given direction leads to the survival of the variable in that direction. Lamarckian paleontologists are apt to overlook this fact that natural selection produces determinate variation.

19. The transmissionist, fixing his attention first on the modification, and secondly [on] the fact that organic effects similar to those produced by the modification gradually become congenitally stereotyped, assumes that the modification *as such* is inherited.

20. It is here suggested that the modification *as such* is not inherited, but is the condition under which congenital variations are favored and given time to get a hold on the organism, and are thus enabled by degrees to reach the fully adaptive level.

When we remember that plastic modification and germinal variation have been working together all along the line of organic evolution, to reach the common goal of adaptation, it is difficult to believe that they have been all along wholly independent of each other. If the direct dependence advocated by the transmissionists be rejected, perhaps the indirect dependence here suggested may be found worthy of consideration.

Chapter 7:
Canalization of Development and the Inheritance of Acquired Characters

This chapter originally appeared in *Nature* **3811** (November 14, 1942): 563–565. Reprinted by permission.

The battle, which raged for so long between the theories of evolution supported by geneticists on one hand and by naturalists on the other, has in present years gone strongly in favour of the former. Few biologists now doubt that genetical investigation has revealed at any rate the most important categories of hereditary variation; and the classical naturalist theory—the inheritance of acquired characters—has been very generally relegated to the background because, in the forms in which it has been put forward, it has required a type of hereditary variation for the existence of which there was no adequate evidence. The long popularity of the theory was based, not on any positive evidence for it, but on its usefulness in accounting for some of the most striking of the results of evolution. Naturalists cannot fail to be continually and deeply impressed by the adaptation of a an organism to its surroundings and of the parts of the organism to each other. These adaptive characters are inherited and some explanation of this must be provided. If we are deprived of the hypothesis of the inheritance of the effects of use and disuse, we seem thrown back on an exclusive reliance on the natural selection of merely chance mutations.

It is doubtful, however, whether even the most statistically minded geneticists are entirely satisfied that nothing more is involved than the sorting out of random mutations by the natural selective filter. It is the purpose of this short communication to suggest that recent views on the nature of the developmental process make it easier to understand how the genotypes of evolving organisms can respond to the environment in a more co-ordinated fashion.

It will be convenient to have in mind an actual example of the kind of difficulties in evolutionary theory with which we wish to deal. We may quote from Robson and Richards (1936): "A single case will make the difficulty clear. Duerden (1920) has shown that the sternal, alar, etc., callosities of the ostrich, which are undoubtedly related to the crouching position of the bird, appear in the embryo. The case is analogous to the thickening of the soles of the feet of the human embryo attributed by Darwin (1901) 'to the inherited effects of pressure.' As Detlefsen (1925) points out, this would have to be explained on selectionist grounds by the assumption that it was of advantage to have the callosities, as it were, preformed at the place at which they are required in the adult. But it is a large assumption that variations would arise at this place and nowhere else."

In this case we have an adaptive character (the callosities) of a kind which it is known can be provoked by an environmental stimulus during a single lifetime (since skin very generally becomes calloused by continued friction) but which is in this case certainly inherited. The standard hypotheses which come in question are the two considered by Robson and Richards: the Lamarckian explanation in terms of the inheritance of the effects of use, which they cannot bring themselves to support at all strongly, and the 'selectionist' explanation, which, in the form in which they understand it, leaves entirely out of account the fact that callosities may be produced by an environmental stimulus and postulates the occurrence of a gene with the required developmental effect. A third possible type of explanation is to suppose that in earlier members of the evolutionary chain, the callosities were formed as responses to external friction, but that during the course of evolution the environmental stimulus has been superseded by an internal genetical factor. It is an explanation of this kind which will be advanced here.

The first step in the argument is one which will scarcely be denied but is perhaps often overlooked. The capacity to respond to an external stimulus by some developmental reaction, such as the formation of a callosity, must itself be under genetic control. There is little doubt, though no positive evidence in this particular case so far as I know, that individual ostriches differ genetically in the responsiveness of their skin to friction and pressure. If we suppose then, that in the early ostrich ancestors callosities were formed by direct response to external pressure, there would be a natural selection among the birds for a genotype which gave an optimum response.

The next point to be put forward is the one which is, perhaps, new in such discussions, and which therefore requires the most careful scrutiny. It is best considered as one general thesis and one particular application of it.

The main thesis is that developmental reactions, *as they occur in organisms submitted to natural selection,* are in general canalized. That is to say, they are adjusted so as to bring about one definite end-result regardless of minor variations in conditions during the course of the reaction.

The evidence for this comes from two sides, the embryological and genetical. In embryology we have abundant evidence of canalization on two scales. On the small scale of single tissues, one may direct attention to the obvious but not unimportant fact that animals are built up of sharply defined different tissues and not of masses of material which shade off gradually into one another. Similarly, from the experimental point of view, it is usual to find that, while it may be possible to steer a mass of developing tissue into one of a number of possible paths, it is difficult to persuade it to differentiate into something intermediate between two of the normal possibilities. Passing from the scale of tissues to that of organs, it is not too much to claim it as a general rule that there is some stage in every life-history (though it may be an extremely early and short stage) when minor variations in morphology become 'regulated' or regenerated; and that is, again, a tendency to produce the standard end-product. Of course neither of these types of canalization is absolute. Morphological regulation may fail if the abnormalities are too great or occur too late in development; and intermediate types of tissue can occasionally be found, particularly in pathological conditions.

The limitations on canalization which are important for our present purposes can better be seen when the problem is viewed from the other, genetical, side. The canalization, or perhaps it would be better to call it the buffering, of the genotype is evidenced most clearly by constancy of the wild type. It is a very general observation to which little attention has been directed (but see Huxley, 1942; Plunkett, 1932; Ford, 1940) that the wild type of an organism, that is to say, the form which occurs in Nature under the influence of natural selection, is much less variable in appearance than the majority of the mutant races. In *Drosophila* the phenomenon is extremely obvious; there is scarcely a mutant which is comparable in constancy with the wild type, and there are very large numbers whose variability, either in the frequency with which the gene become[s] expressed at all or in the grade of expression, is so great that it presents a considerable technical difficulty. Yet the wild type is equally amazingly constant. If wild animals of almost any species are collected, they will usually be found 'as like as peas in a pod.' Variation there is, of course, but of an altogether lesser order than that between the different individuals of a mutant type.

The constancy of the wild type must be taken as evidence of the buffering of the genotype against minor variations not only in the environment in which the animals developed but also in its genetic make-up. That is to say, the genotype can, as it were, absorb a certain amount of its own variation without exhibiting any alteration in development. Considerable stress has been laid in recent years on certain aspects of this buffering. Fisher (1928) and many authors following him have discussed 'the evolution of dominance,' by which the genotype comes to be able to produce the standard developmental effects even when certain genes have been replaced by others of less efficiency. Again, Stern (1929) and Muller (1932) directed attention to the phenomenon of 'dosage compensation,' by which it comes about that a single dose of a sex-linked gene in the heterogametic sex has the same developmental effect as a double dose in the homogametic. These two processes are part of the larger phenomenon which we have called the canalization of development. This also includes other, at first sight unrelated, features of the genotypic control of development. For example, attention has been directed (Waddington, 1940a) to genes which cause certain regions of developing tissue to take an abnormal choice out of a range of alternative possible paths; Mather and de Winton (1941) have recently spoken of such genes as 'switch genes.' Finally, Goldschmidt has shown that environmental stimuli may, by switching development into a path which is usually only followed under the influence of some particular gene, produce what he has called a 'phenocopy' of a previously known mutant type.

There seems, then, to be a considerable amount of evidence from a number of sides that development is canalized in the naturally selected animal. At the same time, it is clear that this canalization is not a necessary characteristic of all organic development, since it breaks down in mutants, which may be extremely variable, and in pathological conditions, when abnormal types of tissue may be produced. It seems, then, that the canalization is a feature of the system which is built up by natural selection; and it is not difficult to see its advantages, since it ensures the production of the normal, that is, optimal, type in the face of the unavoidable hazards of existence.

The particular application of this general thesis which we require in connexion with 'the inheritance of acquired characters' is that a similar canalization will occur when natural selection favours some characteristic in the development of which the environment plays an important part. It is first necessary to point out the ways in which the environment can influence the developmental system. If we conceptually rigidify such a system into a definite formal scheme, we can think of it as a set of alternative canalized paths; and the environment can act either as a switch, or as a factor involved in the system of mutally interacting processes to which the buffering of the paths is due. This is, of course, too dead and formal a scheme to be a true picture of development as it actually occurs. In so far as it is always to some extent, but not entirely, a matter of convenience what we decide to call a complete organ, so far will it be a matter of convenience what we consider to be different alternative paths; and the question of whether a given influence is thought of as a switch mechanism or a modification of a path will depend on how we choose our alternatives. There are some cases, however, in which the alternatives are very clearly defined. Thus it is commonly assumed that the evolution of sexuality passed through a stage in which, as in Bonellia, the environment acted as a switch between two well-defined alternatives; later, genetic factors arose which superseded the environmental determination by an internal one.

More commonly, however, the original environmental effect will be to produce a modification of an already existent developmental path. Thus in the case of the ostrich ancestors, the formation of callosities following environmental stimulation is a response by a developmental system which is normally present in vertebrates. This system must, in all species, be subject to natural selection; outside certain limits, too great or too low a reactivity of the skin would be manifestly disadvantageous. If we suppose that the callosities, when they were first evolved, were dependent on the environmental stimulus, then the evolution appears as a readjustment of the reactivity of the skin to such a degree that a just sufficient thickening is produced with the normally occurring stimulus.

There would appear to be two possible ways in which such a development might be organized. It might on one hand remain uncanalized, the formation of the thickening in each individual depending on the reception of the adequate stimulus, to which the response remained strictly proportional. If this possibility was realized, the well-known difficulty of accounting for the hereditary fixation of the character remains unimpaired. The alternative is that the development does become canalized, to a greater or lesser extent. In that case, the magnitude of the response would not be proportional to that of the stimulus; there would be a threshold of stimulus, above which the optimum (that is, naturally selected) response would be formed. In so far as the response became canalized, the environment would be acting as a switch.

Systems of either type can be built up by natural selection, and one can point to examples of them in animals at the present day. The reaction of the patterns on Lepidopteran wings (for example, in Ephestia, Kühn, 1936) to temperature during the sensitive period scarcely seems to involve thresholds, while the metamorphosis of the axolotl, for example, clearly does. In general, it seems likely that the optimum response to the environment will involve both some degree of proportionality and some restriction of this by canalization. The most favourable mixture of the two tendencies will presumably differ for different characters. It is easy to see why a much sharper distinction between alternatives is generally evolved in connexion with sex differences than with the degree of muscular development, for example; but even the former is to some extent modifiable by extreme and specialized environmental disturbances (heavy and early hormone treatment), and even the latter has some degree of genetic determination.

The canalization of an environmentally induced character is accounted for if it is an advantage for the adult animal to have some optimum degree of development of the character irrespective of the exact extent of stimulus which it has met in its early life; if, for example, it is an advantage to the young ostrich going out into the hard world to have adequate callosities even if it were reared in a particularly soft and cosy nest. Now in so far as the development of the character becomes canalized, the action of the external stimulus is reduced to that of a switch mechanism, simply in order that the optimum response shall be regularly produced. But switch mechanisms may notoriously be set off by any of a number of factors. The choice between the alternative developmental pathways open to *gastrula ectoderm*, for example, may be made by the normal evocator or by a number of other things (the mode of action of which may be through the release of the normal evocator (cf. Waddington, 1940b, 1940c), but which remain different to the normal evocator nevertheless). Again, we know many instances in which several different genes, by switching development into the same path, produce similar effects; and attention has already been directed to the 'phenocopying' of a gene by a suitable environmental stimulus. Thus once a developmental response to an environmental stimulus has become canalized, it should not be too difficult to switch development into that track by mechanisms other than the original external stimulus, for example, by the internal mechanism of a genetic factor; and, as the canalization will only have been built up by natural selection if there is an advantage in the regular production of the optimum response, there will be a selective value in such a supersession of the environment by the even more regularly acting gene. Such a gene must always act before the normal time at which the environmental stimulus was applied, otherwise its work would already be done for it, and it could have no appreciable selective advantage.

Summarizing, then, we may say that the occurrence of an adaptive response to an environmental stimulus depends on the selection of a suitable genetically controlled reactivity in the organism. If it is an advantage, as it usually seems to be for developmental mechanisms, that the response should attain an optimum value more or less independently of the intensity of stimulus received by a particular animal, then the reactivity will become canalized, again under the influence of natural selection. Once the developmental path has been canalized, it is to be expected that many different agents, including a number of mutations available in the germplasm of the species, will be able to switch development into it; and the same considerations which render the canalization advantageous will favour the supersession of the environmental stimulus by a genetic one. By such a series of steps, then, it is possible that an adaptive response can be fixed without waiting for the occurrence of a mutation which, in the original genetic background, mimics the response well enough to enjoy a selective advantage.

REFERENCES

Darwin, C. *The Descent of Man and Selection in Relation to Sex.* London, 1901.

Detlefsen, J. A. "The Inheritance of Acquired Characters." *Physiol. Rev.* **5** (1925): 244.

Duerden, J. E. "The Inheritance of the Callosities in the Ostrich." *Amer. Nat.* **54** (1920): 289.

Fisher, R. A. "The Possible Modification of the Response of the Wild Type to Recurrent Mutations." *Amer. Nat.* **62** (1928): 115.

Ford, E. B. "Genetic Research in the Lepidoptera." *Ann. Eugen.* **10** (1940): 227.

Huxley, J. S. *Evolution: The Modern Synthesis.* London, 1942.

Kühn, A. "Versuche über die Wirkungsweise der Erbanlagen." *Naturwiss.* **2** (1936): 1.

Mather, K., and D. de Winton. "Adaptation and Counter-Adaptation of the Breeding System in Primula." *Ann. Bot.* **5** (1941): 297.

Muller, H. J. "Further Studies on the Nature and Causes of Gene Mutations." *Proc. 6th Int. Congr. Gen.* **1** (1932): 213.

Plunkett, C. C. "Temperature as a Tool in Research in Phenogenetics." *Proc. 6th Int. Congr. Gen.* **2** (1932): 158.

Robson, G. C., and O. W. Richards. *The Variation of Animals in Nature.* London, 1936.

Stern, C. "Uber die additive Wirkung multipler Allele." *Biol. Zbl.* **49** (1929): 231.

Waddington, C. H. *Growth* Suppl., (1940a): 37.

Waddington, C. H. *Organisers and Genes.* Cambridge, 1940b.

Waddington, C. H. *Genes as Evocators in Development.* 1940c.

George Gaylord Simpson

Chapter 8:
The Baldwin Effect

This chapter originally appeared in *Evolution* **7** (June 1953): 110–117. Reprinted by permission.

HISTORICAL INTRODUCTION

Characters individually acquired by members of a group of organisms may eventually, under the influence of selection, be reenforced or replaced by similar hereditary characters. That is the essence of the evolutionary phenomenon here called "the Baldwin effect."

The possibility of such an effect was noted independently and almost simultaneously by J. M. Baldwin[1] (1896), Lloyd Morgan (1896), and H. F. Osborn (1896). Lloyd Morgan (e.g., 1900) and Osborn (e.g., 1897a and b) made occasional later references to the effect (or factor, principle, or hypothesis, as you will). Baldwin (especially 1902) followed it up in greater detail. E. B. Poulton early joined the discussion, as did several others in the period 1896–1905, approximately.

That three workers independently thought of the Baldwin effect at the same time demonstrates that the idea was in the air, that it was an inevitable outgrowth of the intellectual atmosphere of the time. That time was at the height of the neo-Darwinian *versus* neo-Lamarckian controversy and shortly before the rediscovery of Mendelism gave a radically different turn to biological thought. There was a sharp issue, still familiar to all of us. Organism and environment obviously interact and obviously are closely fitted, that is, adapted to each other. Yet, as was already clear in the 1890's, it is improbable (to say the least) that the effects of the interaction can become heritable directly and in the same form. The Baldwin effect ostensibly provides a reconciliation between neo-Darwinism and neo-Lamarckism. To the extent that it may really occur, it provides a mechanism that is capable of making acquired characters hereditary—or of seeming to do so. Baldwin, Lloyd Morgan, and Osborn all explicitly postulated the Baldwin effect as a way out of the neo-Darwinian–neo-Lamarckian dilemma.

Mendelism and later genetic theory so conclusively ruled out the extreme neo-Lamarckian position that reconciliation came to seem unnecessary. After general acceptance of Mendelism and before clear statement of the modern synthesis of evolutionary theory, the Baldwin effect was seldom discussed in detail, although it continued to be mentioned under various names in reviews of evolutionary theory (e.g., Delage and Goldsmith, 1912; Lull, 1917; Herbert, 1919). Huxley (1942) brought the "unduly neglected" Baldwin effect into the synthetic theory as a subsidiary factor. It is so recognized by most followers of the synthetic theory (e.g., Mayr, 1951), although it is seldom assigned a major role in evolution.

In the meantime the notorious conflict about theories of genetics and of evolution was developing in the U.S.S.R. Until 1948 the most active and able Soviet biologists were contributing substantially to the synthesis of genetics and what was elsewhere usually called neo-Darwinism. There was, however, increasing ideological opposition to Mendelism and to theories of evolution that include random or indeterminate processes, as neo-Darwinism was accused of doing. The result was a conflict which because of its political apriorism must be called pseudo-scientific but

---

[1]James Mark Baldwin was born in Columbia, South Carolina, in 1861 and was educated at Princeton and under Wundt in Germany. He taught at several universities, including Princeton in 1893–1903 (the period of work on the Baldwin effect) and the Johns Hopkins, 1903–1908. After 1908 he spent some time in Mexico and finally settled in France, where he died in 1934. He was a pioneer in experimental psychology and an indefatigable writer and editor. His interest in evolution was part of a widespread effort to develop an evolutionary psychology. (See Boring, 1950; Langfeld, 1944.) Lloyd Morgan and Osborn are more familiar to evolutionists.

which in form closely paralleled the genuinely scientific disagreement of the neo-Darwinians and neo-Lamarckians in the 1890's. In the U.S.S.R. the conflict was between Mendelism (associated with neo-Darwinism through the synthetic theory) and Michurinism (essentially the same as neo-Lamarckism). The similar situation had a similar result: Soviet students independently thought of the Baldwin effect. First was apparently Lukin around 1936,[2] and others followed up and strongly emphasized this trend of thought, notably Kirpichnikov (e.g., 1947), Gause (e.g., 1947), and Schmalhausen (e.g., 1949).

As everyone knows, Michurinism was triumphant in the U.S.S.R., and Mendelism and neo-Darwinism were, in effect, outlawed in 1948. It is significant that particularly virulent attack was made on Schmalhausen's work, which embodied the Baldwin effect. The Baldwin effect could, as its earlier proponents had suggested, be considered as a compromise between the opposing schools, and compromise was ideologically even less acceptable than Mendelism.

Recourse has further been made to the Baldwin effect in still a third context: the conflicting philosophies of finalism (generally associated with vitalism) and materialism. As in neo-Darwinism *vs.* neo-Lamarckism and Mendelism *vs.* Michurinism, the evolutionary issue here centers on the problem of adaptation. Adaptation *seems* to be purposeful. The finalist view is that it is purposeful, in fact, but materialists generally rule out purpose as a possible factor in evolution. There are several possible materialistic explanations for seemingly purposeful adaptation. Among them is the Baldwin effect, which conceivably could, for instance, account for the genetic fixation of purposeful (or pseudo-purposeful) individual activities. The controversy between finalistic and materialistic philosophies of evolution has been especially active in France, and there Hovasse (1943) proposed what is essentially the Baldwin effect to account for adaptations considered finalistic by Cuénot (1941).[3] Hovasse (1950) later wrote a short book on the subject.

TERMINOLOGY AND DEFINITION

Baldwin called the effect in question "organic selection" and defined it as follows (Baldwin, 1902):

> *Organic Selection:* The process of individual accommodation considered as keeping organisms alive, and so, by also securing the accumulation of variations, determining evolution in subsequent generations.

---

[2] *Fide* Gause and Schmalhausen. I have not read Lukin's publications, which are in Ukrainian and Russian.

[3] Cuénot's final (posthumously published) word on the subject (1951) was less decisively finalistic.

In Baldwin's usage an "accommodation" was non-hereditary, *i.e.*, it was an acquired character, while a "variation" was (by definition) hereditary, *i.e.*, genetic. Thus his definition, somewhat ambiguous by present-day usage, designates a sequential process in which acquired characters are replaced by genetic characters. This is clear in Baldwin's discussion of the matter, which also brings in natural selection as the mechanism of replacement. Osborn and Lloyd Morgan accepted Baldwin's term with essentially his definition, and so have several later students, notably Gause. Others have used the term but not precisely with its original meaning. For example Lutz (1948) defines organic selection as "selection of the environment by the organism," a definition radically different from that of Baldwin and those who have followed him more closely.[4] The term is in any case misleading. "Organic selection" is no more organic than any other sort of selection. Moreover, the phenomenon discussed by Baldwin is not directly or solely selection anyway. It is a complex process in which selection, strictly speaking, is only one of several factors, or it is an effect that is postulated as a result of selection.

Osborn used the term "coincident selection" alternatively as a synonym of "organic selection," the significance being that the "germinal variations" selected coincide with adaptive individual modifications. The same thought underlies Hovasse's preference for "parallel selection." Both terms are again misleading in that the process involves or results from but is not as a whole equivalent to selection, and "parallel selection" is liable to confusion with the wholly different process of parallel evolution (in two or more lineages) under the influence of natural selection. Schmalhausen and some others (mostly Russian) speak of "stabilizing selection." The term is sometimes equated with Baldwin's "organic selection," but the equation is misleading. "Stabilizing selection" applies literally and in Schmalhausen's usage to any mechanism tending to fix an adaptive type and to bring it under more rigid genetic control. The Baldwin effect is one such mechanism, but not the only one and not (even in Schmalhausen's opinion) the most important.

Those ambiguities and the difficulty of finding an apt descriptive term for so complex a process led Huxley (1942) to speak of the "Baldwin and Lloyd Morgan principle,"[5] and Hovasse (1950) calls it "Baldwin's principle." That usage seems to me the simplest way toward a term both brief and unambiguous. Whether the mechanism or process in question is really a "principle" remains debatable, and I prefer the expression "Baldwin effect."

From Baldwin to Hovasse all those who have discussed the Baldwin effect under any name make it clear that what is meant is a complex sequence of events. The effect may be analyzed as involving three distinct (but partly simultaneous) steps:

---

[4]Selection of the environment may, however be a first step in the Baldwin effect, a possibility stressed by Thorpe (1945a).

[5]Even this length term fails to credit the simultaneous expressions of the principle, for Osborn thought of it as early as Lloyd Morgan and independently. When the coincidence was discovered, both Osborn and Lloyd Morgan deferred to Baldwin.

1. Individual organisms interact with the environment in such a way as systematically to produce in them behavioral, physiological, or structural modifications that are not hereditary as such but that are advantageous for survival, *i.e.*, are adaptive for the individuals having them.
2. There occur in the population genetic factors producing hereditary characteristics similar to the individual modifications referred to in (1), or having the same sorts of adaptive advantages.
3. The genetic factors of 2 [above] are favored by natural selection and tend to spread in the population over the course of generations. The net result is that adaptation originally individual and non-hereditary becomes hereditary.

That description of the Baldwin effect is also a more precise definition of the term. At this point it need not be taken for granted that the effect actually occurs or has an essential role in evolution. It may be taken as a hypothesis subject to investigation.

SUPPOSED EXAMPLES OF THE BALDWIN EFFECT

The three processes involved in the Baldwin effect are all known to occur separately. The development of adaptive individual modifications or accommodations (sometimes called "somations") is widespread, a matter of common observation sufficiently established by the banal example of the strengthening of muscles by use. Partial or even complete correspondence between the effects of non-heritable modification and heritable mutations (broadly speaking) is also well-established in some instances. The phenocopies of Goldschmidt (1938) conclusively demonstrate this phenomenon. The existence of phenocopies, copying genetic effects without change in heredity, implies equally the existence of genocopies (Hovasse's apt term), copying non-genetic effects by change in heredity. That genetic effects, therefore also genocopies, can be spread through populations by natural selection requires no further substantiation at this late date.

Thus each process necessary for the Baldwin effect does factually occur. There is no reason to doubt that they could occur together, in the stated sequence, and so produce the Baldwin effect. There is even some probability that they must have produced that effect sometimes. Nevertheless two points remain decidedly questionable: whether the Baldwin effect does in fact explain particular instances of evolutionary change, and the extent to which this effect has been involved in evolution or can explain the general phenomenon of adaptation. Basis for judgment on these points is provided by brief review of some supposed examples.

Both Baldwin and Lloyd Morgan considered the Baldwin effect as a way in which, without transmission of acquired characters, habits and other learned behavior could become instincts, *i.e.*, inherited behavior. Their examples included the instinct (if such it be) of chicks to drink by throwing the head in the air, the instinct[6] in primates to grasp with the thumb opposed to the other fingers, and other examples of a similar sort. Osborn went so far as to suppose that an arboreal race of man could be developed by rearing infants in trees, where they would adaptively accommodate to the environment and where accommodation would eventually be replaced by "congenital variations." The soberer examples of Baldwin, Lloyd Morgan, and Osborn are usually open to the objection that when the characters in question are demonstrated to be hereditary there is no evidence whatever that they had occurred as accommodations before they became hereditary.

Somewhat more impressive is the example of bird song, already discussed at length by Lloyd Morgan (1896) and frequently mentioned by later authors. The characteristic song of some species of birds is learned by imitation. In other species the song is innate, hence presumably genetically determined, and in still others the situation is intermediate. It does seem possible, at least, that in some instances a learned song has become innate through the Baldwin effect. More direct evidence seems to be quite lacking, but that hypothesis has been accepted by Huxley (1942), among others.

Huxley also maintains that the Baldwin effect is usually involved in early stages of biological differentiation, that is, in the origin of races and eventually of species characterized by preferences for different hosts or food plants. Examples of such races are numerous as are also races distinguished in part by other behavioral and ecological preferences (see especially Thorpe, 1930, 1939, 1940, 1945a, 1945b). Thorpe transferred insect larvae to new food plants, which the insects thereafter preferred, thus demonstrating that the preference may be caused by early conditioning. Other experiments, notably one by Harrison (1927), strongly suggested that natural selection of genetic variation was responsible for development of a strain with changed preferences: there was high initial mortality and only slow establishment of a population adapted to a new host plant. In that example individual modification perhaps also occurred, but the adaptation was mainly genetical. In sum the experiments by Thorpe and Harrison certainly show that adaptation to a new host may occur either by individual acquisition or by genetic selection, but they do not conclusively prove that hereditary adaptation *replaced* non-hereditary.

The most extensive experimental work on the subject is that of Gause (1947 and earlier work there cited). Asexually reproducing clones of the ciliate *Euplotes* showed individual accommodation by decrease of size when transferred from 2.5 to 5% salinity, and populations of ex-conjugants subjected to genetical selection under the same conditions showed a similar but more extreme response. Parallel results were obtained in adaptations of *Paramecium* to different temperatures. The results confirmed the possibility of the Baldwin effect, but I cannot agree that they demonstrated its occurrence. In fact the Baldwin effect did not occur in these

---

[6]There are problems here as to whether these are instincts or reflexes, if there is a proper distinction between the two, and also as to whether the "instincts" of Baldwin and Lloyd Morgan were inherited as such and not at all learned.

experiments: they show, again, that similar adaptation may be produced either by non-hereditary modification or by genetical selection, but they do not show the latter replacing the former.

Gause's experiments brought out other interesting points. In a medium of 1% salinity *Euplotes* was viable but showed no individual modifications. In the same medium genetical selection among ex-conjugants resulted in increase in size. In 7% salinity pure, non-conjugating clones failed to accommodate and eventually died out, but selection produced strains genetically adapted to high salinity. In these experiments adaptation occurred, but only by genetical mechanisms. The Baldwin effect was definitely ruled out as even a possibility. The results bear on Hovasse's opinion that the Baldwin effect is a usual or necessary part of adaptation.

Hovasse (1950) goes so far as to say that "the application of this principle [the Baldwin effect] can lead to a general explanation of adaptation." The path to this extreme is acceptance at face value of all the criticisms of natural selection advanced on one hand by the neo-Lamarckians and on the other by the finalists. Each of those schools claims a general explanation of adaptation, and Hovasse accepts the generality but substitutes the Baldwin effect as mechanism. That is certainly going much too far. It seems to me to require no argument now that most of the neo-Lamarckian examples supposed to show inheritance of acquired characteristics are fully explained by ordinary natural selection without invoking the Baldwin effect or any other additional principle. Yet in a few cases, the prize exhibits of the neo-Lamarckians, the Baldwin effect does provide a plausible explanation not alternative to that of natural selection but showing how natural selection can have produced the observed result.

An example of this sort is the classic one of callosities still sometimes claimed as conclusive evidence for inheritance of acquired characters (e.g., Wood Jones, 1943). Many vertebrates form calluses where the skin is habitually rubbed. The calluses are protective and may therefore be considered adaptive.

They may be caused entirely by individual modification and are usually intensified, at least, by such accommodation. It has, however, long been observed that in some instances the calluses begin to appear in the embryo, for example, plantar calluses in man (Darwin, 1871), sternal calluses in the rhea (Cuénot, 1951), or elbow calluses in the wart hog (Leche, 1902). The Baldwin effect could explain this phenomenon. Such examples are perhaps most suggestive of a real role for the Baldwin effect in evolution, but they do seem too trivial to establish that role as universal or even particularly important.

Among the adaptive phenomena similarly considered crucial evidence by the finalists what Cuénot (1941) called "coaptations" are perhaps the most striking. In "coaptations d'accrochage" two parts of an organism arise separately in the embryo and subsequently fit together and have a single function. An example is the femoral groove of the mantis, into which the tibia is folded. (Numerous other examples are given by Cuénot, 1941;. Corset, 1931; Hovasse, 1951.) To Cuénot coaptations were irrefutable proof of finalism. Hovasse agrees that they are inexplicable by mutation and natural selection and maintains that only "parallel selection," that is, the Baldwin effect, can explain them without recourse to finalism. The Baldwin effect may, indeed, be involved, but few will agree that it is the only possible non-finalistic explanation and there is a glaring weakness in the argument. Again there is no good evidence that the coaptations (to the extent that they are hereditary; they are not invariably or wholly so) did really begin as accommodations.

Hovasse goes still further. He attempts to meet all criticisms of natural selection by substitution of the Baldwin effect and so ends by making the Baldwin effect virtually all-powerful in adaptation. To give one more example, he cites the butterfly *Kallima*, which so closely mimics a leaf, and accepts the criticism that natural selection cannot have developed that mimicry because its advantage is all-or-none, not favored by selection until it is already established. Hovasse's own interpretation is that the ancestors of *Kallima* were changeable in color and pattern, that they actively copied leaves, and that this variable accommodation was finally fixed genetically by the Baldwin effect. This and some similar arguments seem to me so wildly improbable or, at best, so completely lacking in evidence that they merely weaken the whole case for the over-all importance of the Baldwin effect.

STATUS OF THE BALDWIN EFFECT IN EVOLUTIONARY THEORY

The Baldwin effect is fully plausible under current theories of evolution. Yet a review of supposed examples and of pertinent experiments reveals no instance in which it indubitably occurred, no observations explicable only in this way, and few that seem better explained in this way than in some other. It probably has occurred, but there is singularly little concrete ground for the view that it is a frequent and important element in adaptation.

From 1896 up to now, everyone who has discussed it at any length has taken the position that the Baldwin effect is something distinct from natural selection acting on genetical variation and that its real importance is in meeting or explaining away the criticisms leveled at natural selection by, especially, the neo-Lamarckians, the Michurinists, and the finalists. The Baldwin effect is both possible and probable, but assignment to it of that role in evolutionary theory seems to me fallacious.

As an alternative to neo-Lamarckism or Michurinism, the Baldwin effect supposes that accommodation (adaptive somation) is paralleled by genetic changes with similar results. Actually this is no alternative at all and still leaves the basic decision to be made. If the Baldwin effect occurs, either there is or is not a causal connection between an individual accommodation and subsequent genetic change in a population. If there is no such connection, then the truly genetic change must occur wholly by mutation, reproduction, and natural selection, and the accommodation may be irrelevant. If there is a causal connection, the neo-Lamarckian argument is as much supported as supplanted. Indeed the claim (as by Hovasse) that the Baldwin effect is usual in adaptive evolution could be taken as an argument in favor of neo-Lamarckism: frequent coincidence of somation and mutation might suggest that one causes the other. Nor is the Baldwin effect an adequate answer to the arguments of the finalists, who can as readily see directive purpose in somation as in mutation.

The synthetic theory rests on grounds that have essentially nothing to do with the Baldwin effect. Occurrence of the Baldwin effect is nevertheless consistent with that theory and (if, indeed, it does occur) is an interesting but, I would judge, relatively minor outcome of the theory. It is simply one way in which natural selection may sometimes affect populations, and clearly it is not a factor either contradictory or additional to natural selection.

Of all those who have discussed the Baldwin effect, Schmalhausen seems to me to have most nearly placed it in true perspective within modern theories of evolution. Its place, not precisely in Schmalhausen's terms, may be summarized as follows. Genetical systems do not directly and rigidly determine the characteristics of organisms but set up reaction ranges within which those characteristics develop. An "acquired character" or specifically an adaptive modification (that is, an accommodation) necessarily occurs within a genetically determined reaction range. The range may be relatively broad or extremely narrow. In any case an accommodation has genetical limits and develops only in the framework of the genetical system, but in a labile reaction range the particular form taken by a developing organism depends also on interaction with the environment. The genetical system evolves and the reaction range correspondingly changes. The range may come to cover different possibilities or it may become broader or narrower. If it becomes narrower, the possibilities for individual modification of characteristics become fewer. An accommodation that in a broader range occurs only as a specific response to a particular interaction with the environment may as the range narrows become the only developmental possibility. Then the Baldwin effect may occur: a response formerly dependent on a combination of genetical and environmental variables may become relatively or even absolutely invariable. It is not putting the matter in the right terms to say, as has usually been done, that this contrast is between "acquired" and "inherited" characters.

The ability to "acquire" a character has, in itself, a genetical basis. Selection acts (with some exceptions) on the phenotype, so that it is valid to say that selection is actually not on genetical characters but on the ability to acquire characters. This point is emphasized by an example recently discussed by Waddington (1952). The phenomenon involves, again, a broader principle of which the Baldwin effect may be considered a special case. The Baldwin effect would ensue when selection for the ability to acquire an adaptive character so narrowed the developmental range that the character would usually or invariably appear. There is, further, no evident reason why such selection might not act on genetical variation tending to push back appearance of the character into earlier developmental stages. Another aspect of the matter is that the genetical system producing such an effect may well differ in different instances. The absence, on both sides, of one-to-one correspondence of phenotype and genotype is well known. There is therefore wide possible scope for the Baldwin effect or especially for the phenomena of specialization and adaptation of which that effect is one aspect. All this does not seem to me to support the view of 'Espinasse (1952), who in discussion of Waddington's results concludes from the absence of such correspondence that the characters and adaptive changes of populations cannot usefully be interpreted in genetical terms. It only enriches and makes more widely explanatory such interpretation.

There is, finally, as Schmalhausen also pointed out, a certain balance between lability and stability of developmental ranges and norms in evolution. Wide ranges, with labile development, permit individual adjustment to the immediate environment and to short-term vicissitudes. Narrower ranges promote more highly specific and long-continued adaptation, usually advantageous under relatively constant conditions. Narrowing of the reaction range thus exchanges short-term and more plastic for long-term and more rigid adaptation. It is one of the aspects of specialization, and an important special case in evolutionary stabilization. Such a sequence need not precisely correspond with the Baldwin effect, but it includes that effect among the possibilities.

Seen in a modern context, the Baldwin effect helps to focus attention on a host of problems, especially in developmental (or physiological) genetics, well worthy of further study. It does not, however, seem to require any modification of the opinion that the *directive force* in adaptation, in the Baldwin effect or in any other particular way, is natural selection.

LITERATURE CITED

Baldwin, J. M. "A New Factor In Evolution." *Amer. Nat.* **30** (1896): 441–451, 536–553.

Baldwin, J. M. *Development and Evolution.* New York: Macmillan, 1902.

Boring, E. G. "The Influences of Evolutionary Theory Upon American Psychological Thought." In *Evolutionary Thought in America*, edited by S. Person, 268–298. New Haven, CT: Yale University Press, 1950.

Corset, J. "Les coaptations chez les insectes." *Bull. Biol. suppl.* **13** (1931): 1–337.

Cuénot, L. *Invention et finalite en biologie.* Paris: Flammarion, 1941.

Cuénot, L. *L'evolution biologique.* Paris: Masson, 1951.

Darwin, C. *The Descent of Man.* London: Murray, 1871.

Delage, Y. and M. Goldsmith. *The Theories of Evolution.* New York: Huebsch, 1912.

'Espinasse, P. G. "Selection of the Genetic Basis for an Acquired Character." *Nature* **170** (1952): 71.

Gause, G. F. "Problems of Evolution." *Trans. Connecticut Acad. Arts, Sci.* **73** (1947): 17–68.

Goldsmith, R. *Physiological Genetics.* New York: McGraw-Hill, 1938.

Harrison, J. W. H. "Experiments on the Egg-Laying Habits of the Sawfly *Pontania salias* Chr. ..." *Proc. Roy. Soc. London (B)* **101** (1927): 115–126.

Herbert, S. *The First Principles of Evolution,* 2nd ed. London: Black, 1919.

Hovasse, R. *De l'adaptation á l'évolution par la sélection.* Paris: Hermann, 1943.

Hovasse, R. *Adaptation et evolution.* Paris: Hermann, 1950.

Huxley, J. S. *Evolution The Modern Synthesis.* London: Allen and Unwin, 1942; New York: Harper, 1942.

Kirpichnikov, V. S. "The Problem of Non-Hereditary Adaptive Modifications." *Jour. Genetics* **48** (1947): 164–175.

Langfeld, H. S. "James Mark Baldwin." *Dict. Amer. Biog.* **21** (1944): 49–50.

Leche, W. "Ein Fall von Vererbung erworbener Eigenschaften." *Biol. Zentralbl.* **22** (1902): 79–92.

Lull, R. S. *Organic Evolution.* New York: Macmillan, 1917.

Lutz, B. "Ontogenetic Evolution in Frogs." *Evolution* **2** (1948): 29–39.

Mayr, E. "Speciation in Birds." *Proc. Xth Int. Ornith. Congress* (1951): 91-131.

Morgan, C. Lloyd. *Habit and Instinct.* London: Arnold, 1896.

Morgan, C. Lloyd. *Animal Behaviour.* London: Arnold, 1900.

Osborn, H. F. "A Mode of Evolution Requiring Neither Natural Selection nor the Inheritance of Acquired Characters." *Trans. New York Acad. Sci.* **15** (1896): 141–142, 148.

Osborn, H. F. "Organic Selection." *Science* (Oct. 15, 1897a): 583– 587.

Osborn, H. F. "The Limits of Organic Selection." *Amer. Nat.* **31** (1897b): 944–951.

Schmalhausen, I. I. *Factors of Evolution.* (Translated by I. Dordick; edited by Th. Dobzhansky.) Philadelphia: Blakiston, 1949.

Thorpe, W. H. "Biological Races in Insects and Allied Groups." *Biol. Rev.* **5** (1930): 177–212.

Thorpe, W. H. "Further Studies on Pre-imaginal Olfactory Conditioning in Insects." *Proc. Roy. Soc. London (B)* **127** (1939): 424–433.

Thorpe, W. H. "Ecology and the Future of Systematics." In *The New Systematics,* edited by J. S. Huxley, 341–364. Oxford, 1940.

Thorpe, W. H. "The Evolutionary Significance of Habitat Selection." *Jour. Animal Ecol.* **14** (1945a): 67–70.

Thorpe, W. H. "Animal Learning and Evolution." *Nature* **156** (1945b): 46.

Waddington, C. H. "Selection of the Genetic Basis for an Acquired Character." *Nature* **169** (1952): 278.

Wood Jones, F. *Habit and Heritage.* London: Kegan Paul, 1943.

Chapter 9:
The Role of Somatic Change in Evolution[1]

This paper originally appeared in *Evolution* **17** (December, 1963): 529–539.
Reprinted by permission. Footnote in title appeared in the original paper.

All theories of biological evolution depend upon at least three sorts of change:
(a) change of genotype, either by mutation or by redistribution of genes; (b) somatic
change under pressure of environment; and (c) changes in environmental conditions.
The problem for the evolutionist is to build a theory combining these types of
change into an ongoing process which, under natural selection, will account for the
phenomena of adaptation and phylogeny.

Certain conventional premises may be selected to govern such theory building:

[1]This essay is a by-product of research conducted at the Veterans Administration Hospital, Palo
Alto, and sponsored by the Palo Alto Medical Research Foundation under the NIMH Grant OM-
324.

A. THE THEORY SHALL NOT DEPEND UPON LAMARCKIAN INHERITANCE. August Weismann's argument for this premise still stands. There is no reason to believe that either somatic change or changes in environment can, in principle, call (by physiological communication) for appropriate genotypic change. Indeed, the little that we know about communication within the multicellular[2] individual indicates that such communication from soma to gene script is likely to be rare and unlikely to be adaptive in effect. However, it is appropriate to attempt to spell out in this essay what this premise implies:

Whenever some characteristic of an organism is measurably modifiable under measurable environmental impact or under measurable impact of internal physiology, it is possible to write an equation in which the value of the characteristic in question is expressed as some function of the value of the impacting circumstance. "Human skin color is some function of exposure to sunlight," "respiration rate is some function of atmospheric pressure," etc. Such equations are constructed to be true for a variety of particular observations, and necessarily contain subsidiary propositions which are stable (i.e., continue to be true) over a wide range of values of impacting circumstance and somatic characteristic. These subsidiary propositions are of different logical type from the original observations in the laboratory and are, in fact, descriptive not of the data but of *our* equations. They are statements about the form of the particular equation and about the values of the parameters mentioned within it.

It would be simple, at this point, to draw the line between genotype and phenotype by saying that the *forms and parameters* of such equations are provided by genes, while the impacts of environment, etc. determine the actual event within this frame. This would amount to saying, e.g., that the *ability* to tan is genotypically determined, while the amount of tanning in a particular case depends upon exposure to sunlight.

In terms of this oversimplified approach to the overlapping roles of genotype and environment, the proposition excluding Lamarckian inheritance would read somewhat as follows: In the attempt to explain evolutionary process, there shall be no assumption that the achievement of a particular value of some variable under particular circumstances will affect, in the gametes produced by that individual, the form or parameters of the functional equation governing the relationship between that variable and its environmental circumstances.

Such a view is oversimplified, and parentheses must be added to deal with more complex and extreme cases. First, it is important to recognize that the organism, considered as a communicational system, may itself operate at multiple levels of logical typing; i.e., that there will be instances in which what were above called "parameters" are subject to change. The individual organism might as a result of "training" change its ability to develop a tan under sunlight. And this type of change is certainly of very great importance in the field of animal behavior, where "learning to learn" can never be ignored.

---

[2]The problems of bacterial genetics are here deliberately excluded.

Second, the oversimplified view must be elaborated to cover *negative* effects. An environmental circumstance may have such impact upon an organism unable to adapt to it, that the individual in question will in fact produce *no* gametes.

Third, it is expectable that some of the parameters in one equation may be subject to change under impact from some environmental or physiologic circumstance other than the circumstance mentioned in that equation.

Be all that as it may, both Weismann's objection to Lamarckian theory and my own attempt to spell the matter out share a certain parsimony: an assumption that the principles which order phenomena shall not themselves be supposed changed by those phenomena which they order. William of Occam's razor might be reformulated: in any explanation, logical types shall not be multiplied beyond necessity.

B. SOMATIC CHANGE IS ABSOLUTELY NECESSARY FOR SURVIVAL. Any change of environment which requires adaptive change in the species will be lethal unless, by somatic change, the organisms (or some of them) are able to weather out a period of unpredictable duration, until either appropriate genotypic change occurs (whether by mutation or by redistribution of genes already available in the population), or because the environment returns to the previous normal. The premise is truistical, regardless of the magnitude of the time span involved.

C. SOMATIC CHANGE IS ALSO NECESSARY TO COPE WITH ANY CHANGES OF GENOTYPE WHICH MIGHT AID THE ORGANISM IN ITS EXTERNAL STRUGGLE WITH THE ENVIRONMENT. The individual organism is a complex organization of interdependent parts. A mutational or other genotypic change in any one of these (however externally valuable in terms of survival) is certain to require change in many others—which changes will probably not be specified or implicit in the single mutational change of the genes. A hypothetical pre-giraffe, which had the luck to carry a mutant gene "long neck," would have to adjust to this change by complex modifications of the heart and circulatory system. These collateral adjustments would have to be achieved at the somatic level. Only those pre-giraffes which are (genotypically) capable of these somatic modifications would survive.

D. IN THIS ESSAY, IT IS ASSUMED THAT *THE CORPUS OF GENOTYPIC MESSAGES IS PREPONDERANTLY DIGITAL* IN NATURE. In contrast, the soma is seen as a working system in which the genotypic recipes are tried out. Should it transpire that the genotypic corpus is also in some degree analogic—a working model of the soma—premise C (above) would be negated to that degree. It would then be conceivable that the mutant gene "long neck" might modify the message of those genes which affect the development of the heart. It is, of course, known that genes may have pleiotropic effect, but these phenomena are relevant in the present connection only if it can be shown, e.g., that the effect of gene A upon the phenotype and its effect upon the phenotypic expression of gene B are mutually appropriate in the overall integration and adaptation of the organism.

These considerations lead to a classifying of both genotypic and environmental changes in terms of the *price* which they exact of the flexibility of the somatic system. A lethal change in either environment or genotype is simply one which demands somatic modifications which the organism cannot achieve.

But the somatic price of a given change must depend, not absolutely upon the change in question, but upon the range of somatic flexibility available to the organism at the given time. This range, in turn, will depend upon how much of the organism's somatic flexibility is already being used up in adjusting to other mutations or environmental changes. We face an *economics* of flexibility which, like any other economics, will become determinative for the course of evolution if and only if the organism is operating close to the limits set by this economics.

However, this economics of somatic flexibility will differ in one important respect from the more familiar economics of money or available energy. In these latter, each new expenditure can simply be *added* to the preceding expenditures and the economics becomes coercive when the additive total approaches the limit of the budget. In contrast, the combined effect of multiple changes, each of which exacts a price in the soma, will be *multiplicative*. This point may be stated as follows: Let S be the finite set of all possible living states of the organism. Within S, let s_1 be the smaller set of all states compatible with a given mutation (m_1), and let s_2 be the set of states compatible with a second mutation (m_2). It follows that the two mutations in combination will limit the organism to the logical product of s_1 and s_2, i.e., to that usually smaller subset of states which is composed only of members common to both s_1 and s_2. In this way each successive mutation (or other genotypic change) will fractionate the possibilities for somatic adjustment of the organism. And, should the one mutation require some somatic change, the exact opposite of a change required by the other, the possibilities for somatic adjustment may immediately be reduced to zero.

The same argument must surely apply to multiple environmental changes which demand somatic adjustments; and this will be true even of those changes in environment which might seem to benefit the organism. An improvement in diet, for example, will exclude from the organism's range of somatic adjustments those patterns of growth which we would call "stunted" and which might be required to meet some other exigency of the environment.

From these considerations it follows that if evolution proceeded in accordance with conventional theory, its process would be blocked. The finite nature of somatic change indicates that no ongoing process of evolution can result only from successive externally adaptive genotypic changes since these must, in combination, become lethal, demanding combinations of internal somatic adjustments of which the soma is incapable.

We turn therefore to a consideration of other classes of genotypic change. What is required to give a balanced theory of evolution is the occurrence of genotypic changes which shall *increase* the available range of somatic flexibility. When the internal organization of the organisms of a species has been limited by environmental or mutational pressure to some narrow subset of the total range of living states, further evolutionary progress will require some sort of genotypic change which will compensate for this limitation.

We note first that while the results of genotypic change are irreversible within the life of the individual organism, the opposite is usually true of changes which are achieved at the somatic level. When the latter are produced in response to special environmental conditions, a return of the environment to the previous norm is usually followed by a diminution or loss of the characteristic. (We may reasonably expect that the same would be true of those somatic adjustments which must accompany an externally adaptive mutation but, of course, it is impossible in this case to remove from the individual the impact of the mutational change.)

A further point regarding these reversible somatic changes is of special interest. Among higher organisms it is not unusual to find that there is what we may call a "defense in depth" against environmental demands. If a man is moved from sea level to 10,000 feet, he may begin to pant and his heart may race. But these first changes are swiftly reversible: if he descends the same day, they will disappear immediately. If, however, he remains at the high altitude, a second line of defense appears. He will become slowly acclimated as a result of complex physiological changes. His heart will cease to race, and he will no longer pant unless he undertakes some special exertion. If now he returns to sea level, the characteristics of the second line of defense will disappear rather slowly and he may even experience some discomfort.

From the point of view of an economics of somatic flexibility, the first effect of high altitude is to reduce the organism to a limited set of states (s_1) characterized by the racing of the heart and the panting. The man can still survive, but only as a comparatively inflexible creature. The later acclimation has precisely this value: it corrects for the loss of flexibility. After the man is acclimated he can use his panting mechanisms to adjust to *other* emergencies which might otherwise be lethal.

A similar "defense in depth" is clearly recognizable in the field of behavior. When we encounter a new problem for the first time, we deal with it either by trial and error or possibly by insight. Later, and more or less gradually, we form the "habit" of acting in the way which earlier experience rewarded. To continue to use insight or trial and error upon this class of problem would be wasteful. These mechanisms can now be saved for *other* problems (Bateson, 1960).

Both in acclimation and in habit formation the economy of flexibility is achieved by substituting a deeper and more enduring change for a more superficial and more reversible one. In the terms used above in discussing the anti-Lamarckian premise, a change has occurred in the parameters of the functional equation linking rate of respiration to external atmospheric pressure. Here it seems that the organism is behaving as we may expect any ultrastable system to behave. Ashby (1945, 1952) has shown that it is a general, formal characteristic of such systems that those circuits controlling the more rapidly fluctuating variables act as balancing mechanisms to protect the ongoing constancy of those variables in which change is normally slow and of small amplitude; and that any interference which fixes the values of the changeful variables must have a disturbing effect upon the constancy of the normally steady components of the system. For the man who must constantly pant at high altitudes, the respiration rate can no longer be used as a changeable quantity in the maintaining of physiological balance. Conversely, if the respiration rate is to become available again as a rapidly fluctuating variable, some change must occur among the more stable components of the system. Such a change will, in the nature of the case, be achieved comparatively slowly and be comparatively irreversible.

Even acclimation and habit formation are, however, still reversible within the life of the individual, and this very reversibility indicates a lack of communicational economy in these adaptive mechanisms. Reversibility implies that the changed value of some variable is achieved by means of homeostatic, error-activated circuits. There must be a means of detecting an undesirable or threatening change in some variable, and there must be a train of cause and effect whereby corrective action is initiated. Moreover, this entire circuit must, in some degree, be available for this purpose for the entire time during which the reversible change is maintained—a considerable using up of available message pathways.

The matter of communicational economics becomes still more serious when we note that the homeostatic circuits of an organism are not separate but complexly interlocked, e.g., hormonal messengers which play a part in the homeostatic control of organ A will also affect the states of organs B, C, and D. Any special ongoing loading of the circuit controlling A will therefore diminish the organism's freedom to control B, C, and D.

In contrast, the changes brought about by mutation or other genotypic change are presumably of a totally different nature. Every cell contains a copy of the new genotypic corpus and therefore will (when appropriate) behave in the changed manner, without any change in the messages which it receives from surrounding tissues or organs. If the hypothetical pre-giraffes carrying the mutant gene "long neck" could also get the gene "big heart," their hearts would be enlarged without the necessity of using the homeostatic pathways of the body to achieve and maintain this enlargement. Such a mutation will have survival value not because it enables the pre-giraffe to supply its elevated head with sufficient blood, since this was already achieved by somatic change—but because it increases the overall flexibility of the organism, enabling it to survive *other* demands which may be placed upon it either by environmental or genotypic change.

It appears, then, that the process of biological evolution could be continuous if there were a class of mutations or other genotypic changes which would simulate Lamarckian inheritance. The function of these changes would be to achieve by genotypic fiat those characteristics which the organism at the given time is already achieving by the uneconomical method of somatic change.

Such a hypothesis, I believe, conflicts in no way with conventional theories of genetics and natural selection. It does, however, somewhat alter the current conventional picture of evolution as a whole, though related ideas were put forward over sixty years ago. Baldwin (1897) suggested that we consider not only the operation of the external environment in natural selection but also what he called "organic selection" in which the fate of a given variation would depend upon its physiologic viability. In the same article, Baldwin attributes to Lloyd Morgan the suggestion that there might exist "coincident variations" which would simulate Lamarckian inheritance (the so-called "Baldwin effect").

According to such a hypothesis, genotypic change in an organism becomes comparable to legislative change in a society. The wise legislator will only rarely initiate a new rule of behavior; more usually he will confine himself to affirming in law that which has already become the custom of the people. An innovative rule can be introduced only at the price of activating and perhaps overloading a large number of homeostatic circuits in the society.

It is interesting to ask how a hypothetical process of evolution would work *if* Lamarckian inheritance were the rule, i.e., if characteristics achieved by somatic homeostasis were inherited. The answer is simple: *it would not work*, for the following reasons:

1. The question turns upon the concept of economy in the use of homeostatic circuits, and it would be the reverse of economical to fix by genotypic change *all* the variables which accompany a given desirable and homeostatically achieved characteristic. Every such characteristic is achieved by ancillary homeostatic changes all around the circuits, and it is most undesirable that these ancillary changes should be fixed by inheritance, as would logically happen according to any theory involving an indiscriminate Lamarckian inheritance. Those who would defend a Lamarckian theory must be prepared to suggest how in the genotype an appropriate selection can be achieved. Without such a selection, the inheritance of acquired characteristics would merely increase the proportion of non-viable genotypic changes.

2. Lamarckian inheritance would disturb the relative timing of the processes upon which evolution must—according to the present hypothesis—depend. It is essential that there be a time lag between the uneconomical but reversible somatic achievement of a given characteristic and the economical but more enduring alterations of the genotype. If we look upon every soma as a working model which can be modified in various ways in the workshop, it is clear that sufficient but not infinite time must be given for these workshop trials before the results of these trials are incorporated into the final blueprint for mass production. This delay is provided by the indirection of stochastic process. It would be unduly shortened by Lamarckian inheritance.

The principle involved here is general and by no means trivial. It obtains in all homeostatic systems in which a given effect can be brought about by means of a homeostatic circuit, which circuit can, in turn, be modified in its characteristics by some higher system of control. In all such systems (ranging from the house thermostat to systems of government and administration) it is important that the higher system of control *lag behind* the event sequences in the peripheral homeostatic circuit.

In evolution two control systems are present: the homeostases of the body which deal with tolerable internal stress, and the action of natural selection upon the (genetically) non-viable members of the population. From an engineering point of view, the problem is to *limit* communication from the lower, reversible somatic system to the higher irreversible genotypic system.

Another aspect of the proposed hypothesis about which we can only speculate is the probable relative frequency of the two classes of genotypic change: those which initiate something new and those which affirm some homeostatically achieved characteristic. In the Metazoa and multicellular plants, we face complex networks of multiple interlocking homeostatic circuits, and any given mutation or gene recombination which initiates change will probably require very various and multiple somatic characteristics to be achieved by homeostasis. The hypothetical pre-giraffe with the mutant gene "long neck" will need to modify not only its heart and circulatory system but also perhaps its semicircular canals, its intervertebral discs, its postural reflexes, the ratio of length and thickness of many muscles, its evasive tactics *vis-à-vis* predators, etc. This suggests that in such complex organisms, the merely affirmative genotypic changes must far outnumber those which initiate change, if the species is to avoid that *cul de sac* in which the flexibility of the soma approaches zero.

Conversely, this picture suggests that most organisms, at any given time, are probably in such a state that there are multiple possibilities for affirmative genotypic change. If, as seem probable, both mutation and gene redistribution are in some sense random phenomena, at least the chances are considerable that one or other of these multiple possibilities will be met.

Finally, it is appropriate to discuss what evidence is available or might be sought to support or disprove such a hypothesis. It is clear at the outset that such a testing will be difficult. The affirmative mutations upon which the hypothesis depends will usually be *invisible*. From among the many members of a population which are achieving a given adjustment to environmental circumstances by somatic change, it will not be possible immediately to pick out those few in which the same adjustment is provided by the genotypic method. In such a case, the genotypically changed individuals will have to be identified by breeding and raising the offspring under more normal conditions.

A still greater difficulty arises in cases where we would investigate those homeostatically acquired characteristics which are achieved in response to some innovative genotypic change. It will often be impossible, by mere inspection of the organism, to tell which of its characteristics are the primary results of genotypic change and which are secondary somatic adjustments to these. In the imaginary case of the pre-giraffe with a somewhat elongated neck and an enlarged heart, it may be easy to *guess* that the modification of the neck is genotypic while that of the heart is somatic. But all such guesses will depend upon the very imperfect present knowledge of what an organism can achieve in way of somatic adjustment.

It is a major tragedy that the Lamarckian controversy has deflected the attention of geneticists away from the phenomenon of somatic adaptability. After all, the mechanisms, thresholds, and maxima of individual phenotypic change under stress must surely be genotypically determined.

Another difficulty, of rather similar nature, arises at the population level, where we encounter another "economics" of potential change, theoretically distinguishable from that which operates within the individual. The population of a wild species is today conventionally regarded as genotypically heterogeneous in spite of the high degree of superficial resemblance between the individual phenotypes. Such a population expectably functions as a storehouse of genotypic possibilities. The economic aspect of this storehouse of possibilities has, for example, been stressed by Simmonds (1962). He points out that farmers and breeders who demand 100% phenotypic uniformity in a highly select crop are in fact throwing away most of the multiple genetic possibilities accumulated through hundreds of generations in the wild population. From this Simmonds argues that there is urgent need for institutions which shall "conserve" this storehouse of variability by maintaining unselected populations.

Lerner (1954) has argued that self-corrective or buffering mechanisms operate to hold constant the composition of these mixtures of wild genotypes and to resist the effects of artificial selection. There is therefore at least a presumption that this economics of variability within the population will turn out to be of the multiplicative kind.

Now, the difficulty of discriminating between a characteristic achieved by somatic homeostasis and the same characteristic achieved (more economically) by a genotypic shortcut is clearly going to be compounded when we come to consider populations instead of physiologic individuals. All actual experimentation in the field will inevitably work with populations, and, in this work, it will be necessary to discriminate the effects of that economics of *flexibility* which operates inside the individuals from the effects of the economics of *variability* which operates at the population level. These two orders of economics may be easy to separate in theory, but to separate them in experimentation will surely be difficult.

Be all that as it may, let us consider what evidential support may be available for some of the propositions which are crucial to the hypothesis:

1. *That the phenomena of somatic adjustment are appropriately described in terms of an economics of flexibility.* In general, we believe that the presence of stress A may reduce an organism's ability to respond to stress B and, guided by this opinion, we commonly protect the sick from the weather. Those who have adjusted to the office life may have difficulty in climbing mountains, and trained mountain climbers may have difficulty when confined to offices; the stresses of retirement from business may be lethal; and so on. But scientific knowledge of these matters, in man or other organisms, is very slight.

2. *That this economics of flexibility has the logical structure described above—each successive demand upon flexibility fractionating the set of available possibilities.*

The proposition is expectable, but so far as I know there is no evidence for it. It is, however, worthwhile to examine the criteria which determine whether a given "economic" system is more appropriately described in additive or multiplicative terms. There would seem to be two such criteria:

a. A system will be additive insofar as the units of its currency are mutually interchangeable and, therefore, cannot meaningfully be classified into sets such as were used earlier in this paper to show that the economics of flexibility must surely be multiplicative. Calories in the economics of energy are completely interchangeable and unclassifiable, as are dollars in the individual budget. Both these systems are therefore additive. The permutations and combinations of variables which define the states of an organism are classifiable and—to this extent—non-interchangeable. The system is therefore multiplicative. Its mathematics will resemble that of information theory or negative entropy rather than that of money or energy conservation.

b. A system will be additive insofar as the units of its currency are mutually independent. Here there would seem to be a difference between the economic system of the individual, whose budgetary problems are additive (or subtractive) and those of society at large, where the overall distribution or flow of wealth is governed by complex (and perhaps imperfect) homeostatic systems. Is there, perhaps, an economics of economic flexibility (a meta-economics) which is multiplicative and so resembles the economics of physiological flexibility discussed above? Notice, however, that the units of this wider economics will be not dollars but patterns of distribution of wealth. Similarly, Lerner's "genetic homeostasis," insofar as it is truly homeostatic, will have multiplicative character.

The matter is, however, not simple and we cannot expect that every system will be either totally multiplicative or totally additive. There will be intermediate cases which combine the two characteristics. Specifically, where several *independent* alternative homeostatic circuits control a single variable, it is clear that the system may show additive characteristics—and even that it may pay to incorporate such alternative pathways in the system provided they can be effectively insulated from each other. Such systems of multiple alternative controls may give survival advantage insofar as the mathematics of addition and subtraction will pay better than the mathematics of logical fractionation.

3. *That innovative genotypic change commonly makes demands upon the adjustive ability of the soma.* This proposition is orthodoxly believed by biologists but cannot in the nature of the case be verified by direct evidence.

4. *That successive genotypic innovations make multiplicative demands upon the soma.* This proposition (which involves *both* the notion of multiplicative economics of flexibility and the notion that each innovative genotypic change has its somatic price) has several interesting and perhaps verifiable implications.

a. We may expect that organisms in which numerous recent genotypic changes have accumulated (e.g., as a result of selection, or planned breeding) will be delicate, i.e., will need to be protected from environmental stress. This sensitivity to stress is to be expected in new breeds of domesticated animals and plants and experimentally produced organisms carrying either several mutant genes or unusual (i.e., recently achieved) genotypic combinations.

b. We may expect that for such organisms further genotypic innovation (of any kind other than the affirmative changes discussed above) will be progressively deleterious.

c. Such new and special breeds should become more resistant both to environmental stress and to genotypic change, as selection works upon successive generations to favor those individuals in which "genetic assimilation of acquired characteristics" (Waddington, 1953, 1957) is achieved (Proposition 5).

5. *That environmentally induced acquired characteristics may, under appropriate conditions of selection, be replaced by similar characteristics which are genetically determined.* This phenomena has been demonstrated by Waddington (1953, 1957) for the *bithorax* phenotypes of *Drosophila*. He calls it the "genetic assimilation of acquired characteristics." Similar phenomena have also probably occurred in various experiments when the experimenters set out to prove the inheritance of acquired characteristics but did not achieve this proof through failure to control the conditions of selection. We have, however, no evidence at all as to the frequency of this phenomenon of genetic assimilation. It is worth noting, however, that, according to the arguments of this essay, it may be impossible, in principle, to exclude the factor of selection from experiments which would test "the inheritance of acquired characteristics." It is precisely my thesis that the *simulation* of Lamarckian inheritance will have survival value under circumstance of *undefined* or multiple stress.

6. *That it is, in general, more economical of flexibility to achieve a given characteristic by genotypic than by somatic change.* Here the Waddington experiments do not throw any light, because it was the experimenter who did the selecting. To test this proposition, we need experiments in which the population of organisms is placed under double stress: (a) that stress which will induce the characteristic in which we are interested, and (b) a second stress which will selectively decimate the population, favoring, we hope, the survival of those individuals whose flexibility is more able to meet this second stress after adjusting to the first. According to the hypothesis, such a system should favor those individuals which achieve their adjustment to the first stress by genotypic process.

7. Finally, it is interesting to consider a corollary which is the converse of the thesis of this essay. It has been argued here that simulated Lamarckian inheritance will have survival value when the population must adjust to a stress

which remains constant over successive generations. This case is in fact the one which has been examined by those who would demonstrate an inheritance of acquired characteristics. A converse problem is presented by those cases in which a population faces a stress which changes its intensity unpredictably and rather often—perhaps every two or three generations. Such situations are perhaps very rare in nature, but could be produced in the laboratory.

Under such variable circumstances, it might pay the organisms in survival terms to achieve the *converse* of the genetic assimilation of acquired characteristics. That is, they might profitably hand over to somatic homeostatic mechanisms the control of some characteristic which had previously been more rigidly controlled by the genotype.

It is evident, however, that such experimentation would be very difficult. Merely to establish the genetic assimilation of such characteristics as *bithorax* requires selection on an astronomical scale, the final population in which the genetically determined *bithorax* individuals can be found being a selected sample from a potential population of something like 10^{50} or 10^{60} individuals. It is very doubtful whether, after this selective process, there would still exist in the sample enough genetic heterogeneity to undergo a further converse selection favoring those individuals which still achieve their *bithorax* phenotype by somatic means.

Nevertheless, though this converse corollary is possibly not demonstrable in the laboratory, something of the sort seems to operate in the broad picture of evolution. The matter may be presented in dramatic form by considering the dichotomy between "regulators" and "adjusters" (Prosser, 1955). Prosser proposes that where internal physiology contains some variable of the same dimensions as some external environmental variable, it is convenient to classify organisms according to the degree to which they hold the internal variable constant in spite of changes in the external variable. Thus, the homoiothermic animals are classified as "regulators" in regard to temperature while the poikilothermic are "adjusters." The same dichotomy can be applied to aquatic animals according to how they handle internal and external osmotic pressure.

We usually think of regulators as being in some broad evolutionary sense "higher" than adjusters. Let us now consider what this might mean. If there is a broad evolutionary trend in favor of regulators, is this trend consistent with what has been said above about the survival benefits which accrue when control is transferred to genotypic mechanisms?

Clearly, not only the regulators but also the adjusters must rely upon homeostatic mechanisms. If life is to go on, a large number of essential physiological variables must be held within narrow limits. If the internal osmotic pressure, for example, is allowed to change, there must be mechanisms which will defend these essential variables. It follows that the difference between adjusters and regulators is a matter of *where*, in the complex network of physiologic causes and effects, homeostatic process operates.

In the regulators, the homeostatic processes operate at or close to the input and output points of that network which is the individual organism. In the adjusters, the environmental variables are permitted to enter the body and the organism must then cope with their effects, using mechanisms which will involve deeper loops of the total network.

In terms of this analysis, the polarity between adjusters and regulators can be extrapolated another step to include what we may call "extraregulators" which achieve homeostatic controls *outside* the body by changing and controlling the environment—man being the most conspicuous example of this class.

In the earlier part of this essay, it was argued that in adjusting to high altitude there is a benefit to be obtained, in terms of an economics of flexibility, by shifting from, e.g., panting to the more profound and less reversible changes of acclimation; that habit is more economical than trial and error; and that genotypic control may be more economical than acclimation. These are all *centripetal* changes in the location of control.

In the broad picture of evolution, however, it seems that the trend is in the opposite direction: that natural selection, in the long run, favors regulators more than adjusters, and extraregulators more than regulators. This seems to indicate that there is a long time evolutionary advantage to be gained by *centrifugal* shifts in the locus of control.

To speculate about problems so vast is perhaps romantic, but it is worth noting that this contrast between the overall evolutionary trend and the trend in a population faced with constant stress is what we might expect from the converse corollary here being considered. If constant stress favors centripetal shift in the locus of control, and variable stress favors centrifugal shift, then it should follow that in the vast spans of time and change which determine the broad evolutionary picture, centrifugal shift of control will be favored.

SUMMARY

In this essay the author uses a deductive approach. Starting from premises of conventional physiology and evolutionary theory and applying to these the arguments of cybernetics, he shows that there must be an *economics of somatic flexibility* and that this economics must, in the long run, be coercive upon the evolutionary process. External adaptation by mutation or genotypic reshuffling, as ordinarily thought of, will inevitably use up the available somatic flexibility. It follows—if evolution is to be continuous—that there must also be a class of genotypic changes which will confer a bonus of somatic flexibility.

In general, the somatic achievement of change is uneconomical because the process depends upon homeostasis, i.e., upon whole circuits of interdependent variables. It follows that inheritance of acquired characteristics would be lethal to the evolutionary system because it would *fix* the values of these variables all around the circuits. The organism or species would, however, benefit (in survival terms) by genotypic change which would *simulate* Lamarckian inheritance, i.e., would bring about the adaptive component of somatic homeostasis without involving the whole homeostatic circuit. Such a genotypic change (erroneously called the "Baldwin effect") would confer a bonus of somatic flexibility and would therefore have marked survival value.

Finally, it is suggested that a contrary argument can be applied in those cases where a population must acclimate to *variable* stress. Here natural selection should favor an anti-Baldwin effect.

REFERENCES

Ashby, W. R. "The Effect of Controls on Stability." *Nature* 155 (1945): 242.

Ashby, W. R. *Design for a Brain.* New York: Wiley, 1952.

Baldwin, J. M. "Organic Selection." *Science* 5 (1897): 634.

Bateson, G. "Minimal Requirements for a Theory of Schizophrenia." *AMA Arch. Gen. Psychiatry* 2 (1960): 447.

Lerner, I. M. *Genetic Homeostasis.* Edinburgh: Oliver and Boyd, 1954.

Prosser, C. L. "Physiological Variation in Animals." *Biol. Rev.* 30 (1955): 22–262.

Simmonds, N. W. "Variability in Crop Plants, Its Use and Conservation." *Biol. Rev.* 37 (1962): 422–462.

Waddington, C. H. "Genetic Assimilation of an Acquired Character." *Evolution* 7 (1953): 118.

Waddington, C. H. *The Strategy of the Genes.* London: Allen and Unwin, 1957.

New Work

Richard K. Belew

Preface to Chapter 10

A central premise wedding individual cognition to population evolution is that the more an individual "knows" about its environment, the more appropriately it will behave and hence the higher its selective fitness. One common statement of the advantage of maintaining an internal "model" goes back to early cybernetics:

> If the organism carries a 'small-scale model' of external reality and of its own possible actions within its head, it is able to try out various alternatives, conclude which is the best of them, react to future situations before they arise, utilize the knowledge of past events in dealing with the present and future, and in every way to react in a much fuller, safer, and more competent manner to the emergencies which face it (Craik, 1968, p.290).

There are, of course, many aspects of an environment that an organism might perceive, remember, and exploit in appropriate circumstances. Common to almost all of these is the simple ability to *predict*: to hazard a guess about the state of the environment at some point in the future.

By focusing exclusively on this problem, Zhivotovsky, Bergman, and Feldman (Chapter 10) are able to develop a simple model of individual predictive behavior and relate it to the genetics of a population. The environment every organism faces is imagined to be in a sequence of discrete states, with transitions from one state in time to the next controlled by a random Markov process. Markov processes are an extremely well studied class of dynamical systems in which the probabilistic transition to environmental state ε_t^i at time t depends on what state ε_{t-1}^j the system was in the previous time step and the conditional probability $Pr(\varepsilon^j|\varepsilon^i)$ of this particular state transition occurring. For example, we could imagine the weather as being in one of three states, Sunny, Cloudy, or Rainy. The behavior of this "weather process" is then fully described[1] in the matrix below, showing the probability $Pr(\varepsilon^j|\varepsilon^i)$ of making the state transition from row i to column j:

$Pr(\varepsilon^i|\varepsilon^j)$	Sunny	Cloudy	Rainy
Sunny	0.6	0.4	0.0
Cloudy	0.2	0.6	0.2
Rainy	0.1	0.4	0.5

An "mth-order Markov process" is one in which the environmental state transitions depend not only on ε_{t-1} but also $\varepsilon_{t-2}, \varepsilon_{t-3}, \ldots, \varepsilon_{t-m}$; i.e., the stochastic changes are functionally dependent on a longer history of prior states. With this model of an environment fluctuating according to an mth-order Markov process, Zhivotovsky et al. (Chapter 10) then use reasoning like Craik's above to associate the fitness of an individual with its ability to predict the next environmental state. An individual is assumed to live for N time steps and its total fitness is the sum (or the product, in Section 5) over this lifetime.

Just what might an individual "know" that might help it better predict and survive in such an environment? First, it could have a memory of recent prior states of the environment. Since we know the environment is a Markov process sensitive to these prior states, an individual with an accurate memory of them can be expected to do better at predicting future environmental states than one without such memory. Zhivotovsky et al. (Chapter 10) call such memory "*a posteriori* information" since it is knowledge of the environment's states after the fact of their occurrence. They are especially interested in studying the individual's behavior as the depth k of an individual's memory is varied, from no memory up to the full m states on which the environment itself depends.

The second form of knowledge an individual might have is insight into the transition probabilities $Pr(\varepsilon^j|\varepsilon^i)$ themselves. The assumption that individuals possess such "*a priori*" insight into the environmental process certainly makes stronger cognitive demands on the individual than simply remembering its past. But a great deal of psychological research shows that many organisms can in fact induce such probabilities given sufficient experience with a process. Further, machine learning research has recently proposed computational methods (including neural networks) for accomplishing exactly this task. While this chapter does not address the acquisition of the individual's *a priori* knowledge through learning, the connections it

---

[1]To be precise, a vector of initial probabilities is also required.

suggests between this active area of machine learning research and the genetics of populations of such predictors is of major significance.

A subtle but important distinction (which Zhivotovsky et al. address in their conclusion) is that selection is based on the "strategy" an individual's uses as it "...prepares itself to live in environmental state ε_j," rather than on simply a rule for predicting that next state. The authors are able to derive an optimal strategy given limited *a priori* and *a posteriori* information. A somewhat surprising result is that the optimal strategy is "locally deterministic," meaning that all transition probabilities are either zero or unity rather than any intermediate values (e.g., such as those used in the weather example above). Zhivotovsky et al. believe that an individual's behavior might nevertheless appear probabilistic (due to changes in *a posteriori* information as it learns about its environment) but on its face such slavish, deterministic behavior seems counterintuitive.

The mathematical effort required to establish these results is significant, Zhivotovsky et al. do not shy away from any of the details, and the paper will therefore be heavy-going for many of this book's readers. The fact that the model is capable of generating strong hypotheses, counterintuitive or not, is impressive indeed, and the connection that this paper establishes between predictive learning methods and population genetics will amply reward the reader's effort.

REFERENCES

Craik, K. J. "Hypotheses on the Nature of Thought." In *Thinking and Reasoning*, edited by P. C. Wason and P. N. Johnson-Laird, 283–291. 1968.

Lev A. Zhivotovsky, Aviv Bergman, and Marcus W. Feldman

Chapter 10:
A Model of Individual Adaptive Behavior in a Fluctuating Environment

Individual behavioral strategies that use conditional probabilities for future environments and information about past environments are studied. The environments are random and Markovian. The individual uses the available information to prepare for the next environmental state in order to increase its fitness. The fitness depends on the discrepancy between the realized environment and that for which the individual is prepared. Additive and multiplicative combinations of the fitnesses accruing to the individual at each environmental epoch are studied. A semi-optimal strategy is found, which maximizes individual fitness given the depth of information about the environment available to the individual. The effects of randomly varying fitnesses and errors in the individual's perception of the environmental parameters are also analyzed.

1. INTRODUCTION

For many biological organisms the process of adaptation is one of survival and reproduction in an uncertain environment. Two kinds of adaptations are important

Adaptive Individuals in Evolving Populations, Ed. R. K. Belew & M. Mitchell,
SFI Studies in the Sciences of Complexity, Vol. XXVI, Addison-Wesley, 1996   **131**

to distinguish. The first might be viewed as occurring at the population level and envisages an array of different genotypes or phenotypes, each adapted to a characteristic range of environmental conditions. Taken as an ensemble, this array permits the *population* to adapt to changing conditions. This situation is usually modeled in terms of the evolution of the frequencies of the types under natural selection with emphasis on between-generation changes in environmental conditions (Lewontin and Cohen, 1969; Gillespie, 1973; Hartl and Cook, 1973; Karlin and Liberman, 1974; Stephens, 1991; Bergman and Feldman, 1994), or by making use of optimality reasoning where the distinction between generations is often blurred (Cohen, 1966, 1993; Harley, 1981; Harley and Maynard Smith, 1983; Houston and Sumida, 1987; McNamara and Houston, 1987).

A second kind of adaptation may occur when *individuals* exhibit plasticity that permits them to respond to environmental conditions in a manner that enhances their survival. Thus, individuals seek a general strategy that permits them to learn or seek a behavior that increases fitness (Shettleworth, 1984). For this response to be appropriate, the individual should possess some "information" about the future environment, it should have stored information about previous environments (i.e., memory), and it should be able to predict and prepare for pending environments. These three properties contribute to the individual's ability to survive in an environment that changes *within* its lifetime.

Both kinds of adaptations, population level and individual level, may have occurred in the process of evolution. Both may involve selection on *behavioral* differences between the types in a population.

Ecological analyses of evolution in changing environments typically assume that individual behaviors are strategies in a game against the environment (Maynard Smith, 1982). This framework assumes that individuals have no knowledge about their environments and cannot make predictions. In such situations a minimax strategy is a convenient way for an individual to counter the worst possible environmental conditions. In reality, however, the environment does not play a game with individuals, and individuals may have some *a priori* and *a posteriori* information about their environment on a short-term time scale during their lifetime.

Many models of behavior based on individual learning couch the learning process in terms of changes in the probabilities with which a response is chosen due to previously acquired information about stimuli and responses (e.g., Bush and Mosteller, 1955). Here the environment is represented as a set of stimuli; and the conceptualization of environmental changes remains somewhat vague. In this paper we investigate how such adaptive behavior depends on the extent of historical knowledge about the environment, on the amount of prior knowledge about the probabilistic law that governs environmental change, and whether the individual's behavior is intrinsically probabilistic or deterministic for a given set of information. To address these questions, we consider an environment with several possible different conditions, called *states*, whose temporal pattern follows given conditional probabilities. The individual possesses some information about both these probabilities and the history of previous environmental outcomes. The individual chooses which next (unknown) state to prepare for in order to increase its fitness, which depends on the discrepancy between the expected and realized environmental states.

2. THE ENVIRONMENT

Suppose that at each discrete point in time, t, the environment is a random variable $\mathcal{E}_t$ which can take values from the set $\varepsilon_1, \varepsilon_2, \ldots, \varepsilon_n$. The time series $\{\varepsilon_t\}$ constitutes an mth-order Markov process. These random variables are described by the conditional probabilities

$$P\{i|s_m\} = P\{\mathcal{E}_t = \varepsilon_i | \mathcal{E}_{t-1} = \varepsilon_{i_1}, \mathcal{E}_{t-2} = \varepsilon_{i_2}, \ldots, \mathcal{E}_{t-m} = \varepsilon_{i_m}\} \tag{1}$$

where s_m can be abbreviated by the m-vector $(i_1, i_2, \ldots, i_m)$, with each coordinate representing one of the environmental states $1, 2, \ldots, n$. Thus, if $m = 4$, for example, and $s_m = (2, 3, 1, 7)$, the process took the values ε_2 at $t - 1$, ε_3 at $t - 2$, ε_1 at $t - 3$, and ε_7 at $t - 4$. Such an environment will be defined to have *depth* m. In the case $m = 0$, temporal changes in the environment are completely independent and, at any time, the values $\varepsilon_1, \varepsilon_2, \ldots, \varepsilon_n$ occur with probabilities $P_1, P_2, \ldots, P_n$. If $m = 1$, the environment is a standard Markov process in which only the previous state affects the present. In all cases the environment will be assumed to be stationary. That is, for each $i = 1, 2, \ldots, n$, the unconditional probability that the environment is ε_i is

$$P_i = \sum_{s_m} P\{i|s_m\} P(s_m), \tag{2}$$

where $P(s_m)$ is the stationary probability of the sequence of m environmental states, assumed to be independent of where in time the sequence occurs. We shall be concerned with properties of the sequence length, m. In general, s_p is the p-vector in which each coordinate represents one of the environmental states. We write $\sum_{s_p} f(s_p)$ to denote the summation of the values of function f where summation is taken over all n^p possible sequences $(i_1, i_2, \ldots, i_p)$. If $p > q$, we write $s_p \subset s_q$ for the case where the first q coordinates of s_p coincide with the q-vector s_q. For example, if $q = 3$ and $s_q = (\varepsilon_4, \varepsilon_1, \varepsilon_3)$, then for $p = 5$, sequences s_p with $s_p \subset s_q$ are represented by the set of $n^{p-q} = n^2$ vectors of length 5: $(\varepsilon_4, \varepsilon_1, \varepsilon_3, \varepsilon_i, \varepsilon_j)$, where i and j are arbitrary $(i, j = 1, 2, \ldots, n)$. The notation $\sum_{s_p \subset s_q} f(s_p)$ is used for the sum of values of $f(s_p)$ over all n^{p-q} sequences, s_p, whose first q coordinates are identical to s_q.

3. INDIVIDUAL FITNESS

To an individual, environmental changes may be due either to actual temporal effects described by the conditional probabilities (1), or they may appear to occur because the individual migrates across a spatially varying habitat. In any case, we shall assume that the environment experienced by the individual is as described in the previous section by relations (1) with depth m.

We assume that individuals are able to predict the environment and are able to prepare in some way for that predicted environment. Suppose that an individual predicts that the environment at time t will be in state ε_j, and that the environment at time t actually takes state ε_i. In this case, a fitness E_{ij} will be assigned to that individual. It is natural to assume $E_{ii} > E_{ij}$ if $j \neq i$; that is, correct prediction of the environment entails higher fitness than incorrect, although in general we do not use these inequalities.

During an individual's lifetime, N environmental changes occur, of which N_{ij} are such that the individual predicted state ε_j but the actual environmental state was ε_i. Thus

$$N = \sum_{i=1}^{n} \sum_{j=1}^{n} N_{ij}. \tag{3}$$

The *total normalized additive fitness* of such an individual is

$$W = \frac{1}{N} \sum_{i=1}^{n} \sum_{j=1}^{n} N_{ij} E_{ij} \tag{4}$$

$$= \sum_{i=1}^{n} \sum_{j=1}^{n} n_{ij} E_{ij} \tag{5}$$

where $n_{ij} = N_{ij}/N$ is the frequency that ε_j is predicted and ε_i occurs. Later, in Section 6, we introduce a multiplicative fitness formulation.

4. INDIVIDUAL STRATEGIES

Suppose that just prior to the realization of the environment at time t an individual knows the realizations of the k previous environments $t - 1$, $t - 2, \ldots, t - k$. We say that this·individual possesses *a posteriori information* of *depth* k, or has a *memory of depth* k. Let $Q\{j|s_k\}$ be the probability that such an individual prepares itself to live in environmental state ε_j, knowing the previous k realizations of the environment. Of course, $\sum_{j=1}^{n} Q(j|s_k) = 1$ for each s_k. It should be understood that $Q\{j|s_k\}$ is not a rule for predicting the environmental process; rather, as we shall discuss later, these values represent an individual's choices of a strategy based on its previous history in past environments.

Assume that the number of environmental changes, N, that occur during the lifetime of an individual is large enough that model (5) may be regarded in terms of infinitesimal changes as $N \to \infty$. This assumption ensures that the contribution of each environmental change is sufficiently small (i.e., $o\,(1/N)$). The individual lives through the first k steps of its life without using information about the environment (an assumption that could be altered, for example, in the presence of cultural transmission). After this stage, it activates a memory of depth k.

An individual characterized by $Q\{\cdot|s_k\}$ in an environment characterized by the rule $P\{\cdot|s_k\}$ will obtain fitness E_{ij} with probability

$$\pi_{ij}(s_k) = P(s_k)P\{i|s_k\}Q\{j|s_k\}, \tag{6}$$

given s_k and $k \le m$, where $s_m \subset s_k$, because the individual's knowledge of environmental history is independent of the actual realization of the next environment. Hence, for the additive model, the total fitness of such an individual is

$$W(Q) = \sum_{s_k}\sum_{i=1}^{n}\sum_{j=1}^{n} \pi_{ij}(s_k)E_{ij}. \tag{7}$$

5. A SEMI-OPTIMAL STRATEGY

An individual's total fitness depends on both the probability law of the environment, $P\{i|s_m\}$, and that of its preparation, $Q\{j|s_k\}$. We seek a good strategy $Q\{\cdot|s_k\}$, namely, one that gives rise to high individual fitness. Suppose that the individual knows the conditional probability law for the environment given the previous ℓ environments, with $\ell \le m$; i.e., it knows $P\{i|s_\ell\}$. Such an individual will be said to possess *a priori information* of depth ℓ. Since the environment is stationary,

$$P\{i|s_\ell\} = \sum_{s_m \subset s_\ell} \frac{P(s_m)P\{i|s_m\}}{P(s_\ell)} \text{ and}$$
$$P(s_\ell) = \sum_{s_m \subset s_\ell} P(s_m). \tag{8}$$

Given *a priori* information of depth ℓ, and memory of depth k, we define a *semi-optimal strategy*, $\hat{Q}$, as one which maximizes $W(Q)$:

$$\hat{W} = W(\hat{Q}) = \max_{Q} W(Q). \tag{9}$$

The term semi-optimal is used here because there exist "best" strategies that can be determined with complete knowledge of the environmental law and the previous states. (See Note 3 following Result 3.)

Since Eq. (7) is a linear function of Q, its maximum is reached at 0 or 1 for each s_k. In order to obtain the solution, suppose that an individual having *a priori* information of depth ℓ possesses information s_k about the k previous states at times $t - 1, t - 2, \ldots, t - k$ and wants to make the best prediction of the environmental state at time t. Obviously, the individual should choose ε_j if the expected value of E_{ij}, weighted by the probabilities $P\{i|s_k\}$ of the next environments, i.e., $\sum_{i=1}^{n} P\{i|s_k\} E_{ij}$, is greater than each of the expectations for different states ε_{j_0} $(j_0 \neq j)$, namely $\sum_{i=1}^{n} \{P_i|s_k\} E_{ij_0}$. (If there are several environmental states of the same maximal expected fitness value, then any combination of these ε's is equally preferred.) We formulate this in terms of the following:

RESULT 1. Suppose that $k \leq \ell$. Then the semi-optimal strategy for additive fitnesses (5) is

$$
\hat{Q}\{j|s_k\} = \begin{cases} 1 & \text{if } \displaystyle\sum_{i=1}^{n} P\{i|s_k\}(E_{ij} - E_{ij_0}) > 0 \quad \text{for all} \quad j_0 \neq j, \\[4mm] 0 & \text{if } \displaystyle\sum_{i=1}^{n} P\{i|s_k\}(E_{ij} - E_{ij_0}) < 0 \quad \text{for at least one} \quad j_0 \neq j. \end{cases} \tag{10}
$$

If there exists a set J of more than one j for which

$$
\begin{aligned}
&\sum_{i=1}^{n} P\{i|s_k\}(E_{ij} - E_{ij_0}) > 0 \quad \text{for all} \quad j_0 \notin J, \text{ and} \\
&\sum_{i=1}^{n} P\{i|s_k\}(E_{ij_1} - E_{ij_2}) = 0 \quad \text{for} \quad j_1, j_2 \in J,
\end{aligned} \tag{11}
$$

then $\hat{Q}\{j|s_k\}$, $j \in J$, are arbitrary within the simplex $\sum_{j \in J} \hat{Q}\{j|s_k\} = 1$, $\hat{Q}\{j|s_k\} \geq 0$.

With this strategy

$$
\hat{W} = \sum_{s_k} P(s_k) \max_{1 \leq j \leq n} \sum_{i=1}^{n} P\{i|s_k\} E_{ij}. \tag{12}
$$

The same result holds where there is deeper *a posteriori* information, $k > \ell$:

RESULT 2. Suppose that $k > \ell$. Then, for each $s_k \subset s_\ell$, the semi-optimal strategy is

$$\hat{Q}\{j|s_k\} = \begin{cases} 1 & \text{if} \quad \sum_{i=1}^{n} P\{i|s_\ell\}(E_{ij} - E_{ij_0}) > 0 \quad \text{for all} \quad j_0 \neq j, \\[4mm] 0 & \text{if} \quad \sum_{i=1}^{n} P\{i|s_\ell\}(E_{ij} - E_{ij_0}) < 0 \quad \text{for at least one} \quad j_0 \neq j. \end{cases} \tag{13}$$

If there exists a set J of more than one j for which

$$\sum_{i=1}^{n} P\{i|s_\ell\}(E_{ij} - E_{ij_0}) > 0 \quad \text{for all} \quad j_0 \notin J \text{ and}$$
$$\sum_{i=1}^{n} P\{i|s_\ell\}(E_{ij_1} - E_{ij_2}) = 0 \quad \text{for all} \quad j_1, j_2 \in J, \tag{14}$$

then $\hat{Q}\{j|s_k\}, j \in J$, are arbitrary within the simplex $\sum_{j \in J} \hat{Q}\{j|s_k\} = 1, \hat{Q}\{j|s_k\} \geq 0$.

Under this semi-optimal strategy

$$\hat{W} = \sum_{s_\ell} P(s_\ell) \max_{1 \leq j \leq n} \sum_{i=1}^{n} P\{i|s_\ell\}E_{ij}. \tag{15}$$

Appendix A contains the proofs of both Results 1 and 2.

COROLLARY 1. Both results show that, for given depths k of memory and ℓ of *a priori* information, the semi-optimal strategy for model (7) depends only on the conditional probabilities $P\{i|s_k\}$ or $P\{i|s_\ell\}$ (from Eqs. (10) and (13)) and not on the unconditional probabilities of the environment, $P(s_k)$. This is true even though the total fitness actually depends on both conditional and unconditional probabilities, as in Eqs. (12) and (15). This means that in preparing for future environments, estimates of these conditional probabilities are the most valuable information, assuming, of course, that the individual knows the relevant values of E_{ij} in the computation of the semi-optimal strategy.

Results 1 and 2 show that the semi-optimal strategy is importantly affected by both the knowledge about the environmental process with depth ℓ and the depth of an individual's memory, k. Result 1 says that the semi-optimal strategy is restricted by the depth of information, (memory) k, if the depth of environmental knowledge is greater, i.e., if $\ell > k$. In this case, the additional knowledge about $P\{i|s_\ell\}$ has no value. Further, if the depth of information is greater than that of the environmental knowledge, i.e., $k > \ell$, the additional information about the previous environments at times $t - \ell + 1, \ldots, t - k$ also has no value at time t. The effective parameter, therefore, is $\min(k, \ell)$.

COROLLARY 2. If there is a cost for the acquisition of deeper information, k, or for more advanced information about the environment, ℓ, then evolution tends to decrease $|\ell - k|$.

It seems clear that deeper knowledge and information should increase fitness. In fact, we have the following:

RESULT 3. Let k', ℓ', k'' and ℓ''' be such that $\min(k', \ell') > \min(k'', \ell'')$, and suppose that $\hat{W}'$ and $\hat{W}''$ are the fitnesses corresponding to the semi-optimal strategies for these pairs of parameters from Result 1. Then $\hat{W}' \geq \hat{W}''$.

The proof of Result 3 is provided in Appendix B.

Formally we may consider the case $k > m$; i.e., an organism's memory extends into the past further than the environment's. This may appear to be unrealistic. But imagine that an individual came from, or was adapted to, an environment of depth k and now performs in another environment with m much less than k, even zero. This individual will, at least in the beginning, act according to its former model Q corresponding to depth k, even though this is not optimal. As a result, its fitness would, at least temporarily, be lower in the environment of depth m.

If $k > m$, then $P\{i|s_k\} = P\{i|s_m\}$ by definition. It follows from Eq. (7) that $W(Q)$ may be rewritten as

$$W(Q) = \sum_{s_m} \sum_{s_k \subset s_m} \sum_{i=1}^{n} \sum_{j=1}^{n} \pi_{ij}(s_k) E_{ij}$$

$$= \sum_{s_m} \sum_{i} \sum_{j} \pi_{ij}^{+}(s_m) E_{ij}, \tag{16}$$

where

$$\pi_{ij}^{+}(s_m) = P(s_m) P\{i|s_m\} Q^{+}\{j|s_m\}, \tag{17}$$

with

$$Q^{+}\{j|s_m\} = \sum_{\substack{s_k \\ s_k \subset s_m}} \frac{P(s_k)}{P(s_m)} Q\{j|s_k\}. \tag{18}$$

We may also consider cases where $\ell > m$.

Note 1. Usually the inequality $\min(k'\ell') > \min(k'', \ell'')$ implies $\hat{W}' > \hat{W}''$ if $\min(k'', \ell'') < m$. $\hat{W}' = \hat{W}''$ only in the special case described in Appendix A.

Note 2. The fitness $\hat{W}$, determined by the semi-optimal strategy, increases as a function of $\min(k, \ell)$ only if $\min(k, \ell) \leq m$; $\hat{W}$ reaches its greatest possible value at $\min(k, \ell) = m$. Hence, increasing the depth of information (k) or the depth of knowledge (ℓ) beyond the depth of the environmental law (m) has no effect. For this reason we consider only the case $\min(k, \ell) \leq m$.

COROLLARY 3. If $k < \ell$, that is, the depth of information, k, is less than the depth of knowledge, ℓ, the fitness is an increasing function of k until k exceeds ℓ (or m, if $\ell > m$). On the other hand, the fitness remains constant if ℓ increases but k is fixed.

COROLLARY 4. If $k > \ell$, fitness is an increasing function of ℓ until $\ell \geq k$ (or until $\ell \geq m$, if $k > m$) while it does not increase in k for fixed ℓ.

Note 3. Denote by $\hat{Q}_{k\ell}$ the semi-optimal strategy defined for given depths, k, of memory and *a priori* information, ℓ. Since the fitness corresponding to $\hat{Q}_{k\ell}$ reaches its maximum as a function of k and ℓ at $k = \ell = m$, and then remains constant for $k \geq m$, $\ell \geq m$, only the strategy $\hat{Q}_{m,m}$ should be called *optimal*. Within the constraint of limited information, $\hat{Q}_{k\ell}$ is the best strategy. Increasing k and ℓ improves the semi-optimal strategy by increasing the fitness of the individual at the semi-optimal strategy.

6. MULTIPLICATIVE FITNESSES

The total normalized multiplicative fitness is defined for positive E_{ij} by

$$W = \left(\prod_{i=1}^{n} \prod_{j=1}^{n} E_{ij}^{N_{ij}} \right)^{1/N} . \tag{4'}$$

The multiplicative and additive models both express independence of the contributions from future environments to the total fitness. They are related since Eq. (4') is equivalent to

$$\ln W = \sum_{i=1}^{n} \sum_{j=1}^{n} n_{ij} \ln E_{ij}. \tag{19}$$

All of the above results which hold for additive fitnesses are also valid for multiplicative fitnesses just by changing E_{ij} to $\ln E_{ij}$. In spite of this close relationship, in some complex situations, the two models produce qualitatively different results (see Section 7).

For the multiplicative model, the fitness of an individual with strategy $Q\{j|s_k\}$ is obtained using Eq. (6) as

$$W(Q) = \prod_{s_k} \prod_{i=1}^{n} \prod_{j=1}^{n} E_{ij}^{\pi_{ij}(s_k)},$$

so that

$$\ln W(Q) = \sum_{s_k} \sum_{i=1}^{n} \sum_{j=1}^{n} \pi_{ij}(s_k) \ln E_{ij}. \tag{7'}$$

Since Eq. (7') has absolutely the same form as for the additive model (7), the results for the additive case allow us to infer the following:

RESULT 1'. Suppose that $k \leq \ell$. Then, the semi-optimal strategy for multiplicative fitnesses (4') is

$$\hat{Q}\{j|s_k\} = \begin{cases} 1 & \text{if } \sum_{i=1}^{n} P\{i|s_k\} \ln \dfrac{E_{ij}}{E_{ij_0}} > 0 \quad \text{for all} \quad j_0 \neq j, \\[3mm] 0 & \text{if } \sum_{i=1}^{n} P\{i|s_k\} \ln \dfrac{E_{ij}}{E_{ij_0}} < 0 \quad \text{for at least one} \quad j_0 \neq j. \end{cases} \tag{10'}$$

If there exists a set J of more than one j for which

$$\sum_{i=1}^{n} P\{i|s_k\} \ln \frac{E_{ij}}{E_{ij_0}} > 0 \quad \text{for all} \quad j_0 \notin J$$

$$\text{and} \tag{11'}$$

$$\sum_{i=1}^{n} P\{i|s_k\} \ln \frac{E_{ij_1}}{E_{ij_2}} = 0 \quad \text{for all} \quad j_1, j_2 \in J,$$

then $\hat{Q}\{j|s_k\}, j \in J$, are arbitrary within the simplex $\sum_{j \in J} \hat{Q}\{j|s_k\} = 1, \hat{Q}\{j|s_k\} \geq 0$.

Under the semi-optimal strategy,

$$\ln \hat{W} = \sum_{s_k} P(s_k) \max_{1 \leq j \leq n} \sum_{i=1}^{n} P\{i|s_k\} \ln E_{ij}. \tag{12'}$$

RESULT 2'. Suppose that $k > \ell$. Then in the case of multiplicative fitnesses, for all $s_k \subset s_\ell$,

$$
\hat{Q}\{j|s_k\} = \begin{cases} 1 & \text{if} \quad \displaystyle\sum_{i=1}^{n} P\{i|s_\ell\} \ln \frac{E_{ij}}{E_{ij_0}} > 0 \quad \text{for all} \quad j_0 \neq j, \\[4ex] 0 & \text{if} \quad \displaystyle\sum_{i=1}^{n} P\{i|s_\ell\} \ln \frac{E_{ij}}{E_{ij_0}} < 0 \quad \text{for at least one} \quad j_0 \neq j. \end{cases} \tag{13'}
$$

If there exists a set J of more than one j for which

$$
\sum_{i=1}^{n} P\{i|s_\ell\} \ln \frac{E_{ij}}{E_{ij_0}} > 0 \quad \text{for all} \quad j_0 \notin J
$$
$$
\text{and} \tag{14'}
$$
$$
\sum_{i=1}^{n} P\{i|s_\ell\} \ln \frac{E_{ij_1}}{E_{ij_2}} = 0 \quad \text{for all} \quad j_1, j_2 \in J,
$$

then $\hat{Q}\{j|s_k\}, j \in J$, are arbitrary within the simplex $\sum_{j \in J} \hat{Q}\{j|s_k\} = 1, \hat{Q}\{j|s_k\} \geq 0$.

Under the semi-optimal strategy

$$
\ln \hat{W} = \sum_{s_\ell} P(s_\ell) \max_{1 \leq j \leq n} \sum_{i=1}^{n} P\{i|s_\ell\} \ln E_{ij}. \tag{15'}
$$

Result 3 is also valid for multiplicative fitnesses, if $\hat{W}$ takes the form (12') or (15'). All of the corollaries of Results 1–3 also hold with multiplicative fitnesses.

---

7. RANDOMLY VARYING FITNESSES

So far we have assumed that the parameters of the model are deterministic. In reality, however, they may be subject to individual estimation error, or be influenced by intrinsic factors. Thus, in this section we consider fitnesses E_{ij} which vary during an individual's lifetime, the changes occurring when the environment changes. We take E_{ij} to be continuous random variables such that, at each new environmental state, E_{ij} are independent of the previous fitnesses and are distributed with joint density function $f(E_{11}, E_{12}, \ldots, E_{nn})$.

Denote the marginal density function of E_{ij} by $f_{ij}(E_{ij})$ with

$$f_{ij}(E_{ij}) = \underset{\substack{\text{all } E_{k\ell} \\ k\ell \neq ij}}{\int \int \cdots \int} f(E_{11} \cdots E_{ij} \cdots E_{nn}) \, dE_{11} \underset{\text{not } ij}{\cdots} dE_{nn}$$

with mean $\bar{E}_{ij}$ and variance V_{ij} given by

$$\bar{E}_{ij} = \int x f_{ij}(x) dx, \tag{20}$$

and

$$V_{ij} = \int (x - \bar{E}_{ij})^2 f_{ij}(x) dx, \tag{21}$$

respectively. Also, define the logarithmic mean

$$\overline{\ln E}_{ij} = \int \ln x f_{ij}(x) dx. \tag{22}$$

We have:

RESULT 4. Suppose that the random fitnesses E_{ij} are independent of the changing environments that occur during the lifetime of an individual. Then for the infinitesimal case

$$W(Q) = \sum_{s_k} \sum_{i=1}^{n} \sum_{j=1}^{n} \pi_{ij}(s_k) \bar{E}_{ij} \tag{23}$$

for the additive case (7), and

$$W(Q) = \prod_{s_k} \prod_{i=1}^{n} \prod_{j=1}^{n} \hat{E}^{\pi_{ij}(s_k)} \tag{23'}$$

with multiplicative fitnesses (7'), where

$$\hat{E}_{ij} = \exp\left[\overline{\ln E}_{ij}\right]. \tag{24}$$

To prove this result for the multiplicative case, note that the analogy to the fitness (4') in the present case is

$$W = \left[\left(E_{11}^{(1)} \cdots E_{11}^{(N_{11})} \right) \left(E_{12}^{(1)} \cdots E_{12}^{(N_{12})} \right) \cdots \left(E_{nn}^{(1)} \cdots E_{nn}^{(N_{nn})} \right) \right]^{1/N},$$

where E_{ij} is the infinitesimal fitness of the individual at the event when the predicted environmental state is ε_j and its realization is ε_i. Thus

$$W = \prod_{i=1}^{n} \prod_{j=1}^{n} w_{ij}$$

with

$$w_{ij} = \left[E_{ij}^{(1)} \cdots E_{ij}^{(N_{ij})} \right]^{1/N} = \exp \left\{ \frac{1}{N} \sum_{v=1}^{N_{ij}} \ln E_{ij}^{(v)} \right\}$$

$$= \exp \left\{ \frac{N_{ij}}{N} \cdot \frac{1}{N_{ij}} \sum_{v=1}^{N_{ij}} \ln E_{ij}^{(v)} \right\}$$

$$\to \exp \left\{ \pi_{ij} \, \overline{\ln E_{ij}} \right\} = \hat{E}_{ij}^{\pi_{ij}}$$

as $N \to \infty$. This proves the result in the multiplicative case. For the additive model, the proof is similar.

Result 4 confirms that the general conclusions about semi-optimal strategies obtained in Results 1–3 remain valid for randomly distributed fitnesses. The only difference is that individuals know and use $\bar{E}_{ij}$ instead of E_{ij} for random additive fitnesses, and $\overline{\ln E_{ij}}$ instead of $\ln E_{ij}$ for random multiplicative fitnesses. The fact that $\overline{\ln E_{ij}}$ is used instead of $\ln E_{ij}$ has qualitative implications. Since for x close to $\bar{E}_{ij}$ we can write

$$\ln x = \ln \bar{E}_{ij} + (x - \bar{E}_{ij}) \frac{d \ln x}{dx} \bigg|_{x=\bar{E}_{ij}} + \frac{1}{2}(x - \bar{E}_{ij})^2 \frac{d^2 \ln x}{dx^2} \bigg|_{x=\bar{E}_{ij}} + \cdots,$$

we have

$$\overline{\ln E_{ij}} \cong \ln \bar{E}_{ij} - \frac{1}{2} C_{ij}^2, \tag{25}$$

where C_{ij} is the coefficient of variation of E_{ij}:

$$C_{ij} = V_{ij}^{1/2} / \bar{E}_{ij}.$$

Hence, unlike the random additive model, in the random multiplicative model, optimal strategies depend on both the mean and variation (and other statistics) of the fitness distribution.

8. SPECIAL CASE: TWO ENVIRONMENTAL STATES

Consider an environment with two states ε_0 and ε_1 with corresponding fitness values E_{00}, E_{01}, E_{10}, and E_{11} as defined above. Thus, for example, E_{10} is the fitness when the current state of the environment is ε_1, but the state for which the individual is prepared is ε_0. Assume $E_{11} > E_{10}$ and $E_{00} > E_{01}$. Consider first additive fitnesses. From Result 1, a semi-optimal strategy $\hat{Q}\{0|s_k\}$, with $\hat{Q}\{1|s_k\} = 1 - \hat{Q}\{0|s_k\}$, depends on the ratio of $P\{0|s_k\}(E_{00} - E_{01})$ to $P\{1|s_k\}(E_{11} - E_{10})$. Define Z by

$$Z = \frac{E_{11} - E_{10}}{E_{11} + E_{00} - E_{01} - E_{10}}. \tag{26}$$

Then we have the following corollaries to Result 1:

RESULT 5. Suppose $k \leq \ell$. Then the semi-optimal strategy $\hat{Q}$ satisfies

$$\hat{Q}\{0|s_k\} = \begin{cases} 1 & \text{if} \quad P\{0|s_k\} > Z, \\ 0 & \text{if} \quad P\{0|s_k\} < Z, \end{cases} \tag{27}$$

with an arbitrary $\hat{Q}(0 \leq \hat{Q} \leq 1)$ if $P\{0|s_k\} = Z$.

RESULT 6. Suppose $k > \ell$. For every $s_k \subset s_\ell$, the semi-optimal strategy is

$$\hat{Q}\{0|s_k\} = \begin{cases} 1 & \text{if} \quad P\{0|s_\ell\} > Z, \\ 0 & \text{if} \quad P\{0|s_\ell\} < Z, \end{cases} \tag{28}$$

with an arbitrary $\hat{Q}$ $(0 \leq \hat{Q} \leq 1)$ if $P\{0|s_\ell\} = Z$ (for $s_k \subset s_\ell$).

Results 5 and 6 constitute a very simple algorithm for obtaining the semi-optimal strategy. The decision rule is simply to compare the conditional probabilities of the environmental law, $P\{0|s_k\}$, with a single number Z for each s_k. These results also hold for the multiplicative fitness model with Z defined by

$$Z = \frac{\ln(E_{11}/E_{10})}{\ln(E_{11}E_{00}/E_{10}E_{01})}. \tag{26'}$$

The semi-optimal strategies for the multiplicative case are the same as those for the additive model, namely Eqs. (27) and (28).

9. ERRORS IN INFORMATION ABOUT PARAMETERS

We have seen that partial knowledge about the depth of the environment leads to inferior strategies and reduced individual fitness. It seems clear that errors in the knowledge of other parameters of the models should also weaken the strategy and decrease individual fitness. We now analyze this effect quantitatively in the special case of two environmental states ε_0 and ε_1. We assume for simplicity that $k \leq \ell$.

Suppose that an individual considers $\tilde{P}\{i|s_k\}$ and $\tilde{E}_{ij}$ (which, in general, differ from $P\{i|s_k\}$ and E_{ij}), as exact parameters and behaves according to the strategy $\tilde{Q}\{j|s_k\}$ specified by

$$\tilde{Q}\{0|s_k\} = \begin{cases} 1 & \text{if } \tilde{P}\{0|s_k\} > \tilde{Z}, \\ 0 & \text{if } \tilde{P}\{0|s_k\} < \tilde{Z}, \end{cases} \tag{29}$$

where

$$\tilde{Z} = \frac{\tilde{E}_{11} - \tilde{E}_{10}}{\tilde{E}_{11} + \tilde{E}_{00} - \tilde{E}_{01} - \tilde{E}_{10}} \tag{30}$$

for the additive model, and

$$\tilde{Z} = \frac{\ln(\tilde{E}_{11}/\tilde{E}_{10})}{\ln(\tilde{E}_{00}\tilde{E}_{11})/(\tilde{E}_{01}\tilde{E}_{10})} \tag{26'}$$

for the multiplicative model (see Result 5). Define S_1 as the set of s_k such that $\tilde{Q}\{0|s_k\} = 0$ and $\hat{Q}\{0|s_k\} = 1$, and S_2 as the set of s_k with $\tilde{Q}\{0|s_k\} = 1$ and $\hat{Q}\{0|s_k\} = 0$. In other words, the set $S_0 = S_1 \cup S_2$ contains s_k for which the adopted strategy (29) is wrong.

Take the additive model and denote by $\tilde{W}$ the total fitness for the adopted strategy:

$$\tilde{W} = \sum_{s_k} P(s_k) \sum_{i=1}^{n} \sum_{j=1}^{n} P\{i|s_k\}\tilde{Q}\{j|s_k\}E_{ij}. \tag{31}$$

Recall the total fitness $\hat{W}$ defined by (15). Then

$$\begin{aligned}
\hat{W} - \tilde{W} = &\sum_{s_k \in S_1} P(s_k)\Big\{ P\{0|s_k\}E_{00} + P\{1|s_k\}E_{10} \\
&- P\{0|s_k\}E_{01} - P\{1|s_k\}E_{11}\Big\} \\
&+ \sum_{s_k \in S_2} P(s_k)\Big\{ -P\{0|s_k\}E_{00} - P\{1|s_k\}E_{10} \\
&+ P\{0|s_k\}E_{01} + P\{1|s_k\}E_{11}\Big\}.
\end{aligned}$$

After some rearrangement,

$$\hat{W} - \tilde{W} = (E_{00} + E_{11} - E_{01} - E_{10})$$
$$\times \left\{ \sum_{s_k \in S_1} P(s_k) [P\{0|s_k\} - Z] + \sum_{s_k \in S_2} P(s_k) [Z - P\{0|s_k\}] \right\}.$$

By definition, $\hat{Q}\{0|s_k\} = 1$, and from Result 5, $P\{0|s_k\} \geq Z$ for $s_k \in S_1$. Similarly $P\{0|s_k\} \leq Z$ for $s_k \in S_2$. Hence, we have:

RESULT 7. For the two-state environment, the difference between the fitnesses of the semi-optimal and adopted strategies is

$$(E_{00} + E_{11} - E_{01} - E_{10}) \sum_{s_k \in S_0} P(s_k)|P\{0|s_k\} - Z| \tag{32}$$

for the additive model, and

$$\ln \frac{E_{00} E_{11}}{E_{01} E_{10}} \sum_{s_k \in S_0} P(s_k)|P\{0|s_k\} - Z| \tag{32'}$$

for the multiplicative model.

This result shows that the greater the difference between the fitness consequences of accurate and inaccurate prediction, namely $E_{00} - E_{01}$ and $E_{11} - E_{10}$ ($\ln E_{00} - \ln E_{01}$ and $\ln E_{11} - \ln E_{10}$ for the multiplicative model), the greater is the fitness reduction. The more frequently states s_k occur where such errors are made, the greater is the fitness reduction. On the other hand, Eqs. (32) and (32') show that the consequences of inaccurate knowledge are relatively minor near the zone of unreliable prediction, namely $P\{0|s_k\} \approx Z$, where even small errors in parameters produce, e.g., a strategy $\tilde{Q}\{j|s_k\} = 0$ instead of the best $\hat{Q}\{j|s_k\} = 1$, etc. (see Results 5 and 6).

10. CONCLUDING REMARKS

In the study of animal behavior, it is widely assumed that individuals act during their lifetime to improve their fitness (Shettleworth, 1984). Measures of individual fitness are difficult to design, however, and even in the study of economic and artificial systems the notion of an optimum may be difficult to make precise (Holland, 1975). Several mutually incompatible cost functions may be reasonable in any one system. Nevertheless, the definition of a "good" strategy may assist in the development of a qualitatively useful set of principles concerning the evolution of learning and adaptation in fluctuating environments.

In order to adapt to a fluctuating environment, an organism must have some information about the environment. Otherwise, in totally unpredictable conditions, the best it can do is adopt a minimax strategy that minimizes fitness losses under the worst environmental conditions; the evolution of life history strategies in stochastic environments has been widely discussed (see, e.g., Tuljapurkar, 1990; Stearns, 1992; Yoshimura and Clark, 1993). If, however, the individual has some knowledge about the environment, its strategy can be significantly improved. For example, such knowledge may improve foraging ability, and thereby improve fitness.

The properties of E_{ij} are important to consider. In principle they may be arbitrary, although it is more convenient in practice to assume that $E_{ij} \geq 0$ for the additive models and $E_{ij} \geq 1$ for the multiplicative case. E_{ij} may be allowed to be of arbitrary sign in the additive model, and merely positive in the multiplicative case, if we assume that $W = W_0 + \sum_i \sum_j n_{ij} E_{ij}$ in the former and $W = W_0 \prod_i \prod_j E_{ij}^{n_{ij}}$ in the latter, where W_0 is an individual's "initial fitness." This fitness is augmented or decreased during the individual's lifetime according to its success in predicting coming environments.

A second issue concerning the E_{ij} values relates to what might be an appropriate metric for E_{ij}. If the environment is one-dimensional, then the states may be ordered with respect to their similarity, say, as $\varepsilon_1, \varepsilon_2, \varepsilon_3, \cdots, \varepsilon_n$ with ε_i closer to ε_j than to ε_u if $i < j < u$ or if $u < j < i$. In this case, it is natural to assume that

$$E_{ij} > E_{iu} \quad \text{if} \quad i < j < u \quad \text{or} \quad u < j < i$$

and

$$E_{ij} > E_{vj} \quad \text{if} \quad v < i < j \quad \text{or} \quad j < i < v.$$

These are minimal metric conditions. Convexity or concavity of E_{ij}'s, or other properties, such as $E_{ii} > E_{ij}$ mentioned above, may be used to develop a corresponding metric space. It is of interest to extend such considerations to multidimensional environments.

In our model, the environment appears to the individual to be a multiorder Markov process. This may be the result of actual temporal fluctuations with a "memory" or be due to the individual's movement through a spatially varying habitat. In our model, the individual possesses two kinds of knowledge: *a priori* knowledge which contains information on the conditional probabilities of the environmental outcomes, and *a posteriori* knowledge which gives information about some previous environmental outcomes. The depth of *a posteriori* information, i.e., the depth of memory, k, and the depth, ℓ, of the *a priori* information, determines semi-optimal strategies that maximize individual fitness under these limited amounts of information. An optimal strategy would be obtained if the individual knew the probability law of the environmental stochastic process, and all necessary information about the realizations of previous environments.

We understand "environment" to consist of those external factors that influence the performance of a given individual in specific ecological or social circumstances at a definite stage of its life. The set of "essential" environmental states, ε_i, their number, n, their frequencies, P_i, the environmental depth, m, and transition probabilities $P\{i|s_m\}$ may all be different for different individuals, different populations, and under different ecological conditions. The amount of knowledge about the environment, the depths k and ℓ of this knowledge and the fitnesses E_{ij} may vary among individuals. Moreover, all of these parameters may be sensitive to life-cycle stage in the same individual. Knowledge that might be important for foraging might be irrelevant with respect to mating. Some knowledge important at one life stage may be irrelevant at later stages. Here we have been concerned only with stages of life that are in some sense stationary, that is, lasting long enough that they can be described in terms of stationary stochastic processes. The alternative would require consideration of nonstationary processes and different sets of fitnesses at different life stages.

In this paper, we have not addressed the mechanisms by which an individual acquires prior and posterior information about its environment and fitness, nor what causes the fitnesses E_{ij} in these environments, nor how strategies might be behaviorally altered. Such mechanisms are assumed to be intrinsic to the individual. Their ontogeny is, of course, important, but outside the scope of this analysis.

A further development of our analysis would permit different relative contributions of earlier environmental states, say, those at times $t - m$, $t - m + 1$, etc., from later ones, say, at $t - 1$, $t - 2$, etc., to the environmental transition probabilities $P\{i|s_m\}$ and to the corresponding strategy function $Q\{j|s_k\}$. For example, there might be a recency bias in the environment, in which case the contribution of recent events would be more important than those of the distant past. Here, the individual's memory depths, k and ℓ, should not be large compared to the corresponding values which would be appropriate to the case of primary bias, where early environmental events are more important.

How good the semi-optimal strategies are depends on $\min(k, \ell)$. The larger this value is, the greater is the fitness at the semi-optimal strategy. It follows that in our model, differences between the depths of *a priori* and *a posteriori* knowledge are not desirable in an evolutionary sense. For example, if the *a priori* knowledge is greater than *a posteriori*, i.e., $\ell > k$, increasing ℓ without a corresponding increase in k does not produce an increase in fitness at the semi-optimal strategy. Of course, if increasing ℓ incurs a cost, the fitness may even decrease under these conditions. The best outcome for the model is agreement between k and ℓ, and these should be as large as possible until they exceed m.

A semi-optimal strategy is not a predictive function. It might better be regarded as a preparation function: an individual prepares to accrue a maximal expected contribution to its fitness in the pending environment. For example, in extreme cases, it may choose a constant strategy, preparing itself for the same environmental state if this occurs sufficiently often, and provided that large enough fitness increments accrue to those who choose it. When the fitness parameters E_{ij} include a random component, then differences emerge between the semi-optimal strategies that apply in additive and multiplicative cases. Thus in the additive case, it is sufficient for the individual to know the mean of the variable fitnesses. For the multiplicative case, however, at least the variance of these fitnesses must be known in addition to the mean.

It is important that in our model, with independent additive or multiplicative contributions to fitness, a semi-optimal strategy is locally deterministic. By this we mean that at the time points of environmental changes, the probabilities that an individual prepares for the various environmental options take values 0 and 1 rather than intermediate values. Thus, the individual knows for sure which of the possible future environmental states it must prepare for, using its *a priori* and *a posteriori* knowledge. At the same time, however, throughout an individual's lifetime, its behavior appears to be probabilistic because the individual can choose different environmental states depending on the changing *a posteriori* information it acquires as the environment changes during its life. Although the strategic choices are deterministic, viewed on a longer time scale they might appear to be extremely random. This is reminiscent of animal behavior experiments in which the subject responded quite differently to the same stimulus, to the extent that the behavior was described as "spontaneous" (Manning, 1979).

It should be emphasized that in the presence of errors in the *a priori* or *a posteriori* information, the semi-optimal strategy remains locally deterministic. The only consequence of erroneous or incomplete information is a decrease in the total fitness of an individual.

Although our model is presented in terms of *a priori* and *a posteriori* information, and their use in the search for a semi-optimal strategy, we do not address how this information is acquired. This important issue, namely the role of learning in the process of preparation for environmental fluctuation, is currently under investigation.

ACKNOWLEDGMENT

The authors are grateful to Drs. Rik Belew and Peter Todd for their helpful comments on an earlier draft.

APPENDIX A: PROOFS OF RESULTS 1 AND 2

To see that Result 1 is true, note that for any set of nonnegative numbers $c_1, c_2, \ldots, c_n$,

$$\max(c_1, c_2, \ldots, c_n) = \max_{\sum \alpha_j = 1; \alpha_j \geq 0} \sum \alpha_j c_j,$$

and that the maximum is achieved at a boundary of the simplex $\sum \alpha_j = 1$; $\alpha_j \geq 0$. For the multiplicative model, this implies that for this boundary specified by j,

$$\sum_{i=1}^n P\{i|s_k\} E_{ij} \geq \sum_{i=1}^n P\{i|s_k\} E_{ij_0},$$

for every $j_0 (j_0 \neq j)$.

For Result 2, introduce for $k \geq \ell$

$$Q^+\{j|s_\ell\} = \sum_{s_k \subset s_\ell} \frac{P(s_k)}{P(s_\ell)} Q\{j|s_k\} \tag{A1}$$

(cf. (11)). It follows from the linearity of W that $Q^+\{j|s_\ell\}$ has the form (13). Hence, from Eq. (A1), all of the terms $\hat{Q}\{j|s_k\}$ must be 1 or 0 for every $s_k \subset s_\ell$.

APPENDIX B: PROOF OF RESULT 3

For definiteness, suppose $k' \leq \ell'$ and $k'' \leq \ell''$ so that $k' > k''$. Note that for any set of nonnegative values $A_1, A_2, \ldots, A_n$, their maximum may be written

$$\max\{A_1, A_2, \ldots, A_n\} = \max_j A_j = \max_{\sum_{j=1}^n \alpha_j = 1} \sum_{j=1}^n \alpha_j A_j,$$

where α_j are nonnegative and lie on the simplex $\sum_{j=1}^n \alpha_j = 1$. Denote a set of sequences of possible environmental events of length k' by S' and let S'' be those of length k''. If $k'' < k'$, then

$$\sum_{s \in S'} P(s) = \sum_{s'' \in S''} \sum_{\substack{s \subset s'' \\ s \in S'}} P(s).$$

Since the sum of the maximums is greater than or equal to the maximum of the sums, we have from Result 1 that

$$\hat{W}' = \sum_{s_{k'} \in S'} P(s_{k'}) \max_{j} \sum_{i=1}^{n} P\{i|s_{k'}\} E_{ij}$$

$$= \sum_{s_{k'}} P(s_{k'}) \max_{\sum \alpha_j = 1} \sum_{j=1}^{n} \sum_{i=1}^{n} \alpha_j P\{i|s_{k'}\} E_{ij}$$

$$\geq \sum_{s_k'' \in S''} \max_{\sum \alpha_j = 1} \sum_{j=1}^{n} \sum_{i=1}^{n} \sum_{s_{k'} \subset s_{k''}} \alpha_j P(s_{k'}) P\{i|s_{k'}\} E_{ij}$$

$$= \sum_{s_{k''}} \max_{\sum \alpha_j = 1} \sum_{j=1}^{n} \sum_{i=1}^{n} \alpha_j P(s_{k''}) P\{i|s_{k''}\} E_{ij}$$

$$= \sum_{s_{k''}} P(s_{k''}) \max_{j} \sum_{i=1}^{n} P\{i|s_{k''}\} E_{ij}$$

$$= \hat{W}''.$$

The equality in the proof occurs if and only if, for every $s_{k''}$ with $s_{k'} \subset s_{k''}$, $\sum_j \sum_i \alpha_j P\{i|s_{k'}\} E_{ij}$ reaches its maximum in the simplex $\sum \alpha_j = 1$, $\alpha_j \geq 0$ on the same boundary. The proof remains the same for the multiplicative case if we replace E_{ij} and $\hat{W}$ by their logarithm.

REFERENCES

Bergman, A., and M. W. Feldman. "On the Evolution of Learning." Working paper 94-02-005, Santa Fe Institute, 1994.

Bush, R. R., and F. Mosteller. *Stochastic Models for Learning.* New York: J. Wiley, 1955.

Cohen, D. "Optimising Reproduction in a Randomly Varying Environment." *J. Theor. Biol.* **12** (1966): 119–129.

Cohen, D. "Optimising Reproduction in a Randomly Varying Environment when a Correlation May Exist Between the Conditions at the Time a Choice has to be Made and the Subsequent Outcome." *J. Theor. Biol.* **16** (1967): 1–14.

Cohen, D. "Fitness in Random Environments." In *Adaptation in Stochastic Environments,* edited by J. Yoshimura and C. W. Clark, 8–25. Berlin: Springer-Verlag, 1993.

Gillespie, J. H. "Polymorphism in Random Environments." *Theor. Pop. Biol.* **4** (1973): 193–195.

Harley, C. B. "Learning the Evolutionarily Stable Strategy." *J. Theor. Biol.* **89** (1981): 611–633.

Harley, C. B., and J. Maynard Smith. "Learning—An Evolutionary Approach." *TINS* (1983): 204–208.

Hartl, B. L., and R. D. Cook. "Balanced Polymorphisms of Quasi-Neutral Alleles." *Theor. Pop. Biol.* **4** (1973): 163–172.

Holland, J. H. *Adaptation in Natural and Artificial Systems.* Ann Arbor: The University of Michigan Press, 1975.

Houston, A. I., and B. H. Sumida. "Learning Rules, Matching, and Frequency Dependence." *J. Theor. Biol.* **126** (1987): 289–308.

Karlin, S., and U. Liberman. "Random Temporal Variation in Selection Intensities: Case of Large Population Size." *Theor. Pop. Biol.* **6** (1974): 355–382.

Levins, R. *Evolution in Changing Environment.* Princeton, NJ: Princeton University Press, 1968.

Lewontin, R. C., and D. Cohen. "On Population in a Randomly Varying Environment." *Proc. Natl. Acad. Sci. USA* **62** (1969): 1056–1060.

Manning, A. *An Introduction to Animal Behavior.* Reading, MA: Addison-Wesley, 1979.

Maynard Smith, J. *Evolution and the Theory of Games.* Cambridge, MA: Cambridge University Press, 1982.

McNamara, J. M., and A. I. Houston. "Memory and the Efficient Use of Information." *J. Theor. Biol.* **125** (1987): 385–395.

Shettleworth, S. J. "Learning and Behavioral Ecology." In *Behavioral Ecology: An Evolutionary Approach,* 2nd ed., edited by J. R. Krebs and N .B. Davies, 170–194. Sunderland, MA: Sinauer, 1984.

Stearns, S. C. *The Evolution of Life Histories.* Oxford: Oxford University Press, 1992.

Stephens, D. W. "Change, Regularity, and Value in the Evolution of Animal Learning." *Behav. Ecology* **2** (1991): 77–89.

Tuljapurkar, S. *Population Dynamics in Variable Environments*. Berlin: Springer-Verlag, 1990.

Yoshimura, J., and C. W. Clark, eds. *Adaptation in Stochastic Environments*. Berlin: Springer-Verlag, 1993.

William E. Hart

Preface to Chapter 11

The following chapter examines the role of learning in a model of the immune system. The evolution of the immune system is modeled with a genetic algorithm that evolves a population of antibodies, and the process of somatic hypermutation is modeled with a simple learning procedure. Non-Lamarckian inheritance is performed, which means that offspring of an individual do not inherit information about the results of the learning process. However, the fitness of the final antibody found during learning is used when selecting antibodies in the genetic algorithm.

The authors measure the evolutionary progress of populations of antibodies, varying the learning rate. They observe two effects, the Baldwin effect and the mastery effect. For small rates of learning, learning improves the evolutionary performance. This is the Baldwin effect, where learning helps the genetic algorithm find solutions that are nearly optimal. When the learning rate becomes sufficiently high, increases in the learning rate decrease the evolutionary performance. This is the mastery effect, where so much learning is allowed that most individuals in the population can learn the optimal solution. Consequently, the relative fitness of near-optimal solutions is masked, and evolution cannot distinguish between them.

Note that the mastery effect does not occur in evolutionary models that use learning with Lamarckian inheritance. When the final antibodies found during learning replace the initial antibodies, then the fitnesses of antibodies are their true fitnesses. Consequently, evolution can distinguish between near-optimal solutions.

The Baldwin and mastery effects have also been observed in models of genetic algorithms and artificial neural networks that describe the evolution of real nervous systems (e.g., Hinton and Nowlan, 1987; Keesing and Stork, 1991). However, the immune system model differs from these models by modeling the direct encoding between the antibody and genetic representation. While real nervous systems arise through a developmental process, models of artificial neural networks have been encoded directly. Because of this difference between methods of encoding in the model and natural system, it is not clear that models of genetic algorithms and artificial neural networks provide reliable predictions about the role of Baldwin effects in real nervous systems. The experimental results for the immune system model show that evolutionary models using direct encodings are valid for at least one natural system.

Baldwin effects in the immune system model suggest that the efficiency of evolutionary algorithms used to solve engineering problems can be improved using mechanisms that perform learning. The main difference between the immune system model and these evolutionary algorithms concerns the cost of learning. In the immune system, there is a limited amount of time to respond to a disease before it overruns the body. Consequently, the learning rate in the immune system model is inversely proportional to the number of steps performed during learning. Hart and Belew (Chapter 27) describe a model of evolution with learning in which each step of learning requires a fixed amount of time. This implies that the time needed to perform learning is variable since the number of steps varies from one learning event to another. Further, this model requires that the total time of learning must be summed over all the component steps. Because the time needed to perform learning in Hart and Belew's model is variable, Hightower et al.'s results are not immediately applicable to Hart and Belew's model. However, Hart and Belew observe that evolutionary algorithms can be more efficient when learning mechanisms are used, which suggests that the dynamics of their model are closely related to the dynamics of the immune system model.

This comparison indicates that Hightower et al.'s results are likely to generalize to biological models for which the learning time can vary. Many biological models of independent organisms in natural environments have this characteristic. The results of Hart and Belew (Chapter 27) are not directly applicable to this type of model because the total time of learning is summed, which is not biologically plausible since the learning of independent organisms is typically performed in parallel.

REFERENCES

Keesing, R., and D. G. Stork. "Evaluation and Learning in Neural Networks: The Number and Distribution of Learning Trials Affect the Rate of Evolution." In *NIPS 3* edited by R. P. Lippmann, J. E. Moody, and D. S. Touretzky, 804–810. 1991.

Hinton, G. E. and S. J. Nowlan. "How Learning Can Guide Evolution." *Complex Systems* 1 (1987): 495–502.

Ron Hightower, Stephanie Forrest, and Alan S. Perelson

Chapter 11:
The Baldwin Effect in the Immune System: Learning by Somatic Hypermutation

1. INTRODUCTION

Almost one hundred years have passed since J. Mark Baldwin (1896) proposed that learned or acquired characteristics could become part of the genetic makeup of succeeding generations without Lamarckian inheritance. Individuals that learn or acquire useful characteristics during their lifetimes tend to survive and, Baldwin claimed, this would cause succeeding generations to have a higher probability of acquiring the same characteristics, even though the characteristics themselves were not genetically propagated. Eventually the learned characteristics would become completely genetically encoded. Baldwin called this "a new factor in evolution," and it has come to be known as the *Baldwin effect*.

In recent years the Baldwin effect has been shown to exist in several experiments involving the simulated evolution of learning systems (for example, Hinton and Nowlan, 1987, and Keesing and Stork, 1991). These simulations have typically used the genetic algorithm (GA) to act the part of evolution and neural networks to play the role of the learning mechanism. The nervous system, however, is not the only part of an organism that is capable of learning. This paper discusses the Baldwin effect with respect to the *immune system*—a learning mechanism that rivals the complexity and computational power of the nervous system.

Adaptive Individuals in Evolving Populations, Ed. R. K. Belew & M. Mitchell,
SFI Studies in the Sciences of Complexity, Vol. XXVI, Addison-Wesley, 1996   **159**

In previous work we used a binary model of the immune system to study the effects of evolution on the genetic representation of antibodies (Hightower et al., 1992). Here we have extended this model to include "clonal selection," which is the learning process used by the immune system. Once this learning mechanism was incorporated into the model, it became possible to observe the Baldwin effect in the evolution of our binary immune system. We found that the Baldwin effect is not a universal relationship between evolution and learning, and that the strength of the effect is sensitive to the shape of the fitness landscape.

2. THE IMMUNE SYSTEM AND CLONAL SELECTION

The human immune system is responsible for recognizing, and defending against, pathogens, toxins, and other foreign molecules, collectively called *antigens*. The immune system produces special molecules, *antibodies*, which bind to antigen and thereby lead to their elimination. Antigen recognition is essentially a form of template matching—when the shape and charge of the antibody and antigen molecules match, in a complementary fashion, the molecules can bind and the antigen is recognized. The closer the match is between antibody and antigen, the stronger the molecular binding and the better the recognition.

Antibodies are produced by the B cells of the immune system. Each B cell produces a specific type of antibody that can recognize only specific types of antigen. The antibodies produced by a B cell are expressed on the surface of the cell as receptors. When the antibody receptors of a B cell recognize an antigen, that B cell is stimulated to reproduce. In the presence of antigen, the daughter cells of a stimulated B cell will also become stimulated and reproduce. Thus, the presence of antigen will cause the proliferation of those B cells best suited to recognizing that antigen. This process is called clonal selection.

The stimulated proliferation of B cells activates a mechanism called *somatic hypermutation* (Kepler, 1993). Somatic hypermutation affects the genes that encode for the antibody molecule (in particular the portion of the antibody that binds to antigen, called the variable region). The daughter cells of a stimulated B cell, therefore, exhibit wide variation in their ability to recognize antigen.

Due to competition for binding antigen, the better B cells with higher affinity antibodies will be stimulated by the antigen and will grow at the expense of B cells expressing "poorer" or lower affinity antibodies. By repeating this process of mutation and selection a number of times, the immune system "learns" to produce higher affinity antibodies for the antigen stimulating the system.[1]

---

[1]It is interesting to note that somatic hypermutation and clonal selection behave like an internal version of evolution that operates within an individual's body.

Learned characteristics in the immune system are not passed directly to off-spring. The DNA in a B cell that is modified during the learning process cannot be spliced into the DNA of an egg or sperm cell, and so this DNA cannot be passed on genetically to the next generation. Any contribution that clonal selection and somatic hypermutation have on the evolutionary process must occur via the Baldwin effect.

3. THE BINARY IMMUNE SYSTEM

Our model of the immune system uses bit strings, each of length 64, to represent the shape of antibody and antigen molecules. In this bit string universe, molecular binding takes place when an antibody bit string and an antigen bit string "match" each other in a complementary fashion. The *match score* between an antibody bit string and an antigen bit string is the number of complementary bits. As shown in Figure 1, the match score can be computed by applying the exclusive-or operator (XOR). The expected match score between two randomly chosen bit strings is 32 (half the length of the 64-bit strings).

The *binding value*, derived from the match score, represents how well two molecules bind. In reality, two molecules must match each other over a sufficiently large surface area before a stable bond can form. In our bit string universe this is emulated by requiring the match score to exceed a certain threshold before binding takes place. The curve in Figure 2 shows the relationship of binding value to match score. For match scores below the threshold (in these experiments the threshold was equal to 45 bits), the binding value is essentially zero. For match scores above the threshold, the binding value is one. Note that the binding value function makes a soft transition on either side of the threshold, rather than an abrupt shift from zero to one.

We chose a nonlinear binding value function for two reasons. First, it is a more accurate description of the actual molecular binding process. Second, it appears to be a necessary condition for the Baldwin effect to occur (as will be discussed later).

```
Antigen:  1100100101101001101010000100001111010011001010101001010100110101
Antibody: 1011010101001010010111000101011101011011100011001100100100100111
XOR:      11111   1  111111 1    1 1 1   1   1 1 11 1 111     1  1
```

FIGURE 1 Computing the match score between binary molecules.

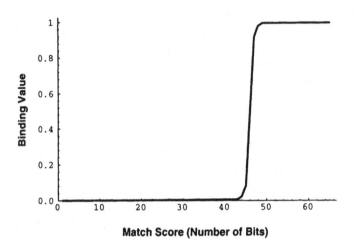

FIGURE 2   Relation of binding value to match score.

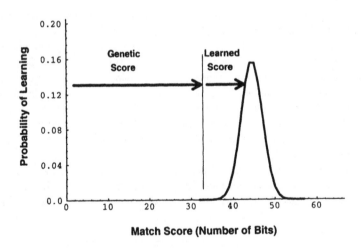

FIGURE 3   Probability of learning, with G = 25 guesses and genetic component of 32.

In our original model of the immune system (Hightower et al., 1992) no learning (somatic hypermutation) took place during fitness evaluation, and the match score was determined entirely by the genes. Here we extend that model and assume that the match score of an antibody has both a genetic component and a learned component, as shown in Figure 3. The genetic component of the match score is computed by comparing an antibody bit string to the antigen before any learning occurs. The antibody is then allowed to make G guesses in an attempt to improve its score. (This learning-as-guessing process follows that used by Hinton and Nowlan [1987].) Each guess has a 50% probability of being correct, so the probability distribution for the learned component is a binomial distribution, with a mean of *Genetic Component + (Allowed Guesses)/2*. Comparing Figure 2 and Figure 3, we see that if the genetic component is insufficient to produce a strong binding value, a small amount of learning may push the match score above the necessary threshold.

4. THE EXPERIMENT

The experiment described in this paper tests for a modified statement of the Baldwin effect. Instead of testing whether "learning guides evolution," the experiment tests whether "learning accelerates evolution." The approach is to vary the learning rate—the number, G, of allowed guesses—and to measure the corresponding amount of evolutionary progress after a fixed amount of time has passed. This is more similar to the experiments of Keesing and Stork (1991) than to the Hinton and Nowlan experiment (Hinton, 1987).

Before looking at the results, we detail the genetic algorithm portion of the experiment. We used a genetic algorithm derived from Grefenstette's GENESIS package (Schraudolph, 1992), with a crossover rate of 0.6 and a mutation rate of 0.0002. The population size was 50. The GA experiments all ran for exactly 1000 generations. Individuals were initially set to all zero bits for the first generation.[2]

An individual in the population was represented as a genome of 512 bits. Using a compression technique found in the real immune system, these 512 bits encode for a total of 4096 antibodies, each of length 64 bits. The details of this genetic representation can be found in Hightower et al. (1992).

The fitness of an individual, as used by the GA, is found by testing all 4096 antibodies against a small number, K, of randomly chosen antigens (here $K = 8$). The K antigen bit strings are randomly selected from a larger set of bit strings called the *antigen universe*. For the experiments described here the antigen universe contained 32 randomly generated antigen bit strings. The 32-bit strings in the antigen universe are generated at the beginning of the experiment and remained fixed for every generation thereafter, whereas a different set of K bit strings is randomly selected every time a fitness evaluation takes place.

---

[2]Initializing the population to all zeros increases the difficulty of the learning task. There is also some biological justification: the genes encoding for antibodies probably arose through a process of gene duplication, so the DNA would have had a high degree of redundancy early in its evolutionary history.

Individual fitness is calculated as follows. For each of the eight antigens, the individual being evaluated selects the single antibody that best matches the antigen. The match score of this antibody with its corresponding antigen is the genetic component of the score for that antigen. The chosen antibody then undergoes the learning process and makes G guesses to determine the learned component of the match score. The genetic component and the learned component are added (see Figure 3), and used to find the corresponding binding value (using the function shown in Figure 2). Thus a binding value is found for each of the $K = 8$ antigens. The fitness of the individual is the average binding value, averaged over the eight randomly chosen antigens. This fitness is used by the GA to determine which individuals will survive into the next generation.

No Lamarckian inheritance is performed in these experiments. The learning that takes place does not modify the genetic character of the individual being evaluated. Therefore the offspring of an individual will inherit no direct information about the results of the learning process.

5. THE RESULTS

In Figure 4 we show the results of these experiments for a number of different learning rates, G, between 0 and 30. Each point is an average over 30 GA runs. The vertical axis shows the genetic component of the match score, averaged for the entire population (and the 30 GA runs). This average genetic match score is essentially the "fitness at birth," before learning takes place, and shows the genetic goodness of the population. The data curve shows that the highest average match scores (around 43 bits) were found by the genetic algorithm when the learning rate was between 8 and 12 guesses per lifetime.

Two things should be pointed out about the scaling of this plot. First, the match score between a randomly chosen antibody and a randomly chosen antigen has an expected value of 32 bits (half the length of the 64-bit strings), and this determined the minimum coordinate for the graph. Second, the binding value for all match scores above 48 is 1.0, so there is no selection pressure for the population to exceed a score of 48. Hence 48 was used as the maximum coordinate of this plot. The plot is scaled between the minimum and maximum obtainable values.

The upward slope on the left side of the curve is due to the Baldwin effect. It shows that increased learning leads to an increase in evolutionary progress during the course of 1000 generations. The explanation for this acceleration of evolution is that learning rewards those individuals that are nearer to the threshold in the Without learning, individuals near the threshold have essentially the same poor fitness as those further away. This is similar to the situation binding value function.

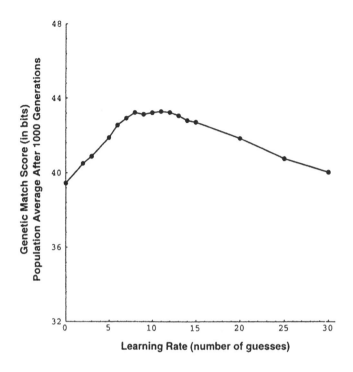

FIGURE 4   The Baldwin effect and the mastery effect. The population average match score at generation = 1000. Thirty experiments per data point. Population size = 50.

described by Hinton and Nowlan (1987) in their experiment where learning put "shoulders" around the single solution, providing fitness information to the GA that would not have been available otherwise.

The ramp on the right side of the curve shows what we call the *mastery effect*. The antigen recognition task has only a finite difficulty, so it is possible for an individual to completely master the task. Given a sufficient amount of learning, even the worst members of the population can learn the correct solution in 1000 generations, which effectively hides their genetic disabilities from the GA. This reduction in information slows the progress of evolution, and the slowing appears to occur linearly with an increase of learning rate. Presumably, an infinite amount of learning would absolutely halt evolution, given a task with finite difficulty. Keesing and Stork (1991) demonstrated a similar effect with a neural network model.

A small modification to this experiment removed the Baldwin effect completely. In Figure 2 we show a binding value function that is nonlinear with respect to the match score. When the binding value function is made to be a linear function of match score, then changes in the learning rate G do not improve evolutionary progress. With a linear binding value function there is no Baldwin effect, and only the mastery effect remains. This makes good sense with respect to Hinton and Nowlan's "shoulders" explanation. Without the nonlinearity, there is nothing for learning to put shoulders around, and therefore no evolutionary advantage to learning.

6. DISCUSSION

We have taken an existing model of the immune system and shown the presence of the Baldwin effect. The explanation of the effect is that learning allows the population to perform local search of the fitness landscape during evolution. As noted by Hinton and Nowlan, this is like putting shoulders around the solution. In these experiments the "shoulders" are put around the nonlinearity in the binding value function. Learning allows evolution to discover which individuals are nearest to the threshold of success.

The Baldwin effect is not a universal relation between evolution and learning, and appears to be sensitive to the shape of the fitness landscape. As we discovered, the Baldwin effect disappears when the nonlinearity in the binding value function is removed. Without the nonlinearity, there is nothing for learning to put shoulders around, so local search (learning) offers no evolutionary advantage. This observation was never made explicit in Hinton and Nowlan's explanation of the Baldwin effect.

Our experiment combines aspects of the work by Hinton and Nowlan and the work by Keesing and Stork. As in the Hinton and Nowlan experiment, our genetic representation is binary and learning is implemented as a process of bit guessing. The experiment itself, however, follows the "learning accelerates evolution" model of Keesing and Stork, and the resulting curve has a shape similar to their experimental results.

The immune system experiment differs, however, in the representational correctness of our model. Learning and evolution operate on the same genetic representation, and this is true for both our model immune system and the real immune system. Contrast this with the nervous system models. Hinton and Nowlan used genes to represent neural connections. Keesing and Stork used genes to represent weights on neural connections. In real nervous systems, there is no direct genetic representation for either weights or connections. Weights and connections derive from a complicated process of growth and development. The point is that evolution and learning operate on different levels of representation for real nervous systems, but in the simulated models, evolution and learning operate on the same level. This may be an important distinction when we begin looking for the specific conditions that promote or inhibit the Baldwin effect.

ACKNOWLEDGMENTS

We thank the Center for Nonlinear Studies, Los Alamos National Laboratory, and the Santa Fe Institute for ongoing support of this project. Forrest also acknowledges the support of the National Science Foundation (grant IRI-9157644). Perelson acknowledges the support of the National Institutes of Health (grant AI28433)

REFERENCES

Baldwin, J. M. "A New Factor in Evolution." *Am. Natur.*, **30** (1896): 441–451.

Hightower, R., S. Forrest, and A. S. Perelson. "The Evolution of Cooperation in Immune System Gene Libraries." Technical Report CS92-14, Department of Computer Science, University of New Mexico, Albuquerque, NM, 1992.

Hinton, G. E., and S. J. Nowlan. "How Learning Can Guide Evolution." *Complex Systems* **1** (1987): 495–502.

Keesing, R., and D. G. Stork. "Evolution and Learning in Neural Networks: The Number and Distribution of Learning Trials Affect the Rate of Evolution." In *Proceedings of Neural Information Processing Systems 3*, edited by R. P. Lippmann, J. E. Moody, and D. S. Touretzky. San Mateo, CA: Morgan Kaufmann, 1991.

Kepler, T. B., and A. S. Perelson. "Somatic Hypermutation in B Cells: An Optimal Control Treatment." *J. Theor. Biol.* **164** (1993): 37–64.

Schraudolph, N. N., and J. J. Grefenstette. "A User's Guide to GA UCSD 1.4." Technical Report CS92-249, Department of Computer Science and Engineering, University of California at San Diego, La Jolla, CA, 1992.

Michael L. Littman and Filippo Menczer

Preface to Chapter 12

A central theme of this volume is identifying constructive connections between the natural sciences and computer science. Work in reinforcement learning is a prime example of the synergy between the two areas. The term "reinforcement" comes from experiments in animal learning in which an animal comes to associate a neutral stimulus such as a flash of light with more meaningful stimuli like the smell of food. In computer science, reinforcement learning refers to a class of solutions to the problem of temporal credit assignment in which a learning agent must decide on an action to take to maximize its long-term reward. The present popularity of reinforcement learning among researchers in Artificial Intelligence is motivated by both its connection to natural systems and more recent mathematical results that provide insight into when such methods can be expected to work.

The following paper by Shafir and Roughgarden is a fascinating new chapter in the reinforcement learning story. The authors, both biologists, propose a reinforcement learning problem directly inspired by the foraging behavior of a particular breed of tropical lizard. At each moment in time, the lizard is presented with a juicy insect at some distance away and it must decide whether to pursue the insect, risking the possibility that another insect will appear at a shorter distance before it returns, or to ignore the insect, risking the possibility of waiting a long time before its next meal. The lizard's decisions are guided by the criterion of maximizing the number of insects consumed per time unit.

Adaptive Individuals in Evolving Populations, Ed. R. K. Belew & M. Mitchell,
SFI Studies in the Sciences of Complexity, Vol. XXVI, Addison-Wesley, 1996   **169**

The authors show that, once the proper assumptions are made, the problem is well-defined and they express the optimal decision policy for the lizard as a foraging distance cutoff that is a function of the insect abundance and the lizard's speed of motion. It is interesting to note that this problem, although well-posed, differs from the type of problems normally studied by reinforcement learning researchers. In one respect, it is harder, since the decision policy must work in the continuous space of insect distances rather than in a discrete set of states. However, unlike traditional applications of reinforcement learning, the lizard's choice of action does not affect the resulting state of the world; instead, the choice is made according to a fixed probability distribution. This simplifies the problem somewhat since the effects of bad decisions, while slowing down the estimation of the unknown probability distribution, do not result in long-term penalties for the lizard.

These differences are interesting because they stem from the authors' interest and experience in natural systems. Consequently, current reinforcement learning techniques cannot be applied to this problem. Instead, the authors propose a novel approach to finding the optimal decision policy. It can be shown (see appendix) that their rule reduces to: "pursue if the time it would take to consume the insect is less than the average time per consumed insect experienced so far." The authors show, by way of a series of simulation experiments, that this decision rule succeeds in quickly identifying the optimal distance cutoff. Furthermore, it is easy to imagine variations of the lizard's foraging task for which the rule would present additional advantages. For example, in a changing environment, such as if the prey distribution were not constant, the decision rule could adapt to the changing conditions.

In spite of the fact that the authors cannot guarantee that this rule converges on the correct behavior, they prefer it to another, complicated equation that expresses optimal behavior given the prey distribution. This brings up an interesting point: although their problem is motivated primarily by biological realism, their solution is justified on computational grounds. An implementation of their learning rule requires less complicated computational machinery than expressing the optimal policy directly.

In fact, we don't know whether neural circuitry for computing a ratio of running sums (as required by the learning rule) is more or less common than that for computing a cube root (as for the optimal distance cutoff). Evolution's "choice" of a decision policy is dictated in part by what decision policies are more fit (an adaptationist perspective) but also by what can be implemented with the available "cognitive hardware" (a computational issue). The field of computer science is deeply concerned with how various calculations can be performed most efficiently for a given model of computation. In this respect, the paper illustrates how computational issues become relevant to biology when the cognitive constraints of the modeled natural system are accounted for.

As a concrete example, computer science (more precisely, stochastic approximation theory) provides the formal machinery to talk about the computation performed by Shafir and Roughgarden's learning rule. The appendix to this preface contains a proof that the decision rule indeed converges to the optimal decision policy, assuming that it does converge. This brings some mathematical underpinning to a computationally inspired solution for a biological problem.

This work also provides valuable insights to computer science. Traditionally, work in machine learning, and more recently in reinforcement learning, has been concerned with finding general purpose algorithms for solving problems. The learning rule here is seemingly novel and fairly specific to a particular survival-oriented task. We need to pay more attention to this type of approach since Artificial Intelligence as a whole has found that most problems are simply too hard to solve in a general way. Special-purpose algorithms and special-purpose learning algorithms are needed to solve real-world problems, just like those in biology.

Sharoni Shafir and Jonathan Roughgarden

Chapter 12:
The Effect of Memory Length on Individual Fitness in a Lizard

The effect of memory length on the ability to solve an optimization problem was studied by simulating foraging by individual *Anolis* lizards having different memory lengths. All individuals employed the same simple decision rule that determined whether the lizard would pursue or ignore an insect that appeared on the ground in front of it. Lizards with no memory made many suboptimal decisions, and their mean rate of net energy gain was almost half of the maximum possible rate in the simulated environment. Initial small increases in memory greatly improved the lizards' performance. As memory was increased further, lizards' net energy gain approached the maximum possible rate, but with diminishing returns.

The reproductive fitness of each lizard was determined based on its energy gain. Lizards that could remember even only the last two prey encounters laid almost twice as many eggs per day as lizards that foraged with no memory. Thus, the evolutionary selective pressure for having at least a short memory is likely to be very strong.

Adaptive Individuals in Evolving Populations, Ed. R. K. Belew & M. Mitchell,
SFI Studies in the Sciences of Complexity, Vol. XXVI, Addison-Wesley, 1996   **173**

1. INTRODUCTION

Memory is central to learning, regardless of whether the learning agent is a human, a robot, an immune system, or a lizard. Under certain conditions it may not be worthwhile to remember the past (Bergman and Feldman, forthcoming; Zhivotovsky et al., Chapter 10.) In predictable environments, the relative weighting of remote and more recent past events that best traces the environment depends on the rate at which the environment is changing (McNamara and Houston, 1987). But what is the price of suboptimal memory? And what is the optimal memory length in a complex environment? Such questions are difficult to address analytically.

Here, we assess the effect of memory length on the ability to solve a complex problem. The model is tailored around an *Anolis* lizard foraging for insects, but the approach may be applied to other biological and computer science paradigms. We simulate real lizards chasing real insects; the result is a simulation, not a realization (cf. Pattee, 1988). We do not intend to create artificial lizards whose behavior resembles that of living ones, but rather to use the simulation to study the effect of memory length on the fitness of real lizards. The success of lizards with different memory lengths is measured according to a metric directly associated with fitness, viz., the number of days it takes a female to produce an egg. Thus, we can determine the cost (in terms of reproductive fitness) of suboptimal memory lengths.

2. THE *ANOLIS* LIZARDS SYSTEM

Anolis is a large genus of insectivorous neo-tropical lizards. They are "sit and wait" predators. Trunk-ground anoles perch on the bottom of tree trunks and scan the ground below them. When an insect appears nearby, they dart in pursuit, then return to their perch.

2.1 A LIZARD'S DILEMMA

Chasing an insect that is a short distance away is energetically less expensive and takes less time than chasing one that is farther away. So to maximize rate of net energy gain, a lizard should chase close insects and ignore distant insects. But what is the cutoff radius beyond which prey should not be pursued?

2.2 THE THEORETICAL SOLUTION

The cutoff radius that maximizes rate of net energy intake, E/T, depends on the abundance of insects in the environment. When insects are rare the optimal cutoff radius is greater than when insects are abundant. In an environment where there is only one type of insect, the cutoff radius, r, that maximizes the rate of net energy intake is:

$$
r = \frac{-\frac{2(e_p - e_w)}{\pi a e} + \left(\frac{3ev}{2\pi a e} + \sqrt{\left(\frac{3ev}{2\pi a e}\right)^2 + \left(\frac{2(e_p - e_w)}{\pi a e}\right)^3} \right)^{2/3}}{\left(\frac{3ev}{2\pi a e} + \sqrt{\left(\frac{3ev}{2\pi a e}\right)^2 + \left(\frac{2(e_p - e_w)}{\pi a e}\right)^3} \right)^{1/3}} \tag{1}
$$

where a is the abundance of prey in units of prey per m^2 per time, e is the insect's energy content, e_w is the energy expended per time while sitting perched and looking for prey, e_p is the energy expended per time while pursuing a prey item, and v is the lizard's speed. The derivation of this solution can be found in Roughgarden's book (1995). The model is also extended to multiple prey types, and to prey that can escape. Here we only consider the simple case of a stochastic, stable environment with one, stationary, prey type.

2.3 THE PROBLEM OF ATTAINABILITY

One of the main criticisms of optimal foraging theory is that attaining the optimal strategy is often well beyond an animal's cognitive abilities (Pierce and Ollason, 1987). It would be argued, for example, that a lizard could not solve cubic equations such as that required by Eq. (1). A growing body of experimental evidence, however, suggests that many animals have a concept of number, time, space, and rate, and that they can perform abstract operations isomorphic to addition, subtraction, and division (Gallistel, 1989, 1990). A simple decision rule proposed by Roughgarden (1995) requires a lizard no more than the ability to perform such basic operations.

In simulations, the average rate of net energy gain of a lizard that follows this rule and that remembers all its prey encounters indefinitely into the past seems to converge on the maximum possible rate. Menzer, Hart, and Littman provide a formal proof for this convergence as an appendix to the chapter.

We begin with a modification of the decision rule using only finite memory length.

2.4 A RULE OF THUMB

When the nth insect appears in the lizard's field of vision the lizard is assumed to make two calculations. It calculates what its long-term mean E/T would be if it pursued the insect ($E/T_p^{[n]}$), and what it would be if it ignored the insect ($E/T_i^{[n]}$). It chooses the option that yields the higher rate.

The net energy intake from the previous m ($m \geq 1$) insect appearances, E^m, is

$$E^m = \sum_{j=1}^{m}(e^{[n-j]} - t_w^{[n-j]}e_w - t_p^{[n-j]}e_p), \tag{2}$$

and the total duration of the previous m insect appearances, T^m, is

$$T^m = \sum_{j=1}^{m}(t_w^{[n-j]} + t_p^{[n-j]}) \tag{3}$$

where t_w is the time waiting for an insect to appear since the last insect appeared, and t_p is the time to pursue the insect and return to the perch. The expected long-term mean E/T is the net energy intake in the past plus the net energy gain (or loss) from pursuing or ignoring the present insect, divided by the total time past plus the time of waiting for the present insect or of waiting and pursuing it. The equations are the following:

$$(E/T)_p^{[n]} = \frac{E^m + e^{[n]} - t_w^{[n]}e_w - t_p^{[n]}e_p}{T^m + t_w^{[n]} + t_p^{[n]}}, \tag{4}$$

and

$$(E/T)_i^{[n]} = \frac{E^m - t_w^{[n]}e_w}{T^m + t_w^{[n]}}. \tag{5}$$

Memory is treated as a window of the last m insect appearances. The larger m is, the greater is the weight given to past episodes relative to the present one. For the special case of $m = 0$, E^m and T^m equal zero.

A memory window results in a step-function weighting of the past. Different weighting functions (e.g., Kacelnik and Krebs, 1985; Todd and Kacelnik, 1993) could be incorporated into the model as a better understanding of how lizard memory is gained.

3. SIMULATION

All the lizards in the simulations foraged according to the above decision rule, and were naive about the state of the environment at the beginning of each simulation, $E^{[0]} = T^{[0]} = 0$. We endowed different individuals with different memory lengths, and at the end of each run we calculated the mean rate of net energy gain of each individual. Values for all the parameters used in the model are available, and are presented by Roughgarden (1995).

3.1 SIMULATION PARAMETERS

Insects were 2 mm long, representing a typical ant or small fly, and appeared at a rate of 60 per m^2 per 12 hours. The lizard could see insects that appeared up to 12 m away, within 180° in front of it.[1] (An insect's energy content is related to its size, and so beyond about 11.5 m, the energy expended pursuing a 2-mm insect is greater than the insect's net energy content. Lizards never chased an insect as far as 12 m). Each simulation ended when a lizard had encountered 500 insects. Each simulation of a lizard with a particular memory length was repeated 400 times.

3.2 SIMULATION RESULTS

The optimal cutoff radius for the parameters used in the simulations is 6.81 m (from Eq. 1). A lizard made a "mistake" when it pursued an insect farther than this distance, or when it ignored an insect that appeared within this radius. In Figure 1 we show the behavior of three lizards. The lizard with memory length of the last 10 episodes made many fewer mistakes than the lizard that only remembered the previous episode. The lizard with a memory window of 100 soon learned not to pursue insects beyond about the optimal cutoff radius and not to ignore insects within it.

The maximum mean rate of net energy intake that a lizard could attain is 0.0726 joules per second. The mean E/T of lizards that had no memory was about half of that (Figure 2). Lizards that remembered even just a few past episodes did much better, but the extra advantage of additional memory showed diminishing returns. Mean E/T continued to slowly approach the maximum rate with increased memory length.

A lizard with no memory that foraged for four hours on an average day, and maintained a resting metabolic rate the rest of the time, would produce an egg every 15.9 days (Figure 3). With a memory of even only two previous episodes, it

---

[1]In the present simulations insects appeared in a rectangle 12 × 24 m.

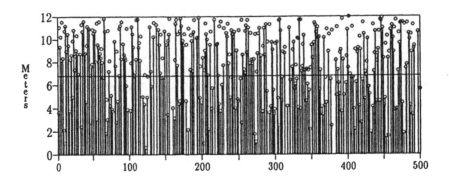

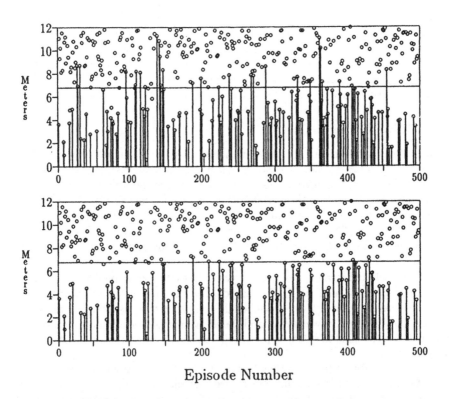

FIGURE 1   The distance insects were pursued by a lizard with a memory length of
1 (top), 10 (middle), and 100 (bottom) last episodes. Each circle represents an insect
appearance. A vertical line to a circle means that the lizard pursued that insect; circles
not attached to a vertical line represent insects that were ignored. The horizontal line is
the optimal cutoff radius.

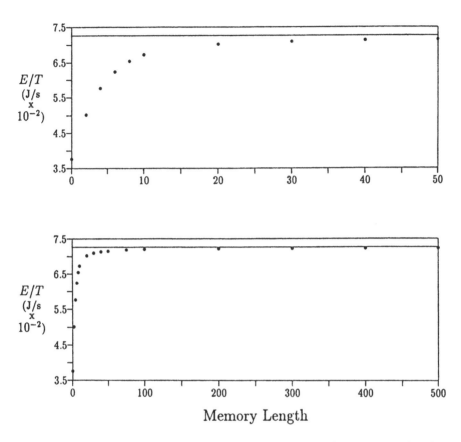

FIGURE 2 The mean rate of net energy gain of lizards with different memory lengths. The top graph emphasizes the shorter memory lengths. For each individual lizard the mean rate was calculated over a foraging bout of 500 episodes. Each data point is the mean of 400 lizards (standard errors are too small to see in the figure). The horizontal line is the maximum possible yield in the simulated environment.

would lay almost twice as many eggs, one every 8.6 days. Fertility rises as memory length increases, approaching the maximum possible yield of an egg every 4.7 days.

The effect of initial small increases in memory length on fertility is more pronounced than on mean rate of net energy intake (compare Figure 3 with top of Figure 2). This is because an individual that has a short memory gains less energy at the end of each day than an individual with a longer memory, and thus it must forage for more days to produce one egg. All lizards spend the same amount of energy each day on self-maintenance during the time that they do not forage.

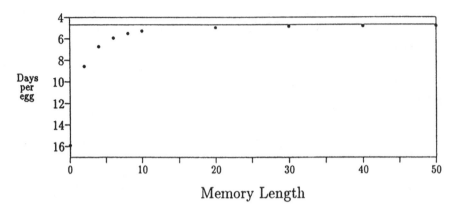

FIGURE 3 The mean number of days to produce one egg for lizards with different memory lengths. Each data point is the mean of 400 lizards. The horizontal line is the minimum number of days to produce an egg in the simulated environment.

Therefore, the more days it takes a lizard to produce an egg, the more total energy that is spent on self-maintenance during the production of that egg. Lizards that forage inefficiently, then, must forage for yet additional days per each egg laid to make up for this extra energy expenditure.

4. DISCUSSION

We have shown that a lizard with a relatively short memory span can make nearly optimal foraging decisions by following a decision rule that requires only simple computational abilities. Many animals are capable of making such computations (Gallistel, 1989, 1990). Anoles, in particular, can quickly learn to modify their foraging behavior in ecologically relevant, instrumental conditioning experiments (Shafir and Roughgarden, 1994).

The penalty for having no memory is very large. A lizard that can remember two previous insect appearances will experience an almost 100% increase in reproductive fitness compared to a lizard that has no memory. The evolutionary selective pressure for initial increase of memory length is likely to be very strong. The selective pressure diminishes as additional increases in memory result in only slight increases in fitness.

Studies of foragers' use of memory have focused on two questions. Theoretical arguments have been used to determine what would be the optimal weighting of past and recent events in particular environments (e.g., McNamara and Houston,

1987). Empirical work has tested how animals actually weight past and recent events when making foraging decisions (e.g., Real et al., 1990; Todd and Kacelnik, 1993; Cuthill et al., 1990). In this paper, by measuring the effect of memory length on an individual's reproductive fitness, we were able to consider a different question: What are the fitness costs of suboptimal memory lengths? In order to understand the evolution of a trait within the whole organism, where there may be tradeoffs between optimizing different traits, it is important to determine both what the optimal condition is, and what is the cost of deviating from it. In general, optimality theory predicts that an animal's memory length should evolve to be more close to the optimal memory the greater the cost of suboptimal memory.

For example, bumble bees (Real et al., 1990), pigeons (Todd and Kacelnik, 1993), and starlings (Cuthill et al., 1990), base their foraging decisions most heavily on the previous one or two flower or prey encounters. Real et al. (1990) have considered some ecological conditions, and possible constraints on learning and memory, under which such sharp discounting of the past could be advantageous. Our results provide a complementary argument, that there would be very weak selection against sharp discounting of the past if the decision rules that these animals followed, like the one presented here, resulted in steep diminishing gains curve with increased memory.

ACKNOWLEDGMENTS

We thank Interval Research Corporation for financial support, and the Santa Fe Institute for supporting our stay in Santa Fe during the PINEP workshop. We thank Michael Littman and Filippo Menczer for very constructive reviews of the manuscript.

REFERENCES

Bergman, A., and M. W. Feldman. "On the Evolution of Learning: Representation of a Stochastic Environment." *Theor. Pop. Biol.* (forthcoming).

Cuthill, I. C., A. Kacelnik, J. R. Krebs, P. Haccou, and Y. Iwasa. "Starlings Exploiting Patches: The Effect of Recent Experience on Foraging Decisions." *Anim. Behav.* **40** (1990): 625–640.

Gallistel, C. R. "Animal Cognition: The Representation of Space, Time and Number." *Ann. Rev. Psychol.* **40** (1989): 155–189.

Gallistel, C. R. *The Organization of Learning.* Cambridge: MIT Press, 1990.

Kacelnik, A., and R. Krebs. "Learning to Exploit Patchily Distributed Food." In *Behavioural Ecology: Ecological Consequences of Adaptive Behaviour*, edited by R. M. Sibly and R. H. Smith, 189–205. Oxford: Blackwell Scientific Publications, 1985.

McNamara, J. M., and A. I. Houston. "Memory and The Efficient Use of Information." *J. Theor. Biol.* **125** (1987): 385–395.

Pattee, H. H. "Simulations, Realizations, and Theories of Life." In *Artificial Life*, edited by C. G. Langton. Santa Fe Institute Studies in the Sciences of Complexity, Proc. Vol. VI, 63–77. Redwood City, CA: Addison-Wesley, 1988.

Pierce, G. J., and J. G. Ollason. "Eight Reasons Why Optimal Foraging Theory is a Complete Waste of Time." *Oikos* **49** (1987): 111–118.

Real, L., S. Ellner, and L. D. Harder. "Short-Term Energy Maximization and Risk-Aversion in Bumble Bees: A Reply to Possingham et al." *Ecology* **71** (1990): 1625–1628.

Roughgarden, J. *Anolis Lizards of the Caribbean: Ecology, Evolution, and Plate Tectonics.* Oxford: Oxford University Press, 1995.

Shafir, S., and J. Roughgarden. "Instrumental Discrimination Conditioning of *Anolis cristatellus* in the Field with Food as a Reward." *Carib. J. Sci.* **30** (1994): 228–233.

Todd, I. A., and A. Kacelnik. "Psychological Mechanisms and The Marginal Value Theorem: Dynamics of Scalar Memory for Travel Time." *Anim. Behav.* **46** (1993): 765–775.

Zhivotovsky, L. A., A. Bergman, and M. W. Feldman. "A Model of Individual Behavior in a Fluctuating Environment." This volume.

Filippo Menczer, William E. Hart, and Michael L. Littman

Appendix to Chapter 12: "The Effect of Memory Length on Individual Fitness in a Lizard"

DEFINITIONS AND SIMPLIFICATIONS

In order to keep the derivation simple, we analyze the special case in which $e_p = e_w = 0$. Since these are constant, the proof of the general case is qualitatively similar. Then Eq. (1) in Shafir and Roughgarden's chapter (this volume) reduces to $r^* = (3v/\pi a)^{1/3}$ where v is the lizard speed and a is the constant rate of prey appearances per unit of area and time.

The decision rule minimizes (T_n/E_n). This ratio is a function of $(T_{n-1}, t_n^w, d_n, E_{n-1})$, where T_{n-1} and E_{n-1} are the memory of the lizard, while t_n^w and d_n are measured for each prey. Prey are points uniformly distributed in a semicircle completely contained in an area A. Rewriting Eqs. (4) and (5) from Shafir and Roughgarden yields:

$$\left(\frac{T_n}{E_n}\right) = \begin{cases} \frac{T_{n-1}+t_n^w}{E_{n-1}} & \text{if } n\text{th prey ignored;} \\ \frac{T_{n-1}+t_n^w+2d_n/v}{E_{n-1}+e} & \text{otherwise.} \end{cases} \tag{A1}$$

Adaptive Individuals in Evolving Populations, Ed. R. K. Belew & M. Mitchell,
SFI Studies in the Sciences of Complexity, Vol. XXVI, Addison-Wesley, 1996   **183**

To start the iteration, $T_0 = E_0 = 0$ and the first point is always pursued. The definitions of E_n and T_n are implicit in Eq. (A1):

$$E_n \equiv e \cdot n \cdot \text{(fraction of pursued points out of } n); \tag{A2}$$

$$T_n \equiv \sum_{i=1}^{n} t_i^w + \frac{2}{v} \sum_{\substack{i=1 \\ i:\text{pursued}}}^{n} d_i. \tag{A3}$$

Using Eq. (A1), the decision rule for the nth point can be rewritten as the following greedy strategy: *pursue if and only if*

$$\frac{T_{n-1} + t_n^w + 2d_n/v}{E_{n-1} + e} < \frac{T_{n-1} + t_n^w}{E_{n-1}}. \tag{A4}$$

With a little algebra we can rewrite Eq. (A4) as

$$d_n < \frac{ev}{2}\left[\left(\frac{T_{n-1}}{E_{n-1}}\right) + \frac{t_n^w}{E_{n-1}}\right] \equiv r_n, \tag{A5}$$

where we have *defined* the critical distance, r_n, estimated implicitly by the decision rule after n points.

THEOREM If the decision rule has a fixed point for (T_n/E_n), i.e., there exists some limit $(T/E)^* \equiv \lim_{n\to\infty}(T_n/E_n)$, then $\lim_{n\to\infty} r_n = \sqrt[3]{3v/\pi a}$.

PROOF We want to prove that the limit of r_n is equal to the optimal foraging distance defined above. From Eq. (A5) this limit is:

$$r^* \equiv \lim_{n\to\infty} r_n = \frac{ev}{2}\left[\lim_{n\to\infty}\left(\frac{T_{n-1}}{E_{n-1}}\right) + \lim_{n\to\infty}\frac{t_n^w}{E_{n-1}}\right]. \tag{A6}$$

Since prey appear with a temporal and spatial distribution that does not change in time, $\forall n : t_n^w < \infty$. Furthermore, from definition (A2) and the convergence hypothesis, it is clear that $E_n \propto_{n\to\infty} n$. Therefore, the second term in Eq. (A6) can be disregarded:

$$r^* = \frac{ev}{2} \lim_{n\to\infty}\left(\frac{T_{n-1}}{E_{n-1}}\right) = \frac{ev}{2}\left(\frac{T}{E}\right)^*. \tag{A7}$$

Using the definitions (A2) and (A3) we can rewrite Eq. (A7):

$$\frac{2r^*}{ev} = \left(\frac{T}{E}\right)^* = \lim_{n\to\infty}\left(\frac{T_n}{E_n}\right)$$

$$= \lim_{n\to\infty}\frac{\sum_{i=1}^n t_i^w + \frac{2}{v}\sum_{\substack{i=1\\i:\text{pursued}}}^n d_i}{E_n} \qquad (A8)$$

$$= \lim_{n\to\infty}\frac{n\langle t^w\rangle_n + \frac{2}{v}\frac{E_n}{e}\langle d_{\text{pursued}}\rangle_n}{E_n}$$

$$= \lim_{n\to\infty}\frac{n\langle t^w\rangle_n}{E_n} + \lim_{n\to\infty}\frac{2\langle d_{\text{pursued}}\rangle_n}{ev}$$

where brackets denote expected values estimated by averages. For the first term of Eq. (A8) we can use the definition (A2) and the central limit theorem to obtain

$$\lim_{n\to\infty}\left(\frac{n}{E_n}\langle t^w\rangle_n\right) = \lim_{n\to\infty}\left(\frac{n}{E_n}\right)\lim_{n\to\infty}\langle t^w\rangle_n$$

$$= \lim_{n\to\infty}\left[\frac{n}{e\cdot n\cdot(\text{fraction of pursued points out of }n)}\right]\cdot\frac{1}{aA}$$

$$= \frac{1}{e\cdot\lim_{n\to\infty}(\text{fraction of pursued points out of }n)}\cdot\frac{1}{aA}$$

$$= \frac{1}{e\cdot\Pr[\forall i: d_i < r_i]}\cdot\frac{1}{aA} \qquad (A9)$$

$$= \frac{2A}{e\cdot\pi(r^*)^2}\cdot\frac{1}{aA} = \frac{2}{e\pi(r^*)^2 a}.$$

Note that in Eq. (A9) we have also used the convergence hypothesis. For the second term of Eq. (A8) we can use Eq. (A5), the known space distribution of prey in the semicircle, and again the central limit theorem and the convergence hypothesis, to obtain

$$
\begin{aligned}
\lim_{n \to \infty} \frac{2 \langle d_{\text{pursued}} \rangle_n}{ev} &= \frac{2}{ev} \lim_{n \to \infty} \langle d_{\text{pursued}} \rangle_n \\
&= \frac{2}{ev} \operatorname{Exp}\left[d_{\text{pursued}} \right] \\
&= \frac{2}{ev} \int_0^{r^*} r \Pr\left[\forall i : r \le d_i < r + dr \right] \\
&= \frac{2}{ev} \int_0^{r^*} r \frac{\pi r \, dr}{\pi (r^*)^2 / 2} \\
&= \frac{4}{ev(r^*)^2} \int_0^{r^*} r^2 \, dr \\
&= \frac{4r^*}{3ev}.
\end{aligned}
\tag{A10}
$$

Combining Eqs. (A8), (A9), and (A10) we finally obtain

$$
\frac{2r^*}{ev} = \frac{2}{e\pi (r^*)^2 a} + \frac{4r^*}{3ev}
\tag{A11}
$$

and, upon solving Eq. (A11) for r^*:

$$
r^* = \sqrt[3]{\frac{3v}{\pi a}}.
\tag{A12}
$$

Thus the critical distance estimated by the decision rule converges exactly to the optimal foraging distance found analytically. Q.E.D.

The hypothesis of the preceding theorem, convergence of (T_n/E_n), cannot be proved in general. We believe, however, that it holds in this case due to the particular function minimized by the optimal foraging distance. In fact, Roughgarden (in press) obtains the critical cutoff radius of Eq. (A12) by minimizing the analytically derived expression

$$
\frac{T}{E} = \frac{1}{e} \left(\frac{2}{a\pi r^2} + \frac{4r}{3v} \right)
\tag{A13}
$$

with respect to r. But clearly the right-hand side of Eq. (A13) has a unique minimum for $r > 0$, and the absence of local optima is a necessary and sufficient condition for a greedy method to converge in the limit of infinite iteration steps.

Jonathan Roughgarden

Preface to Chapter 13

This paper extends artificial life research to a new level of sophistication and at the same time establishes new links between the adaptive computation literature in computer science and the ecology literature in biology. The starting point for the paper is the observation that most implementations of genetic algorithms view an individual as an agent that carries out a fixed program whose fitness is computed by running the program in some computational environment. Over time, mutation, mating, and selection among the various agents' programs cause the population to consist of agents that execute the best programs in the given environment. The genetic algorithm therefore is analogous to the simplest case of evolutionary genetics, where each genotype has a fixed fitness in the given environment. But as Menczer and Belew note, one might also be interested in using the genetic algorithm to construct computer programs that learn how to operate efficiently in their environment. In this context, the fitness measure for an agent and its program is less obvious, and the genetic algorithm may be analogous to density-dependent natural selection in ecology, where the fitnesses are not constants but depend on the quantity of individuals in relation to some finite amount of resource. Indeed, the fundamental problem that Menczer and Belew address is distinguishing whether an agent's adaptive program is better because it actually does a better job or because it is merely offered a richer environment within which to execute.

Adaptive Individuals in Evolving Populations, Ed. R. K. Belew & M. Mitchell,
SFI Studies in the Sciences of Complexity, Vol. XXVI, Addison-Wesley, 1996   **187**

For this reason Menczer and Belew set up a scheme whereby the environment can be configured *a priori* with a desired amount of complexity. Moreover, just how hard it is to live in this environment can also be specified. The agents that must live in the environment are imagined to be objects with two parts, a sensory part and a body, because most programs have both input/output code and a body, or main(), within which the computational work is done. Also, these agents include a neural network to implement each agent's learning during its execution history. Finally, the agents mate, share code, and evolve through generations, as an instance of the genetic algorithm. A particularly important aspect of the way that Menczer and Belew set up the genetic algorithm is that the number of agents is not fixed, but varies in accordance with their success in the environment. Therefore the population size changes, and this suggests how to measure the fitness of an adaptive program.

Menczer and Belew's major conclusion is that the eventual population size is a measure of the fitness of the adaptive programs they are executing. Moreover, by comparing the cleverness of the agents' strategies in the specified environment to that of an agent's random actions, they can distinguish whether a high population size is brought about by an efficient use of the environment or by simply living in a rich environment to begin with. Menczer and Belew conclude by indicating how other algorithmic issues, such as the complexity of the input/output system and the responsiveness of the program to environmental changes, might be investigated in the same way.

Some aspects of this work recapitulate the development of what is called life history theory in ecology. The use of the population size as a fitness measure is called K-selection, where K is the usual symbol to denote the equilibrium population size in a population dynamic model called the logistic equation. It is known that the idea of K-selection as a fitness measure is not valid if the organisms directly interfere with each other; should the agents attack each other (as in a computer virus), then other fitness measures will be needed. There is, however, no counterpart in the ecological literature to the complete environmental description introduced in this paper, and this strikes me as very novel.

My final remark is addressed more to readers who are biologists than those who are computer scientists. It is often very difficult for a scientist to read a paper in the artificial life literature because it is usually not made clear that natural processes are being used as an analogy for a project in which the goal is to develop computer programs. Instead, it may sound as though writers in this area think they are doing natural science. Menczer and Belew refer, for example, to a variable population size in the genetic algorithm as "better founded biologically." What does this mean? It means, I think, that a project whose goal is to evolve computer programs will be more successful if it mimics this aspect of real-life ecology. Still, sometimes a confusion between computer science and natural science seems genuine; for example, the authors state that "LEE can be extended to deal with more complex topics in theoretical biology, such as coevolution of neurosomatic traits, evolution of learning, sexual reproduction, and multiple-species population dynamics." To contribute to ecological theory, except through analogy, requires attention to ecological particulars which seem quite out of place in a project whose goal is to evolve adaptive computational agents.

Chapter 13:
Latent Energy Environments

A novel artificial life (ALife) model, called LEE, is introduced and described. The motivation lies in the need for a measure of complexity across different ALife experiments. This goal is achieved through a careful characterization of environments in which different forms of energy are well-defined. A steady-state genetic algorithm is used to model the evolutionary process. Adaptive organisms in the population are modeled by neural networks with non-Lamarckian learning during life. Behaviors are shown to be crucial in the interactions between organisms and their environment. Some emerging properties of the model are illustrated and discussed.

1. INTRODUCTION

From the tradition of artificial intelligence, the relationship between an adaptive individual and the environment within which it must operate is often viewed in relatively simplistic, stimulus-response terms. The environment provides input on the basis of which the individual then responds. Perhaps because it has grown out of the same tradition, the genetic algorithm (GA) (Holland, 1992) often relies upon an equally simplistic view of the relationship between evolving populations and their environment: the GA generates individuals of a population and the environment must return only the fitness of each of these individuals. The GA can then be viewed as a function optimizer.

In spite of the many reasons to recommend this evolution-as-search perspective, any such model fails to address one of evolution's most obvious characteristics: its *creative* power. One very useful insight arising from an artificial life (ALife) perspective is that the above dichotomy is replaced by a more holistic view of organism and environment as part of a single system to be modeled. In order to allow the necessary evolutionary creativity, the central problem facing every designer of ALife environments is the specification of an artificial task admitting strategies of arbitrary and unanticipated complexity. Unfortunately, this coupling of organisms with environments has created a major methodological problem within ALife research: results that report behaviors of different organisms in different environments are incommensurable, making it very hard to decide whether an observed difference in adaptation is really significant, or is only due to one of the environments being more *complex* (Rössler, 1974; Wilson, 1991; Belew, 1991).

Some models that try to address the environmental complexity issue, such as *Echo* (Holland, 1992), are so inclusive that their results are very hard to interpret. Other recent works have began to analyze simpler environments in open-ended evolutionary simulations (e.g., Todd and Wilson, 1993). The next section will introduce our approach, called "latent energy environments" (LEE), which attempts to extend the characterization of complexity to a more general class of environments, while maintaining the analysis' manageability by carefully enforcing appropriate physical constraints. A view of the model is offered in Figure 1. The arrow labeled "behavior" represents all types of interactions between organisms and environments, described in Section 3. The arrow labeled "evolution" represents, roughly, all genetics-mediated interactions among organisms. In Section 4 we describe the GA used to model such evolutionary processes. In Section 5 we illustrate some

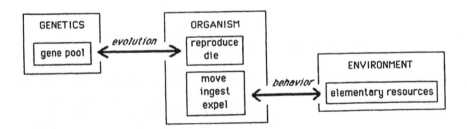

FIGURE 1   LEE model.

	A_1	A_2	A_i	A_t
A_1	0	+1, A_3		+3
A_2		−2		−1, A_t
A_j			E_{ij}, P_{ij}	
A_t				0

FIGURE 2   Example of a reaction table.

general, emerging properties of the model, while in the final section we exemplify some of our objectives for the use of LEE.

2. MODELING ENVIRONMENTAL COMPLEXITY

Our central goal is the specification of a series of ALife environments of graduated complexity. To accomplish this we begin by replacing the standard model of "food" as a spatially localized element of the environment required for survival, with a series of inert environmental "atomic elements" that must be combined by organisms in order to realize the "energy" they require for survival. Consider a simple discrete world with cells placed on a two-dimensional grid. Let the only source of (positive or negative) energy exist via binary reactions; i.e., combining two base elements results in an energy gain or loss, plus some other by-product element(s). Let the base elements belong to an artificial set of types $\{A_1, A_2, \ldots, A_t\}$. Furthermore, let elements of each of these types be generated according to some spatiotemporal distribution. A reaction will be denoted as:

$$A_i + A_j \longrightarrow E_{ij} + P_{ij}$$

where E_{ij} is the resulting energy ($E_{ij} > 0$ for exothermic reactions, $E_{ij} < 0$ for endothermic ones) and P_{ij} indicates zero or more by-products. All the possible reactions can be represented by the entries of a symmetric reaction table like the one in Figure 2. This table can be viewed as that part of the LEE model which is neither directly nor indirectly affected by adaptation, and remains constant throughout the entire evolutionary process. In other words, it represents the laws of physics and chemistry, which by definition do not change over time.

We can now define two critical features of this artificial world. First, let us define a sequence, or chain, of reactions. For any world configuration, there may be many possible combinations of existing elements according to the possible reactions in the table. After one of these reactions takes place, two elements are consumed and some new elements (by-products) may appear, giving rise to a new configuration. Eventually, the chain terminates if no possible reactions exist among the remaining elements. Every sequence of reactions has a corresponding potential energy, given by the arithmetic sum of energies released and/or absorbed by all the reactions in the sequence. For each world configuration, we call latent energy the set of potential energies corresponding to all reaction chains available to organisms starting from that configuration. By controlling the rate at which elements of each type are introduced, we can regulate the amount of potential energy available in its latent form, while actual energy can ultimately be realized only upon dynamically choosing a reaction sequence.

Second, the amount of *work* required to release energy is defined in terms of the distance that one of the elements in a reaction must be moved in order to occupy the cell of the other. This is the correct metric for evaluating organisms that must harvest energy to survive, and whose fundamental behavior, as will be shown in the next section, is movement. By controlling the spatial distributions of element types in the world, we can dynamically regulate the amount of work required to combine them. Similar to the way we have associated potential energy to a reaction sequence, we can also associate to a chain its corresponding work. Subtracting this from latent energy and maximizing over all possible reaction sequences, we can estimate optimal energy efficiencies. In short, we have control and quantitative knowledge of how much latent energy is available in a world, how much work must be done in order to realize this energy, and how difficult therefore it is for an organism (or ecology of organisms) to survive!

These definitions allow us to control parameters of the environment without specifying just how latent energy is to be realized by one or more species. The monitoring of population dynamics makes it possible to compare ecologies in different environments, in terms of how efficiently each exploits the energy latent in its own environment. Different species may exploit different reaction chains in the same environment, but we should expect the most efficient one to prevail if its uses of atomic elements are mutually exclusive.

3. MODELING THE LIFE PROCESS

Each organism in the evolving populations of LEE has a "brain" and a "body." The former interacts with the environment external to the body through a sensory-motor system: it receives input corresponding to the organism's stimuli and produces output controlling its behaviors. The sensory system is composed of a set of sensors with different characteristics. These collect information from either the external world or the internal body of the organism, and map it onto the brain input. Sensors may differ in range, directionality, sensitivity, resolution, specificity,

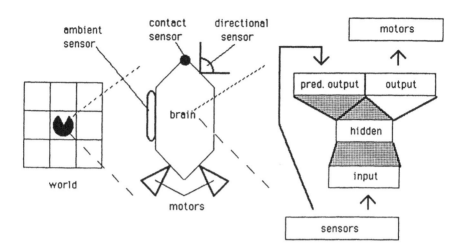

FIGURE 3   A LEE organism. On the left, the organism is displayed situated and oriented in the environment; in the middle, the body of the organism is expanded and different possible sensors and motors are shown; on the right, the architecture of a NNet modeling the organism's brain is illustrated: during prediction learning, the sensory information at a given life cycle is compared with the prediction formulated in the previous cycle.

accuracy, etc. The sensory information is elaborated by the brain to produce an output, interpreted as an action (movement) in the world and executed by a set of motors that make up the motor system. Motors may have different characteristics such as energy cost, power, orientation, accuracy, etc. In Figure 3 we illustrate an organism's body structure with a typical sensory-motor system. The latter provides a strong coupling between environmental complexity and difficulty of the survival task (Menczer and Parisi, 1992).

3.1 BEHAVIOR

One of the fundamental features of LEE is that energy in an environment can be released only through the *behaviors* of organisms. Behaviors induce reaction sequences, thus catalyzing the transformation of latent energy into usable energy. The notion of behavior immediately binds an organism to its environment in a much more intrinsic way than allowed by simple fitness-function GA models.

Ethnologists have long argued that behaviors be treated as first-order phenotypic characteristics of an organism. A second desirable consequence of modeling behaviors, then, is that it demands a more elaborate characterization of the genotype/phenotype distinction (Belew, 1993). It has generally been too easy for

GA modelers to assume a fairly direct correspondence between the genotypic data-structure manipulated by genetic operators (mutation, crossover, etc.) and the phenotype's fitness (i.e., number of offspring) ultimately evaluated by the environment. The appropriateness of behaviors is obviously conditional upon environmental context, and the definition of "adaptive" behavioral features must therefore depend on a complex interplay of genetic and environmental circumstances.

We take *movement* to be our canonical behavior, in part because this overt action is most easily observed. However, certain other actions on the part of the modeled organisms are also part of the model and may be under an organism's discretionary control. One such example is the ingestion of elements in the world: an organism can only carry around a limited number of reaction-causing elements in its "gut," so it may have to discriminate among elements to ingest along its path.

3.2 LEARNING

We wish to allow the behavior of an organism to improve over the course of its lifetime as a consequence of its experience, i.e., to *learn*. Littman (Chapter 26) provides a survey of architectures for learning in evolutionary frameworks. One approach is to use a well-studied type of neural network (NNet) as our computational model of an organism's brain. The LEE context helps to distinguish the particular type of NNet learning of interest to us.

First, we make no assumptions about additional "teaching" input being available to the organism, beyond the stimulus of the organism's environment. This would seem to preclude many successful "supervised" learning techniques, such as backpropagation of error (Rumelhart, Hinton, and Williams, 1986). Such error correction algorithms require that the NNet be given information about the "correct" output/action. Not only are these correct actions unknown, but even attempting such a normative stance violates the spirit of creative discovery that we treat as central.

Fortunately, if the organism is simply forced to *predict* the expected outcome of its actions, differences between its expectations and the actual outcome can generate the same sort of error information, without any additional teacher (Sutton, 1988). Prediction learning has been used successfully in a number of ALife simulations (Nolfi, Parisi, and Elman, 1994; Parisi, Cecconi, and Nolfi, 1990), and the NNets used in LEE will exploit this predictive device. In Figure 3 we show a possible organism's NNet architecture.[1] Note that only a part of the weights of the last connection layer (marked in dark) is learnable through the prediction task.

Besides learning to predict good actions, organisms controlled by NNets must also survive in order for them to reproduce and for the population to evolve. For example, an organism could generate perfect predictions simply by choosing to take

---

[1] Other unsupervised learning architectures have been used in the LEE framework and described elsewhere (Menczer and Belew, 1994).

no action whatsoever (assuming for the moment a static environment in which it is the only actor). The goal of accurate prediction is therefore secondary, subordinate to the organism's primary goal of surviving.

The second important characteristic of our NNet learning situation is that the organism is an *active* selector of its own training experience. One aspect of this problem is to perform "critical experiments" that most rapidly allow the learning system to converge on a consistent explanation of the world. However, an experiment that may immediately help to identify an important pattern may also be deadly to perform. Holland (1992) has characterized this as the "exploitation/exploration" dilemma, where an organism's ability to exploit already identified regularities must be balanced against the acquisition of new information.

4. MODELING THE EVOLUTIONARY PROCESS

The LEE model uses a *steady-state* GA (De Jong and Sarma, 1993) to simulate the evolutionary process. This means that successive generations are interleaved through time, shifting the control of differential reproduction from the experimenter to the adaptive process itself. Rather than being determined by the arbitrary ranking mediation of an explicit fitness measure, selective pressure results directly from intrinsic competition for finite resources within the LEE population. The only currency for such resources in the model is the energy latent in the environment. This characteristic places LEE in the class of *endogenous fitness models* (Mitchell and Forrest, 1994).

For each time step, every organism executes a basic life cycle.[2] The operations performed in one such cycle are outlined in Figure 4. This simple steady-state version of the GA has a few important consequences. The most important one is that the population size does *not* remain constant throughout an experiment. We consider this an important feature of the model, better founded biologically than the fixed-population GA. As shown in the next section, the population size can become stable spontaneously (and quite robustly indeed) when the environmental conditions allow it, without this being imposed externally by the model. Another biologically plausible consequence of having a variable population size is that *extinction* can occur.

Finally, the LEE evolutionary process is *noisy*, for two reasons: the high variance caused by the steady-state GA, and the fact that every parameter for which there is no value motivated by the model is treated as a random variable. Conforming to such a stochastic-oriented principle allows the LEE model to be free of

---

[2]This is an intrinsically parallel process. For sequential machines, however, it can be simulated via sequential calls to the organisms. This is the case in our implementation, where each organism is called in random order so as to minimize the spurious bias of the sequential simulation.

```
for each time cycle {
      for each alive organism {
            sense world;
            feed forward activations;
            move;
            learn;                     /* change phenotype    */
            digest;                    /* catalyze reactions  */
            if (energy > α) {
                  reproduce;           /* copy genotype       */
                  mutate;              /* change new genotype */
                  develop;             /* get new phenotype   */
            }
            else if (energy < ω) die:
      }
      replenish world;
}
```

FIGURE 4   Pseudo-code for the main loop of the LEE GA. The constants α and ω are related to the energy distribution over the population: at initialization, organisms are given random energy uniformly distributed in the interval $[\omega, \alpha]$; at reproduction, energy is conserved by parent and offspring splitting the parent's energy evenly.

unwanted bias from the experimenter. Noise is the cost paid for this: results may be harder to interpret, and adaptation may be slowed down. Furthermore, stochasticity may cause many types of drift in the dynamics of population genetics. Once again, we are satisfied that all these effects are present in nature and thus the LEE model is strengthened by its stochastic features as well.

5. EMERGING FEATURES

While classical GAs constitute the standard paradigm for adaptation viewed as an optimizing process, endogenous fitness models are being used increasingly often as a paradigm for open-ended evolution (Brooks and Maes, 1994). However, the creative power of such models has exposed the lack of appropriate instruments of analysis (Mitchell and Forrest, 1994). In this section we report on some emerging properties of our simulations, to illustrate that the task of analyzing endogenous fitness models is greatly facilitated by relying on well-founded constraints such as the conservation of energy.

5.1 CARRYING CAPACITY

There is a connection between the size of an evolving population and the resources available from its environment. Such resources impose a limit on how many individuals can be sustained: we identify this maximum population size with the *carrying capacity* of the environment. For example, extinction indicates that the carrying capacity of the environment is insufficient to support a population large enough to withstand stochastic fluctuations.

LEE allows us to quantitatively estimate optimality of behaviors by monitoring the population size throughout an experiment. To see how, let us analyze the relationship between latent energy and population size in a particular example. Consider the following reaction energy table for the simple case of an environment with only two types of atomic elements, a and b:

	a	b
a	$-\beta E$	E
b	E	$-\beta E$

$$(0)$$

where

$$E > 0,$$
$$|\beta| < 1, \tag{1}$$

and there are no by-products. Table (0), along with the spatiotemporal distributions of the elements, determines the environmental complexity of the survival task. The minimal set of elements for alternative behavioral strategies is $\{a, a, b, b\}$. In fact, with these elements an organism can catalyze the following sets of reactions:

$$(a + a), (b + b) \longrightarrow -2\beta E, \tag{2}$$
$$(a + b), (a + b) \longrightarrow +2E, \tag{3}$$

where Eq. (3) is clearly a more advantageous strategy than Eq. (2), given conditions (1). If r is the expected rate of replenishment for both a and b elements, i.e.,

$$r_a \equiv \#a/\text{life cycle},$$
$$r_b \equiv \#b/\text{life cycle},$$
$$r_a = r_b \equiv r,$$

then a set $\{a, a, b, b\}$ is produced every $2/r$ life cycles. Thus strategies (2) and (3) produce energy changes

$$\Delta E_2 = \frac{-2\beta E}{2/r} = -r\beta E, \tag{4}$$

$$\Delta E_3 = \frac{2E}{2/r} = rE, \tag{5}$$

per unit time (life cycle), respectively. Let us now consider the situation at equilibrium. On average, two conditions are verified: the population size remains constant, and energy is produced at the same rate at which it is consumed by organisms in the population. Using brackets to indicate time averages, we can write:

$$\langle \text{population size} \rangle = \text{const} \equiv p, \tag{6}$$

$$\langle \Delta E \rangle = 0. \tag{7}$$

Since energy is always conserved, the only consumed energy is that lost in the form of work, that is, used for moving in the world. If we call c the cost per move for any organism and use Eq. (6), the average energy used by the population per unit time is pc. To calculate how much energy is produced, we must know the strategy used on average by the population for combining elements. In other words, we need to determine how efficiently the latent energy contained in the elements is transformed into usable energy. Let us then introduce a probability distribution over strategies. This is quite simple in the present example, with only two strategies: let us call η the probability of strategy (3), so that $(1 - \eta)$ is that of strategy (2). Then the average energy produced per unit time by the population is, using Eqs. (4) and (5),

$$\eta \Delta E_3 + (1 - \eta) \Delta E_2 = r \big[\eta E + (\eta - 1) \beta E \big]$$

so that we can finally rewrite Eq. (7) as

$$\langle \Delta E \rangle = r E (\eta + \eta \beta - \beta) - pc = 0. \tag{8}$$

Equation (8) provides the link between optimality of behavioral strategies, expressed through the probability distribution over catalyzed reactions, and population size. If the former is known, we can solve Eq. (8) for the expected population size p:

$$p = \frac{rE}{c}(\eta + \eta \beta - \beta). \tag{9}$$

In particular, the case of optimal behavior, $\eta = 1$, corresponds to the maximum sustainable population, i.e., the carrying capacity:

$$p_{\text{max}} = \frac{rE}{c}. \tag{10}$$

The converse case is useful for estimating the optimality of a population's behavior at equilibrium, by measuring the population size and solving Eq. (8) for η:

$$\eta = \frac{(pc/rE) + \beta}{1 + \beta}. \tag{11}$$

As an illustration of this simple analysis, in Figure 5 we plot population size versus time in three simulations with different entries in a reaction table like (0). Note the damped oscillations (plus noise), in agreement with population dynamics models. In the steady-state regime, the population size depends on the rate at which energy is introduced into the world. Simulations are labeled by the quantity K, which is the predicted size for a population of random walkers (obtained from Eq. (9) with $\eta = 1/2$):

$$K = p_{\eta=1/2} = \frac{rE}{2c}(1 - \beta). \tag{12}$$

We can use the equations to make predictions about the outcomes of the simulations, and to compare behaviors in the different environments. The smallest K corresponds to a random walk population smaller than the amplitude of the fluctuations, so extinction occurs rapidly. Larger K values result in different stable population levels; the carrying capacities of the two environments, however, are also different (from Eq. (10)). So we can substitute the measured populations into Eq. (11) to find that the two performances are not significantly different ($\eta \sim 1/2$, the random walk value, in both cases). Then the observed difference is to be attributed to the different environments, rather than to different behaviors.

Of course, the linear relation (8) holds only at equilibrium and for this simple example: the more general nonequilibrium case and more complex environments will

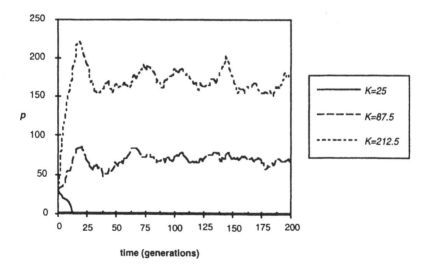

FIGURE 5 Carrying capacity for three simulations with different reaction tables. K is a measure of latent energy: when enough latent energy is supplied by the environment to avoid extinction, an early equilibrium is reached.

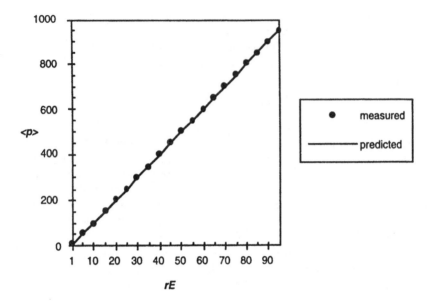

FIGURE 6   Correlation of population size with density-dependent fitness. Measures of population size are time averages in the steady-state regime; predicted population size is from Eq. (13).

yield systems of differential equations that may be too hard to solve analytically. The important point is that carefully designed latent energy environments allow us to maintain an accurate connection between ALife models and simulations.

5.2 FITNESS MEASURES

We have claimed by Eq. (9) that the population size in a LEE simulation is an effective measure of the average optimality of the organisms' behaviors. We want to show how reliable population size really is in estimating population fitness. Since the environmental resources are finite, fitness is affected by the size of the population and is said to be *density dependent*; furthermore, since organisms interact only by indirectly competing for the resources, carrying capacity is the best measure of fitness (Stearns, 1992). Let us consider an environment (simpler than the one described in Table (0)) with just one element, where each atom is associated with energy E (again, no by-products). Equation (6) still holds at equilibrium, and Eq. (7) for energy conservation yields in this case:

$$K \equiv p = \frac{rE}{c} \tag{13}$$

where K is defined as the carrying capacity. The difference between Eqs. (12) and (13) is that, in the former, fitness corresponds to one of the behaviors made possible by the binary reactions, while in the latter, the fitness is purely density dependent; i.e., behavioral effects are negligible with respect to the environmental carrying capacity. Equation (13) predicts perfect correlation between population size and density-dependent fitness. In Figure 6 we illustrate how well the population size measured in LEE simulations with different carrying capacities (determined by rE) matches the predicted fitness.

When evolved behaviors allow organisms to make a more efficient use of the latent energy, an increase in average age and in population size is observed. The situation is illustrated in Figure 7, where measures are from LEE simulations with a reaction table like (0). During these phases, Eq. (6) does not hold, environmental complexity is no longer the sole fitness factor, and thus carrying capacity is neither a sufficient predictor of population size nor a satisfactory fitness measure. Similarly, other biological fitness measures that ignore density dependence, such as *net reproductive rate*, are poorly correlated with population growth in our experiments. Population size and average age, on the other hand, can still be used as fitness

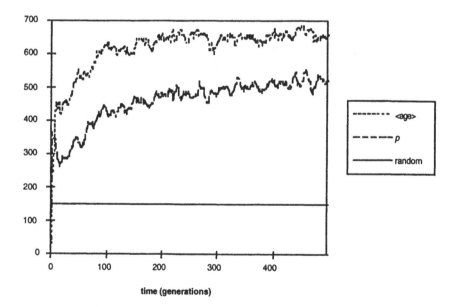

time (generations)

FIGURE 7 Population size beyond density dependence. Here Eq. (10) yields $p_{\max} = 900$, while the size of a population of random walkers (from Eq. (12)) is marked "random." Average age is also shown (units are life cycles).

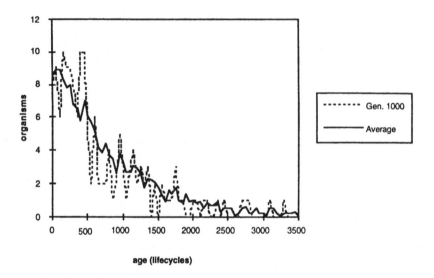

FIGURE 8 Histograms showing age distributions. "Average" refers to a time average of bin values between generations 1,000 and 10,000.

measures at the population level. Their increase is well correlated with *expected reproductive success*, a fitness measure useful in biology with density dependence. Therefore, as long as energy is conserved, density dependence provides a key to analyzing behaviors of evolving populations in endogenous fitness models.

5.3 AGE DISTRIBUTION

Carrying capacity arguments permit us to predict other emerging features of our model. For example, LEE has no control over the age of organisms. It would therefore be possible for organisms to stay alive and keep reproducing forever, introducing a sort of "evolutionary inertia" that would weaken the effect of selective pressure and thus slow down adaptation. The negative evolutionary consequences of immortal organisms have been addressed by Todd (1994). However, we observe that the distribution of age in the LEE population quickly becomes *stable*, according to the Euler-Lotka classic demographic equation. When fitness is density dependent—a consequence, as we have seen, of sharing finite environmental resources—the population reaches zero-growth and the Euler-Lotka equation predicts that the age distribution becomes *stationary*: this is in agreement with the age distributions measured during a LEE simulation and shown in Figure 8. Furthermore, in this situation, the fraction of individuals with age x becomes a direct measure of the probability of survival to age x (Stearns, 1992). The latter by definition is a non-increasing function of age, again in agreement with the measured age distributions (see Figure 8). This explains the fact that, while the average age of the LEE population grows initially, in the course of evolution the organisms begin to die "spontaneously" at a finite age. Therefore we argue that immortals will not evolve in endogenous fitness models, as long as energy is conserved.

6. CURRENT AND FUTURE DIRECTIONS

The LEE model has been implemented to run simulations on UNIX and Macintosh platforms (Menczer and Belew, 1993). In Figure 9 we show a portion of screen in a typical interactive simulation.[3] We believe that LEE represents both a rich theoretical framework and a useful simulation software for the ALife community. This chapter has shown some of its emerging features relevant to population dynamics, aging, and fitness. In the near future we intend to use it to study several other issues.

The first experiment we have proposed for LEE has the objective of studying the evolution of sensory systems. The question is whether efficient encodings of information from sensors can facilitate learning and thus emerge by means of an influence of learning on evolution (Hinton and Nowlan, 1987; Nolfi, Parisi, and Elman, 1994; Belew, 1990). The idea for the experiment comes from a simulation where NNets must look for food while moving in an environment with different zones (Miglino and Parisi, 1991). Analysis of the interaction between evolution and learning in that setting leads to a nonlinear relation between fitness expectancy and sensory information about zones. This inefficient sensory encoding seems to make the task unnecessarily difficult to learn, and suggests that evolution should favor an encoding that would facilitate learning. To test such hypotheses, the simulation with zones has been recast into the LEE framework—this gives good evidence for LEE's flexibility, allowing it to capture different ALife experiments while evaluating their complexity. Organisms in a population adapting in a latent energy environment use a very simple sensory system that is allowed to evolve. Simulation results are in agreement with our hypotheses (Menczer and Belew, 1994).

Another set of experiments (Menczer, 1994) deals more directly with the adaptive advantages of *plasticity*, defined as the capacity for individual changes to occur at the phenotype level. The LEE model and simulator have been used to study the evolutionary interaction between temporal environmental changes and phenotypic plasticity. Both were modeled quite crudely, the former by random changes in the

---

[3]C source code and documentation for release 1.* is available from URL
http://www-cse.ucsd.edu/users/fil/ or ftp://cs.ucsd.edu/pub/LEE/.

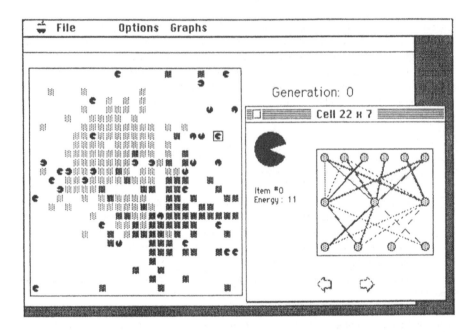

FIGURE 9   Interactive screen during a LEE simulation on the Macintosh.

reaction table and the latter by a form of "developmental noise." The adaptive rate of such noise was found to be correlated with the rate of change in the environment, building partial quantitative support for the claim that plasticity is advantageous for populations evolving in nonstationary environments. Yet another direction of current research is the evolution of reproductive maturity in the presence of cultural transmission by imitation (Cecconi, Menczer, and Belew, 1995).

By including recombination and development into the model, we believe that LEE can be extended to deal with more complex topics in theoretical biology, such as coevolution of neurosomatic traits, evolution of learning, sexual reproduction, and multiple-species population dynamics. It is our hope that the LEE framework can become a useful tool in computational biology.

ACKNOWLEDGMENTS

This project originated from conversations with Stefano Nolfi and Domenico Parisi at the Institute of Psychology of the National Research Center in Rome, Italy. Stefano Nolfi, Jeff Elman, Greg Linden, and Federico Cecconi have assisted in writing the code. Thanks to William Hart, Federico Cecconi, Alex Guazelli, and Jonathan Roughgarden for reviewing earlier drafts of this chapter, Mark Land for discussion on Figure 6, and everyone in the Cognitive Computer Science Research Group for helpful discussions on many issues of the model. Finally, we thank SFI for support during the workshop.

REFERENCES

Belew, R. K. "Evolution, Learning, and Culture: Computational Metaphors for Adaptive Algorithms." *Complex Systems* 4 (1990): 11–49.

Belew, R. K. "Artificial Life: A Constructive Lower Bound for AI." *IEEE Expert* **6-1** (1991): 8–15.

Belew, R. K. "Interposing an Ontogenic Model Between Genetic Algorithms and Neural Networks." In *Advances in Neural Information Processing (NIPS5)*, edited by J. Cowan. San Mateo, CA: Morgan Kaufmann, 1993.

Brooks, R., and P. Maes, eds. *Artificial Life IV*. Cambridge, MA: MIT Press, 1994

Cecconi, F., F. Menczer, and R. K. Belew. "Maturation and the Evolution of Initiative Learning in Artificial Organisms." Technical Report CSE 506, Department of Computer Science and Engineering, University of California at San Diego, La Jolla, CA, 1995.

De Jong, K. A., and J. Sarma. "Generation Gaps Revisited." In *Foundations of Genetic Algorithms 2*, edited by L. D. Whitley. San Mateo, CA: Morgan Kaufmann, 1993.

Hinton, G. E., and S. J. Nowlan. "How Learning Can Guide Evolution." *Complex Systems* 1 (1987): 495–502.

Holland, J. H. *Adaptation in Natural and Artificial Systems*. Cambridge, MA: MIT Press, 1992.

Littman, M. L. "Simulations Combining Evolution and Learning." In this volume.

Menczer, F. "Changing Latent Energy Environments: A Case for the Evolution of Plasticity." Technical Report CS94-336, Department of Computer Science and Engineering, University of California at San Diego, La Jolla, CA, 1994.

Menczer, F., and R. K. Belew. "Evolving Sensors in Environments of Controlled Complexity." In *Artificial Life IV*, edited by R. Brooks and P. Maes, 210–221. Cambridge, MA: MIT Press, 1994.

Menczer, F., and R. K. Belew. "LEE: A Tool for Artificial Life Simulations." Technical Report CS93-301, Department of Computer Science and Engineering, University of California at San Diego, La Jolla, CA, 1993.

Menczer, F., and D. Parisi. "Recombination and Unsupervised Learning: Effects of Crossover in the Genetic Optimization of Neural Networks." *Network* 3 (1992): 423–442.

Miglino, O., and D. Parisi. "Evolutionary Stable and Unstable Strategies in Neural Networks." In *Proc. IJCNN*, Vol. 2, 1448–1453. Piscataway, NJ: IEEE, 1991.

Mitchell, M., and S. Forrest. "Genetic Algorithms and Artificial Life." *Artificial Life* **1:3** (1994): 267–289.

Nolfi, S., D. Parisi, and J. L. Elman. "Learning and Evolution in Neural Networks." *Adaptive Behavior* **3(1)** (1994): 5–28.

Parisi, D., F. Cecconi, and S. Nolfi. "Econets: Neural Networks that Learn in an Environment." *Network* **1** (1990): 149–168.

Rössler, O. E. "Adequate Locomotion Strategies for an Abstract Environment—A Relational Approach." In *Physics and Mathematics of the Nervous System*, edited by M. Conrad, W. Guttinger, and M. D. Cin, 399–418. New York: Springer-Verlag, 1974.

Rumelhart, D. E., G. E. Hinton, and R. J. Williams. "Learning Internal Representations by Error Propagation." In *Parallel Distributed Processing: Explorations in the Microstructure of Cognition*, edited by D. E. Rumelhart and J. L. McClelland, Vol. 1, Ch. 8. Cambridge, MA: MIT Press, 1986.

Stearns, S. C. *The Evolution of Life Histories*, 20–38. New York: Oxford University Press, 1992.

Sutton, R. S. "Learning to Predict by the Methods of Temporal Differences." *Machine Learning* **3** (1988): 9–44.

Todd, P. M. "Artificial Death." In *Pre-proc. Second European Conf. on Artificial Life* (Brussels, Belgium, May 1993). Modified German version to appear in *Jahresring: The German Yearbook of Modern Art*, edited by C. Schneider, 1994.

Todd, P. M., and S. W. Wilson. "Environment Structure and Adaptive Behavior from the Ground Up." In *From Animals to Animats 2: Proc. Second Intl. Conf. on Simulation of Adaptive Behavior*, edited by J.-A. Meyer, H. L. Roitblat, and S. W. Wilson, 11–20. Cambridge, MA: MIT Press, 1993.

Wilson, S. W. "The Animat Path to AI." In *From Animals to Animats: Proc. First Intl. Conf. on Simulation of Adaptive Behavior*, edited by J.-A. Meyer and S. W. Wilson, 15–21. Cambridge, MA: MIT Press, 1991.

PSYCHOLOGY

Peter M. Todd

Chapter 14:
The Causes and Effects of Evolutionary Simulation in the Behavioral Sciences

1. INTRODUCTION

Things change. To prolong their existence in the face of this constant flux (and thereby increase their possibilities for reproduction), organisms must change, too. The behavioral sciences have focused on the ways in which the behavior of organisms changes over time at a variety of scales:

- moment-to-moment decision making and information processing in the face of new environmental inputs, such as deciding whether or not the snake that suddenly crawls out from behind a rock would be a good thing to try to eat or to flee from;

- longer-term alterations of strategies and reactions influenced by learning, such as picking up on the fact after one or two witnessed fatalities that the snakes with diamond patterns on their backs are best avoided;

- lifetime developmental adjustments to changing internal and external circumstances, such as going from only being able to crawl away from dangerous snakes at first, to being able to run at speed; and

Adaptive Individuals in Evolving Populations, Ed. R. K. Belew & M. Mitchell,
SFI Studies in the Sciences of Complexity, Vol. XXVI, Addison-Wesley, 1996 **211**

■   across-lifetime changes in shifting social and cultural structures, which in turn influence individual behavior, such as the snake-worship cult that arises and leads members to try to get closer to diamondbacks rather than fleeing them.

The actions and interactions of these various adaptive processes operating at different time scales are important for our understanding of cognition and behavior.

However, these four adaptive processes do not capture the entire story. Non-living systems can "adapt" at each of these time scales as well: glaciers, for instance, can change in "behavior"—"do different things"—in response to changes in their environment, melting or freezing with instantaneous fluctuations in temperature, advancing or retreating as the climate alters, even affecting the course of future glaciers by carving out valleys and fjords offering preferential paths to take. What many biologists feel distinguishes the adaptability of living things is their capacity to replicate with changes, and to accumulate over time those changes best matched to the changes taking place in the environment—the process of evolution with selection. As John Maynard Smith puts it, "Life should be defined by the possession of those properties which are needed to ensure evolution by natural selection" (Maynard Smith, 1986, p. 7). The great increase in adaptive power that comes when evolution is combined with the other adaptive processes of information processing, learning, development, and culture, is what gives living organisms their unique ability to flourish in the face of ever-shifting adversity.

To fully appreciate the behavior of organisms in their environment, then, we must take into account the evolution of their behavior as well. The behavioral sciences (primarily here psychology, but also linguistics, anthropology, and sociology) are largely only now beginning to do so in earnest. In this chapter, we will briefly explore why evolutionary adaptation has often been left out of the behavioral/cognitive picture, and what is causing it to be painted (back) in now. There is much that remains to be done in this endeavor, and we will discuss the effects this ongoing research will have on the continuing evolution of the behavioral sciences themselves.

2. CAUSES FOR THE ADVENT OF SIMULATION IN STUDYING THE EVOLUTION OF BEHAVIOR

Studying the evolution of behavior and cognition has never been easy. Most of the behavioral sciences have long shown something between reluctance and violent opposition toward considering how evolution can shape behavior, instead relying on notions of all-powerful learning or cultural processes (see Richards, 1987, chapter 11; Tooby and Cosmides, 1992). Even when behavioral scientists have *wanted* to think about the evolved bases of behavior, the tools available, and the results they have so far yielded (both largely borrowed from other fields such as ethology, biology, and paleontology), provide relatively little to work with.

Direct observation of the evolution of behavior is possible only with species that are short-lived enough that we can watch many generations within the limited attention span of an ongoing research project. But fruit flies, nematodes, and microorganisms do not typically exhibit the kinds of cognition that most behavioral scientists are interested in. We can infer the course of past evolution only through the physical traces that are left to us in fossils. But these traces are notoriously spotty and incomplete. In addition, behaviors themselves do not fossilize well, so that another step of indirection must be taken to reconstruct what creatures were doing based on how their remains were affected.

Mathematical models of evolving populations can be constructed using the techniques of theoretical population genetics, but to be tractable these models must be very simple (for instance, involving only two alleles at each of two gene loci). Such necessary simplicity greatly limits what can be said about the often complex processes involved in the evolution of behavior (see Collins and Jefferson, 1992). The comparative cognition approach uses related species in different environments to help determine how environmental features are related to evolved behaviors. For instance, if two related species that live in quite different environments share a particular behavior, this can provide evidence that the behavior evolved in a common ancestor before these two species diverged. But because of the capricious nature of evolution, it can be as difficult to find *living* species in the appropriate environments and with the appropriate behavioral similarities as it is to discover sought-after fossil species.

Finally, evolutionary psychologists have begun to try to "reverse-engineer" the evolved behavioral modules of the (usually human) mind (Cosmides and Tooby, 1987). Researchers in this field proceed by first reconstructing the species' ancestral environment, then inferring the kinds of adaptive challenges that would have faced that species in the environment in which it evolved. Next comes the postulation of adaptive algorithms and mental mechanisms necessary to meet those challenges. Finally, the behavioral implications of those evolved mental mechanisms must be predicted and tested in modern individuals of the species. But here again the information needed to get past the first step—knowledge about the ancient environment—is difficult to come by, relying on the incomplete fossil record. Furthermore, this method, like the previous one, provides only a snapshot of one stage in the evolution of behavior, rather than a picture of its ongoing dynamics.

All of these methods have allowed progress in our understanding of how behavior evolves, and all are useful in providing converging lines of illumination on the central questions in this area. But further tools are obviously still needed. With the recent explosion in computing power available to researchers in this area, another tool has been added to our repertoire: evolutionary computer simulations. In such simulations, as is made clear in the chapters throughout this book, a population of behaving individual organisms is modeled over successive generations, and the ways that their behaviors evolve over time is observed. Each individual "lives its life" in a simulated environment, producing actions in that environment that are influenced by the stimuli it perceives. The synthetic individuals also typically develop and learn, and thereby change their behavioral tendencies over the course of their lifetimes. A fitness function is usually defined in the environment, dependent on such factors as eating food or accumulating energy or finding suitable mates. The higher the fitness an organism attains during its lifetime, the greater number of offspring it gets to contribute to the next generation. The offspring are usually created with behaviors that differ only slightly from those of their parent(s). This fitness-based replication with modification mirrors the logic of evolutionary selection. Like natural evolution, the simulated version serves to produce individuals in successive generations with behaviors that are ever more finely adapted to the environment they inhabit, just as in the natural world.

Evolutionary simulations can be used to address a variety of questions about the evolution of behavior, complementing the investigatory methods listed earlier. First of all, simulation models can provide existence proofs, demonstrating that a certain behavior could in fact evolve through a series of cumulative stages (as opposed to appearing full-blown via some monstrous genetic event). This is the case, for example, in a simulation by Nilsson and Pelger (1994), described by Dawkins (1994), that demonstrates that vision using a spherical, lensed eye can evolve from a flat patch of photoreceptors by a succession of small structural changes, each of which give 1% greater imaging power than the last. This simulation helps answer the perennial question, "what use is half an eye?," and shows that the evolution of such complex traits is possible through small adaptive steps.

Simulations are also one of the best tools for elucidating the dynamics of an evolutionary process, showing what the course of evolution of a certain behavior could have looked like over time. Nowak and Sigmund's (1992) simulation of the evolution of "forgiving" or generous behavior in a population of individuals playing the Iterated Prisoner's Dilemma game is a good example of this kind of study. Their work showed that a possible evolutionary sequence for this behavior could have begun with individuals who always took advantage of each other, and proceeded to individuals who cooperated by reciprocating the behavior of others (the standard Tit-for-Tat strategy). Finally, though, their simulations ended up with individuals who cooperated in a more generous way, overlooking the occasional foibles of their partners. Other evolutionary paths could certainly prove correct, but Nowak and Sigmund's finding provides a useful initial hypothesis.

Third, evolutionary simulations can also be used as an exploratory technique for searching for new behaviors that may or may not exist in nature. Sims's (1994a, 1994b) work on evolving both the bodies and brains of locomoting and competing organisms in a rich and relatively unconstrained environment shows the possibilities for this use of simulations. The variety of behaviors his system discovers or "invents" is remarkable, including snake-like undulating locomotion, paddle-based swimming, and grappling and wrestling for resources. Once found, such artificially evolved behaviors can be used to inspire the search for their counterparts in nature, or, if they do not occur naturally, to delimit the range of natural behavioral possibilities.

In general, then, evolutionary simulations for all three purposes thus can be used to test hypotheses as "idea models" (see Chapter 24) or "runnable models"— instantiated, dynamic thought experiments that can be put in motion within the computer. This provides a means of discovering the implications of ideas that may be too complex to explore purely verbally. These simulations can also be used to generate hypotheses about the evolution of real behaviors, or about the reasons that certain behaviors might not have evolved in our world.

There are several advantages to using evolutionary simulations in these ways to explore behavioral questions. Perhaps the most obvious and important advantage is that these simulations can proceed much more rapidly than natural evolution. This allows the observation of many generations of behavioral adaptation, and, combined with the precise parametric control of simulations, makes it possible to replay the evolutionary movie under different experimental conditions. To make the simulations run quickly, the evolutionary models they instantiate must be made relatively simple and clear, and to run at all, they must be coherent and complete (as in any computer program). This is also an advantage, since it requires the models to be carefully thought out by the researcher and understandable by others.

But evolutionary simulations can also include a degree of complexity much greater than that allowed, for instance, by mathematical modeling (see Collins and Jefferson, 1992). Multiple levels of adaptive processes, from action at the level of DNA, switching regulator genes, and shifting protein complexes up through information processing, learning, development, and evolution, can all be incorporated into the models, adding greatly to their realism and predictive power (see Chapter 22, and Dellaert and Beer, 1994, for models incorporating several adaptive levels). Evolutionary simulations by definition require a population of many individuals, allowing investigation of still higher-level adaptive processes, including kin interactions, social cooperation and competition, and culture (see Section 4). Finally, simulated creatures can be dissected, probed, and prodded in ways that normal animals would not withstand, and a battery of psychological and "neurological" tests can be used to assess their behavior for comparison with living experimental subjects.

Disadvantages also lurk in simulating the evolution of behavior. Because the simulations must be kept relatively simple to run at a useful pace, important complexities of the real world might be left out. In such cases, the predictive fit to reality will be lessened, and the model may be able to tell us little of interest. This will also occur if the model's parameters are set in an unrealistic manner. Both of these problems plague model-building in general though, not just simulations, and they can be avoided by carefully considering the questions being asked and being true to the situation being modeled. More pernicious to the use of evolutionary simulations in particular is their seductive power to sidetrack scientific inquiry. It is very appealing and engaging to watch creatures roaming around in a simulated world on one's screen, behaving more and more adaptively with each passing generation, and it is easy to forget the goal of understanding the natural world in the beguiling presence of an artificial one. Creating enticing entertainments via simulations is an important goal in its own right. But if instead we intend to study the interaction of adaptive processes at multiple levels, we must be careful to build models that are accurate, appropriate, and analyzable.

3. FRUITFUL AREAS FOR EXPLORATION WITH EVOLUTIONARY SIMULATIONS

For any model of the interaction of multiple adaptive processes, including evolution, learning, development, and information processing, we must take into account two major sources of structure affecting those processes. The first is the components that the processes will affect and be implemented by—that is, the internal environment of the organism, incorporating the body, sensors, effectors, and nervous system. The second source of structure is the context in which the adaptive processes operate—that is, the external environment to which the organism becomes adapted. These two factors, internal components and external context, can be thought of as two "pincers" that combine to guide and constrain the evolution of learning and the other adaptive processes working in individuals. (This kind of analysis should be more useful than the traditional "nature/nurture" false dichotomy, because it emphasizes how the external context provides forces of both nature *and* nurture that the internal components adapt to in turn, across different time scales.)

There is much to be understood about the actions of both of these pincers on the behaviors of individuals. A fair amount of work has already been done on how the external context pincer affects the kinds of learning, development, and information processing that might evolve in a given species. Several of the chapters in this book focus on this topic, particularly exploring the effect of the environment on the evolution of learning (see e.g., Littman; Menczer and Belew; Todd; Parisi and Nolfi; [Chapters 26, 13, 21, and 23, respectively]). But many questions remain. A coherent theory of environmental structure needs to be developed to guide research in this area—we must be able to describe the ways in which environments can vary, before we can talk about how their variations affect behavior. Simulation studies are beginning to have an impact here (see Menczer and Belew, Chapter 13; Todd and Wilson, 1993; Todd, Wilson, Somayaji, and Yanco, 1994), but most have focused on relatively fixed environments, such as the spatiotemporal pattern of food in the world. In addition to adaptation to fixed environments, we must consider adaptation to changing environments, including those that the individuals themselves change.

The most complex of such cases is adaptation to the social and cultural environments of reciprocity, communication, memes, and artifacts that populations can construct. The kinds of learning and other processes necessary for individuals to adapt to these shifting environments differ from those that will evolve in response to fixed environments. For example, social and cultural environments can induce the evolution of imitation learning (Cecconi, Menczer, and Belew, in preparation), language (Hurford, 1991; Batali, 1994) and birdsong (Doya, 1994) learning, filial and sexual imprinting (Todd and Miller, 1993; Todd, Chapter 21), social task learning (Hutchins and Hazlehurst, 1991; Hutchins, 1995), and cultural concept learning (Gabora, 1993), among others. This area will greatly benefit from new theories about the construction, structure, and effects of social environments and their corresponding adaptive landscapes (or, perhaps, seascapes, to better capture their continuously rolling nature—see Merrell, 1994).

The internal component pincer, providing constraints on adaptive processes from the internal structure of the adapting organism, falls largely within the domains of neuroscience and development. Here too there are many avenues to be explored, and simulation is proving very useful in this endeavor. Realistic models of neurons and neural circuits are being created at an ever-increasing pace, and are used to investigate how learning and development might be implemented in living organisms. But relatively little work has been done that bridges multiple adaptive processes; in particular, simulations of the evolution and developmental growth of realistic learning neural circuits would be useful.

The vast literature of developmental psychology is also important for helping to guide models of this within-lifetime adaptive process. Simulation work in this area is increasing as well, particularly in language development (Hurford, 1991; Batali, 1994; Bates and Carnevale, 1993), but again, research combining this with other adaptive processes needs to be increased. For instance, simulations could be used to evolve and explore developmental models that grow an increasing memory span across an individual's lifetime, to see if this aids the organism in learning about its environment by first "starting small" and then adding progressively larger amounts of knowledge (Elman, 1993; see also Shafir and Roughgarden, Chapter 12). The evolution of the genotype-phenotype mapping and of the timing of life history stages is also important and readily addressable with simulation models. (It would also be interesting to develop models akin to the genetic algorithm where the genotype affects the phenotype and its behavior throughout the lifespan via the effects of gene switching, protein manufacture, and associated processes responding to the internal and external environment. Of course, such models would require a great amount of complexity in order to map from these effects all the way to behavior.)

The context and component pincers connected to the external and internal environments need not operate in only one direction, always impinging on the individual. Rather, they can also be reversed, causing forces that emanate from the individual to affect both its external surroundings and its own internal structure. When we turn the external context pincer around, it proceeds from individual behavior and learning to exert pressures on the external environment. Such forces for environmental change can in turn possibly change the course of evolution itself. This kind of phenomenon can clearly be seen in simulations of the Baldwin Effect (see Hinton and Nowlan, Chapter 25). In this situation, individuals learn about their environment and thereby change the adaptive landscape their population is situated on. As a result, individual learning can actually speed up an entire population's ability to evolve up a fitness peak in the adaptive landscape.

When individuals learn about their sexual environment—that is, about the other individuals in their own population with whom they can mate—evolution can be affected in other ways, including the creation of new species and the beginnings of runaway selection feedback loops (see Todd and Miller, 1993; Miller and Todd, 1995; Todd, Chapter 21). Similarly, in the shifting social and cultural environments described earlier, learning individuals act as filters and transformers of the sociocultural inputs they receive, as well as generators of new outputs. As a result, individuals create and alter their own sociocultural adaptive landscape and thus their own evolutionary destiny. Lamarckian evolution—the inheritance of acquired cultural traits—can occur in such situations, at a considerably faster rate than traditional Darwinian evolution (see Ackley and Littman, 1994). Investigation is now underway into these and other forms of psychological selection, in which the brains of individuals select the genes of other individuals that will get into the next generation (see Miller, 1993). Multilevel evolutionary simulations will be essential in this area for demonstrating the possible effects of individual behavior on macroevolutionary processes.

Turning the internal component pincer around allows us to consider the impact of learning on an individual's information-processing mechanisms. Phenomena in this category are also commonly studied by neuroscientists, looking at the way neurons and synapses are changed through experience (e.g., synaptic consequences of long-term potentiation). Self-organization of neural circuits into more or less distinct information-processing modules through development and learning also falls in this domain (see Karmiloff-Smith, 1992a; Jordan and Jacobs, 1994), and can have a large impact on the behavior of those circuits. Here again, simulations incorporating different adaptive processes can be useful, helping us to understand how such self-organizing and modularizing processes might evolve (see Kauffman, 1993).

To summarize, simulations of multiple interacting adaptive processes operating in and on individuals can help us discover more about environmental influences on behavior, and behavioral influences on the environment. Both the internal environment (body and brain components) and the external environment (physical, biological, sexual, social, and cultural context) shape the adapting individual and its behavior. Likewise, individual behavior, including development and learning, can affect the individual's internal information-processing machinery; and individual behavior of communicating, choosing mates, helping kin, hunting prey, hiding from predators, and sharing culture can affect the external biological and psychological environment. All these levels and directions of interaction among the various adaptive processes create a complex dynamic web that is difficult to tease apart and study in an isolated reductionist fashion. Complex multilevel simulations, incorporating evolution, learning, development, and information processing, can help us form more complete and realistic models of these phenomena and their effects, and thereby form more accurate hypotheses and conclusions about them.

4. EFFECTS OF SIMULATING THE EVOLUTION OF LEARNING AND OTHER ADAPTIVE PROCESSES

As we have just seen, a variety of factors has brought about the increasing use of simulation models to study the interaction of evolution, development, learning, and behavior. These modeling efforts have, in turn, started to affect the very behavioral sciences giving rise to them. To begin with, simulating the evolution of behavior requires modeling a *population* of individuals changing through time, rather than the intricacies of a *single* individual that are considered in traditional behavioral models. Once this shift in focus has been made from individual to population, it is easy and natural to think about group-oriented behaviors and interactions as well. In contrast to standard cognitive science models of the solitary information-processing behaviors of each individual, evolutionary simulations thus promote an increased emphasis on studying social behaviors, including communication, altruism, cooperation, competition, and the creation and propagation of culture. (Individual differences, rather than similarities, are also of greater interest when we move to a population setting, allowing these simulations to address issues in personality psychology and the study of variation.)

Not only do evolutionary simulations shift attention from individuals to groups, but they also alter how we look at, and model, each individual as well. To create an evolutionary model, we must consider the living and dying, eating and fighting, courting and mating of each individual—the actions involved in accruing energy and propagating genes, the driving forces of evolution. These types of behaviors, and the highly charged ("hot") cognition and emotions underlying them, are very different from the behaviors held paramount from the mind-as-computer viewpoint: gathering, processing, storing, and retrieving "cold" information irrespective of semantic content. These behaviors of importance to evolution are intimately tied up with the body, not segregated off to some disembodied mind. Thus the whole physical context of the body's sensors and effectors, and the environment in which they are embedded and situated, also attain heightened status in an evolutionarily aware cognitive science. This morphology/environment emphasis ties in with the increasing interest in situated cognition and action (Agre, in press; Hutchins, 1995).

The mind-as-computer metaphor, based on serial symbolic computation, is inappropriate in evolutionary simulations for another reason: traditional von Neumann-style computer models of cognition do not evolve as well as other types of models. Imagine that we have written a sequential algorithm to simulate the behavior of some individual, and we want to mutate that program slightly so that the individual's offspring will inherit some variation of their parent's behavior.

As anyone who has ever left a semicolon out of a subroutine knows, small mutations in standard sequential computer code do not usually result in small changes in the code's behavior—rather, the program often fails completely. Other small changes may result in programs that still run, but their behavior is likely to be greatly altered, due to the cascading effect as minor initial changes propagate and grow through the successive lines in the program. Thus, because of the great likelihood of mutations with "lethal" or large effects, evolution through the accrual of small useful mutations will be very slow (if it works at all) when applied to standard sequentially coded algorithms.

To get around this problem, we need a way of modeling behavior for which small changes in structure will result in small changes in behavior. Neural networks, whose behavior is determined by the simultaneous effects of a distributed set of weighted connections between simulated neurons, are one such modeling scheme. Mutating the strength of a connection usually changes the overall performance of a network only slightly. Consequently, the mind-as-neural-network metaphor is widely used in evolutionary simulations of behavior, as the papers in this volume demonstrate. But evolutionary simulations that incorporate neural networks also focus attention on the underlying evolved "innate" structure with which the networks begin learning. This can in turn help neural network research avoid the traps of positing all-powerful learning mechanisms into which behaviorist psychology often fell (see Miller and Todd, 1990; Karmiloff-Smith, 1992b). (Note that other distributed models of cognition, such as classifier systems, can be readily evolved as well—see Wilson, 1994. There has also been a flurry of work recently in the field of genetic programming on how to make traditional programming languages evolvable, by giving up on clean sequential routines and changing how code is represented and altered—see for instance Koza, 1992, 1994.)

Throughout the history of science, the widespread use of new tools has inspired the creation of new theories, from clockwork models of the heavens to computer models of the mind (see Gigerenzer, 1991; Gigerenzer and Goldstein, in preparation). The tools of evolutionary and population thinking, and the simulation techniques used to instantiate them, are no exception: a variety of evolution- and selection-based theories of cognition and culture have also recently been developed. Indeed, entire evolving species can themselves be considered intelligent "organisms" processing information from the environment in ways akin to individual animals (see Schull, 1990). Neural Darwinism (Edelman, 1987) proposes that neural circuits compete and are selected to live based on their computational efficacy. Generate-and-test models of creativity (Perkins, 1994) suggest that novel ideas are selected from a population of possible variations. The variants are thrown forward by an internal generation module, and the most appropriate is then chosen by a "fitness"-testing module in a process akin to Darwinian natural selection. Classifier systems (Wilson and Goldberg, 1989) model adaptive information processing in individuals through the gradual evolution of a population of competing input classifiers and schema processors. In addition, the generation and alteration of culture can be seen as involving the evolution of replicating, mutating ideas or "memes," such as wearing a baseball cap backwards (Dawkins, 1982). These evolutionary theories are unlikely to be the final word in any of these domains. But as the use of evolutionary simulations and models continues to grow, we can expect more such theories inspired by these models to pop up in the future.

Finally, perhaps the most profound effect of evolutionary simulations in the behavioral sciences will be to help overturn the traditional subservience of behavior to evolution. Behaviors evolve; biology dominates psychology. But through evolutionary models, as several of the papers in this volume show, we can come to appreciate and understand how individual adaptive behavior can help alter the very course of evolution itself, via the Baldwin Effect, Lamarckian cultural evolution, sexual selection, behavioral arms races, etc. The behavioral sciences thus emerge as more equal partners with the biological, and individual behavior is seen as both effect and cause of population evolution.

ACKNOWLEDGMENTS

The author thanks the members of the psychology group at the SFI workshop—Liane Gabora, Stefano Nolfi, Domenico Parisi, Jonathan Schull, and Stewart Wilson—for their contributions to this chapter.

REFERENCES

Ackley, D. H., and M. L. Littman. "A Case for Lamarckian Evolution." In *Artificial Life III*, edited by C. G. Langton, 3–10. Santa Fe Institute Studies in the Sciences of Complexity, Proc. Vol. XVII. Reading, MA: Addison-Wesley, 1994.

Agre, P. E. *The Dynamic Structure of Everyday Life*. Cambridge, UK: Cambridge University Press, in press.

Barkow, J. H., L. Cosmides, and J. Tooby, eds. *The Adapted Mind: Evolutionary Psychology and the Generation of Culture*. New York: Oxford University Press, 1992.

Batali, J. "Innate Biases and Critical Periods: Combining Evolution and Learning in the Acquisition of Syntax." In *Artificial Life IV*, edited by R. A. Brooks and P. Maes, 160–171. Cambridge, MA: MIT Press/Bradford Books, 1994.

Bates, E., and G. F. Carnevale. "New Directions in Research on Language Development." *Dev. Rev.* **13** (1993): 436–470.

Belew, R. K., M. Mitchell, and D. H. Ackley. "Computation and the Natural Sciences." This volume.

Cecconi, F., F. Menczer, and R. K. Belew. "Maturation and the Evolution of Imitative Learning in Artificial Organisms." La Jolla, CA: Computer Science and Engineering Department, University of California San Diego, manuscript in preparation.

Collins, R. J., and D. R. Jefferson. "The Evolution of Sexual Selection and Female Choice." In *Toward a Practice of Autonomous Systems: Proceedings of the First European Conference on Artificial Life*, edited by F. J. Varela and P. Bourgine, 327–336. Cambridge, MA: MIT Press/Bradford Books, 1992.

Cosmides, L., and J. Tooby. "From Evolution to Behavior: Evolutionary Psychology as the Missing Link." In *The Latest on the Best: Evolution and Optimality*, edited by J. Dupre, 277–306. Cambridge, MA: MIT Press/Bradford Books, 1987.

Dawkins, R. *The Extended Phenotype: The Gene as the Unit of Selection.* Oxford: W. H. Freeman, 1982.

Dawkins, R. "The Eye in a Twinkling." *Nature* **368** (1994): 690–691.

Dellaert, F., and R. D. Beer. "Toward an Evolvable Model of Development for Autonomous Agent Synthesis." In *Artificial Life IV*, edited by R. A. Brooks and P. Maes, 246–257. Cambridge, MA: MIT Press/Bradford Books, 1994.

Doya, K., and T. J. Sejnowski. "A Computational Model of Song Learning in the Anterior Forebrain Pathway of the Birdsong Control System." *Soc. Neurosci. Abs.* **20** (1994): 166.

Edelman, G. M. *Neural Darwinism: The Theory of Neuronal Group Selection.* New York: Basic Books, 1987.

Elman, J. L. "Learning and Development in Neural Networks: The Importance of Starting Small." *Cognition* **48** (1993): 71–99.

Gabora, L. "Meme and Variations: A Computer Model of Cultural Evolution." In *1993 Lectures in Complex Systems*, edited by D. L. Stein and L. Nadel. Santa Fe Institute Studies in the Sciences of Complexity, Lect. Vol. VI. Reading, MA: Addison-Wesley, 1995.

Gigerenzer, G. "From Tools to Theories: A Heuristic of Discovery in Cognitive Psychology." *Psychol. Rev.* **98** (1991): 254–267.

Gigerenzer, G., and D. G. Goldstein. "Mind as Computer: The Birth of a Metaphor." Chicago: Department of Psychology, University of Chicago, manuscript in preparation.

Hinton, G. E., and S. J. Nowlan. "How Learning Can Guide Evolution." This volume.

Hurford, J. R. "The Evolution of the Critical Period for Language Acquisition." *Cognition* **40** (1991): 159–201.

Hutchins, E. *Cognition in the Wild.* Cambridge, MA: MIT Press, 1995.

Hutchins, E., and B. Hazlehurst. "Learning in the Cultural Process." In *Artificial Life II*, edited by C. G. Langton, C. Taylor, J. D. Farmer, and S. Rasmussen, 689–706. Santa Fe Institute Studies in the Sciences of Complexity, Proc. Vol. XVII. Reading, MA: Addison-Wesley, 1992.

Jordan, M. I., and R. A. Jacobs. "Hierarchical Mixtures of Experts and the EM Algorithm." *Neural Comp.* **6** (1994): 181–214.

Karmiloff-Smith, A. *Beyond Modularity: A Developmental Perspective on Cognitive Science.* Cambridge, MA: MIT Press, 1992a.

Karmiloff-Smith, A. "Nature, Nurture and PDP: Preposterous Developmental Postulates?" *Connection Science* **4** (1992b): 253–269.

Kauffman, S. A. *The Origins of Order: Self-organization and Selection in Evolution.* New York: Oxford University Press, 1993.

Koza, J. R. *Genetic Programming: On the Programming of Computers by Means of Natural Selection.* Cambridge, MA: MIT Press/Bradford Books, 1992.

Koza, J. R. *Genetic Programming II: Automatic Discovery of Reusable Programs.* Cambridge, MA: MIT Press/Bradford Books, 1994.

Littman, M. L. "Simulations Combining Evolution and Learning." This volume.

Maynard Smith, J. *The Problems of Biology.* Oxford, UK: Oxford University Press, 1986.

Menczer, F., and R. K. Belew. "Latent Energy Environments." This volume.

Merrell, D. J. "The Adaptive Seascape: The Mechanism of Evolution." Minneapolis, MN: University of Minnesota Press, 1994.

Miglino, O., S. Nolfi, and D. Parisi. "Discontinuity in Evolution: How Different Levels of Organization Imply Preadaptation." This volume.

Miller, G. F. "Evolution of the Human Brain Through Runaway Sexual Selection." Ph.D. Thesis, Psychology Department, Stanford University, 1993. (To be published by MIT Press/Bradford Books.)

Miller, G. F., and P. M. Todd. "Exploring Adaptive Agency I: Theory and Methods for Simulating the Evolution of Learning." In *Proceedings of the 1990 Connectionist Models Summer School,* edited by D. S. Touretzky, J. L. Elman, T. J. Sejnowski, and G. E. Hinton, 65–80. San Mateo, CA: Morgan Kaufmann, 1990.

Miller, G. F., and P. M. Todd. "The Role of Mate Choice in Biocomputation: Computational Model of Evolution: Sexual Selection as a Process of Search, Optimization, and Diversification." In *Evolution and Biocomputation,* edited by W. Banzhaf and F. H. Eeckman, 169–204. Berlin: Springer-Verlag, 1995.

Nillsson, D. E., and S. Pelger. "A Pessimistic Estimate of the Time Required for an Eye to Evolve." *Proceedings of the Royal Society of London, B* **256** (1994): 53–58.

Nowak, M. A., and K. Sigmund. "Tit For Tat in Heterogeneous Populations." *Nature* **355** (1992): 250–252.

Parisi, D., and S. Nolfi. "The Influence of Learning on Evolution." This volume.

Perkins, D. N. "Creativity: Beyond the Darwinian Paradigm." In *Dimensions of Creativity,* edited by M. A. Boden, 119–142. Cambridge, MA: MIT Press/Bradford Books, 1994.

Richards, R. J. *Darwin and the Emergence of Evolutionary Theories of Mind and Behavior.* Chicago: University of Chicago Press, 1987.

Schull, J. "Are Species Intelligent?" *Behavioral and Brain Sciences* **13** (1990): 63–108.

Shafir, S., and J. Roughgarden. "The Effect of Memory Length on Individual Fitness in a Lizard." This volume.

Sims, K. "Evolving Virtual Creatures." *Computer Graphics, Annual Conference Series* (1994a): 43–50.

Sims, K. "Evolving 3D Morphology and Behavior by Competition." In *Artificial Life IV*, edited by R. A. Brooks and P. Maes, 28–39. Cambridge, MA: MIT Press/Bradford Books, 1994b.

Todd, P. M. "Sexual Selection and the Evolution of Learning." This volume.

Todd, P. M., and G. F. Miller. "Parental Guidance Suggested: How Parental Imprinting Evolves Through Sexual Selection as an Adaptive Learning Mechanism." *Adaptive Behavior* 2 (1993): 5–47.

Todd, P. M., and S. W. Wilson. "Environmental Structure and Adaptive Behavior From the Ground Up." In *From Animals to Animats 2: Proceedings of the Second International Conference on Simulation of Adaptive Behavior*, edited by J.-A. Meyer, H. L. Roitblat, and S. W. Wilson, 11–20. Cambridge, MA: MIT Press/Bradford Books, 1993.

Todd, P. M., S. W. Wilson, A. B. Somayaji, and H. A. Yanco. "The Blind Breeding the Blind: Adaptive Behavior Without Looking." In *From Animals to Animats 3: Proceedings of the Third International Conference on Simulation of Adaptive Behavior*, edited by D. Cliff, P. Husbands, J.-A. Meyer, and S. W. Wilson, 228–237. Cambridge, MA: MIT Press/Bradford Books, 1994.

Tooby, J., and L. Cosmides. "The Psychological Foundations of Culture." In *The Adapted Mind: Evolutionary Psychology and the Generation of Culture*, edited by J. H. Barkow, L. Cosmides, and J. Tooby, 19–136. New York: Oxford University Press, 1992.

Wilson, S. W. "ZCS: A Zeroth Level Classifier System." *Evolutionary Computation* 2 (1994): 1–18.

Wilson, S. W., and D. E. Goldberg. "A Critical Review of Classifier Systems." In *Proceedings of the Third International Conference on Genetic Algorithms*, 244–255. Los Altos, CA: Morgan Kaufmann, 1989.

Reprinted Classics

Peter Godfrey-Smith

Preface to Chapters 15 and 16

Few people had as much influence on the intellectual scene in Victorian England as Herbert Spencer. Few thinkers from any period have fallen from favor so rapidly and comprehensively when the spirit of the times moved on. But Spencer had some interesting things to say about learning and evolution, and his views on this topic were very important in biological, psychological, and philosophical debate in the late nineteenth century.

The most important feature of Spencer's work in this context is his fusing or synthesis of two patterns of explanation: *adaptationism* in biology and *association-ism* in psychology. Spencer recognized that these two frameworks exhibit the same general explanatory structure. For him they describe two different ways in which the internal properties of living systems are brought into relations of "correspondence" with conditions in their environments. Associationism was familiar to the English-speaking philosophical tradition from the eighteenth century. In biology, Spencer was an evolutionist before Darwin's *Origin of Species*. Initially his evolutionary view was based on the inheritance of characteristics acquired by individuals during their lifetimes in response to their environments, in the style of Lamarck (1809). But when Darwin published the *Origin* in 1859, Spencer endorsed natural selection as an important mechanism, especially in evolution's earlier stages.

Adaptive Individuals in Evolving Populations, Ed. R. K. Belew & M. Mitchell,
SFI Studies in the Sciences of Complexity, Vol. XXVI, Addison-Wesley, 1996   **227**

The interesting thing about Spencer is the way he fused these diverse explanatory programs to get a single comprehensive picture of organic change. The diagram (Table 1) in the first reprinted extract, which dates from his *Principles of Biology* (originally published 1864–66), provides a summary of Spencer's view, and it can also be used to explain some of his terminology.

The left-hand side of the diagram represents the environmental causes of biological change. Spencer saw both evolution and individual development as achieving adaptive response or adjustment to environmental conditions. He referred to these processes of adjustment as "equilibration."

Firstly, there is "direct equilibration." This is adaptation in which an environmental condition itself brings about an appropriate organic response. The development of stronger muscles and bones through exercise of those structures is one example. So are various phenomena we would describe as associative learning—learning that a certain sound or shape in the environment tends to be correlated with danger, for example.

Direct equilibration includes both the individual's own adaptation in its lifetime and also the evolutionary consequences of this process. Spencer thought that changes of this type were generally at least partly heritable. The offspring of an individual that has adapted itself to some environmental pattern will typically have a head start, and can take the process of direct equilibration further than its parents did. So this first category includes both individual learning and evolution by the inheritance of acquired characteristics.

Secondly, we have "indirect equilibration." This is Spencer's term for evolution by natural selection, or, to use his more famous coinage, the "survival of the fittest." The environment does not directly bring about an appropriate organic change, but it does act to preserve and accumulate useful changes if they happen to occur for other reasons.

Spencer splits the action of natural selection into three parts in his diagram. Starting from the top, the first type of natural selection recognized is selection for the ability to engage in direct equilibration. That is, natural selection will favor individuals who are sensitive learners or, more generally, who have appropriate properties of what would now be called "phenotypic plasticity." Secondly, there is natural selection for the useful traits themselves, rather than selection for the ability to get useful traits by direct equilibration. Within this category Spencer splits the action of natural selection into "positive" and "negative" aspects—aiding the survival of the good versus killing off the bad. On modern views this is not a significant distinction. Selection simply favors some over others; judgments of fitness are comparative.

Then, at the bottom the diagram, Spencer lists responses to the environment that are properties of populations rather than individuals. There is differentiation of each species into varieties as a consequence of environmental influences, something that happens whether or not the varieties occupy different habitats. I am not sure exactly what is referred to in the last entry at the bottom right of the table. The environment is said to change the structure of the population considered as a whole, by means of changes to the individuals. This may be just an assertion of the fact that changes made to individuals are also changes made to the population comprised of those individuals. Alternatively, it may be change by a process of selection on these local populations, which alters the characteristics of the species by "abstracting a class of its units."

The second extract is from Spencer's *Principles of Psychology* (1855). The passage elaborates on his view of psychological mechanisms. It also shows how Spencer made use of his combination of a broadly empiricist view with an evolutionary perspective to address some common problems for empiricism—the problems of knowledge of supposedly "necessary" truths and the role of concepts such as space and time in the organization of thought. The "universal law" mentioned early in the passage is what Spencer calls "the law of intelligence." This is an associationist principle: the strength of the correlation between any two internal states is proportional to the strength of the correlation between the external conditions which the inner states represent. As the passage shows, Spencer took this law to be applicable to both the experiences of individuals and to the accumulated experiences of populations of individuals. So Spencer had an early version of an idea which has been made popular by evolutionary epistemologists, the view that mental structures or ideas which are prior to experience, or *a priori*, for an individual, are in fact *a posteriori*, or based on experience, with respect to the species (Campbell, 1974).

We have seen that Spencer used associationist learning and evolution by the inheritance of acquired characteristics, and he also embraced the "indirect" mechanism of Darwinian evolution. But there is, of course, a fourth possibility here, which it seems Spencer was well placed to notice. This is the possibility of a within-generation mechanism with a Darwinian or selective nature: trial-and-error learning.

The discussion of intelligence in the pre-Darwin edition of Spencer's *Psychology* (1855) does not recognize the possible role of trial-and-error learning. His "law of intelligence" was the basic mechanism for psychological change in this work, and this is not described as a selection process. In the second edition of the *Psychology* Spencer retained his account of the law of intelligence as before but added a new part to the work in which the physical basis for his psychological principles is discussed (1870, volume 1, part 5). Most of this discussion uses "direct" mechanisms. But when discussing the development of more complex nervous systems, Spencer introduces a trial-and-error principle. If spontaneous, undirected action leads to behavioral success in specific circumstances, a new nervous connection can be established which makes the action more likely to occur in those circumstances in the future. The success produces a "large draught of nervous energy" in the system that somehow flows through and widens the nascent nervous channel responsible for the success. The result is that "what was at first an accidental combination of motions will now be a combination having considerable probability" (1871, p. 545).

So though Spencer did not change his structure of psychological laws and mechanisms, he introduced a selective mechanism into his account of the neural realization of these mechanisms. Robert Boakes claims that Spencer took the idea of a trial-and-error mechanism from Alexander Bain's work without acknowledgment (Boakes, 1984, p. 13). Bain had given what Boakes regards as the first psychological discussions of this mechanism in 1855 and 1859, between the two editions of Spencer's work.

Though Spencer introduced trial-and-error in this belated way, it was not a central part of his psychological picture. William James, in fact, thought that one of Spencer's major oversights was his neglect of spontaneous mental variation and environmental selection as a mechanism for psychological change (James, 1880).

More recently a number of authors have proposed general frameworks for classifying adaptive mechanisms operating at different levels and time scales. For example, Dennett (1975) gives an argument for the necessity of viewing all adaptation and problem solving in terms of different manifestations of the "law of effect," a selective, trial-and-error mechanism. This two-step, generate-and-test model is the basis for many other general discussions of adaptation (Campbell, 1974). Spencer was engaged in a similar type of project, though he made use of different mechanisms. Spencer tried to give a general, environmentalist theory of biological and psychological change in which a central role is given to the older, one-step model in which environmental patterns directly determine the form of organic response.

ACKNOWLEDGMENT

I have benefited from correspondence with Ron Amundson and Daniel Dennett on these issues.

REFERENCES

Boakes, R. *From Darwin to Behaviorism*. Cambridge: Cambridge University Press, 1984.

Campbell, D. T. "Evolutionary Epistemology." In *The Philosophy of Karl Popper*, edited by P. A. Schillp. La Salle: Open Court, 1974.

Dennett, D. C. "Why the Law of Effect Will Not Go Away." 1975. Reprinted in D. C. Dennett, *Brainstorms; Philosophical Essays on Mind and Psychology*. Cambridge, MA: MIT Press, 1978.

James, W. "Great Men and Their Environments." 1880. Reprinted in W. James, *The Will to Believe, and Other Essays in Popular Philosophy*. New York: Longmans, 1897.

Lamarck, J. B. *Zoological Philosophy*. Translated by Hugh Elliot. Chicago: Chicago University Press, 1809, 1984.

Spencer, H. *Principles of Psychology*. London: Longman, Brown and Green, 1855.

Spencer, H. *Principles of Biology*, 2 vols. New York: Appelton. (English edition published 1864–66.)

Spencer, H. *Principles of Psychology*, 2nd ed., 2 volumes. New York: Appelton, 1871. (English edition published 1870.)

Chapter 15:
Excerpts from "The Principles of Biology"

The following excerpts are taken from *The Principles of Biology* by Spencer (New York: Appleton, 1866), first American edition, Vol. 1, Part III, chapter XIII entitled "The Co-operation of the Factors." Reprinted by permission.

§169. Thus the phenomena of organic evolution, may be interpreted in the same way as the phenomena of all other evolution. Those universal laws of the re-distribution of matter and motion, to which things in general conform, are conformed to by all living things; whether considered in their individual histories, in their histories as species, or in their aggregate history.

* * * * * * *

If the forces acting on any aggregate remain the same, the changes produced by them in the aggregate will presently reach a limit, at which the constant outer forces are balanced by the constant inner forces; and thereafter no further metamorphosis will take place. Hence, that there may be continuous changes of structure in organisms, there must be continuous changes in the incident forces. This condition to the evolution of animal and vegetal forms, we find to be fully satisfied. The astronomic, geologic, and meteorologic changes that have been slowly but incessantly going on, and have been increasing in the complexity of their combinations, have been perpetually altering the circumstances of organisms; and organisms, as they have become more numerous in their kinds and higher in their kinds, have been perpetually altering one another's circumstances. Thus, for those progressive modifications upon modifications which organic evolution implies, we find a sufficient cause in the modifications after modifications, which every environment over the Earth's surface has been undergoing, throughout all geologic and pre-geologic times. The progressive inner changes for which we thus find a cause in the continuous outer changes, conform, so far as we can trace them, to that universal law of the instability of the homogeneous, which is manifested throughout evolution in general. We see that in organisms, as in all other things, the exposure of different parts to different kinds and amounts of incident forces, has necessitated their differentiation; and that for the like reason, aggregates of individuals have been lapsing into varieties, and species, and genera, and classes. We also see that in each type of organism, as in the aggregate of types, the multiplication of effects has continually aided this transition from a more homogeneous to a more heterogeneous state. And yet again, we see that that increasing segregation, and concomitant increasing definiteness, which characterizes the growing heterogeneity of organisms, has been insured by the necessary maintenance of them under combinations of forces not greatly unlike preceding combinations—by the continual destruction of those which expose themselves to aggregates of external actions markedly incongruous with the aggregates of their internal actions, and the survival of those subject only to comparatively small incongruities. Finally, we have found that each change of structure, superposed on preceding changes, has been a re-equilibration necessitated by the disturbance of a preceding equilibrium. The maintenance of life being the maintenance of a balanced combination of functions, it follows that individuals and species that have continued to live, are individuals and species in which the balance of functions has not been overthrown. Inevitably, therefore, survival through successive changes of conditions, implies successive adjustments of the balance to the new conditions. This deduction we find to be inductively verified. What is ordinarily called adaptation, is, when translated into mechanical terms, direct equilibration. And that process which, under the name of natural selection, Mr. Darwin has shown to be an ever-acting means of fitting the structures of organisms to their circumstances, we find, on analysis, to be expressible in mechanical terms as indirect equilibration.

The actions that are here specified in succession, are in reality simultaneous; and they must be so conceived before organic evolution can be rightly understood. Some aid towards so conceiving them, will be given by the annexed table, representing the co-operation of the factors.

TABLE 1

on each species : affecting

its individuals,

— immediately through their functions;
- which, partially in the first generation, and completely in the course of generations, are directly equilibrated with the changed agencies.
- which have their direct equilibration with the changed agencies, aided by indirect equilibration, through the more frequent survival of those in which the direct equilibration is most rapid.

— mediately through the aggregate of individuals;
- positively—by aiding the multiplication of those whose moving equilibria happen to be most congruous with the changed agencies: thus, in the course of generations, indirectly equilibrating certain individuals with them.
- negatively—by killing those whose moving equilibria are most incongruous with the changed agencies: thus, in the course of generations, indirectly equilibrating each of its surviving individuals with them.

its aggregate of individuals,

— by acting on it in some parts of the habitat more than in others ; and thus differentiating the species into local varieties.

— by acting differently on slightly-unlike individuals in the same locality;
- and thus causing differentiations of the species into varieties, irrespective of locality.
- and thus causing modification of the species as a whole, by abstracting a certain class of its units,

Factors:

- Astronomic changes
- Geologic changes
- Meteorologic changes

  alter the incidence of inorganic forces.

- Enemies
- Competitors
- Co-operators
- Prey

  varying in number — alter the incidence of organic forces.

- Enemies
- Competitors
- Co-operators
- Prey

  varying in kind

§170. Respecting this co-operation of these factors, it remains only to point out their respective shares in producing the total result; and the way in which the proportions of their respective shares vary as evolution progresses.

At first, changes in the amounts and combinations of external inorganic forces, astronomic, geologic, and meteorologic, were the only causes of the successive modifications undergone by organisms; and these changes have continued, and must still continue, to be causes of such modifications. As, however, through the diffusion of organisms, and the consequent differential actions of inorganic forces on them, there arose unlikenesses among organisms, producing varieties, species, genera, orders, classes, etc.; the actions of organisms on one another became new sources of organic modifications. And as fast as types have multiplied, and become more complex; so fast have the mutual actions of organisms come to be more influential factors in their respective evolutions. Until, eventually, as we see exemplified in the human race, they have come to be the chief factors.

Passing from the external causes of change to the internal processes of change entailed by them, we see that these, too, have varied in their proportions—that which was originally the most important and almost the sole process, becoming gradually less important, if not at last the least important. Always there must have been, and always there must continue to be, a survival of the fittest: natural selection must have been in operation at the outset, and can never cease to operate. While yet organisms had comparatively feeble powers of co-ordinating their actions, and adjusting them to environing actions, natural selection worked almost alone in moulding and re-moulding organisms into fitness for their changing environments; and natural selection has remained almost the sole agency by which plants and inferior orders of animals have been modified and developed. The equilibration of organisms that are comparatively passive, is necessarily effected indirectly, by the action of incident forces on the species as a whole. But along with the gradual evolution of organisms having some activity, there grows up a kind of equilibration that is relatively direct. In proportion as the activity increases, direct equilibration plays a more important part. Until, when the nervo-muscular apparatus becomes greatly developed, and the power of varying the actions to fit the varying requirements becomes considerable, the share taken by direct equilibration rises into co-ordinate importance. We have seen reason to think that as fast as essential faculties multiply, and as fast as the number of organs that co-operate in any given function increases, indirect equilibration through natural selection, becomes less and less capable of producing specific adaptations, and remains fully capable only of maintaining the general fitness of constitution to conditions. Simultaneously, the production of adaptations by direct equilibration, takes the first place—indirect equilibration serving to facilitate it. Until at length, among the civilized human races, the equilibration becomes mainly direct: the action of natural selection being restricted to the destruction of those who are constitutionally too feeble to live, even with external aid. As the preservation of incapables is habitually secured by our social arrangements; and as very few except criminals are prevented by their inferiorities from leaving the average number of offspring (indeed the balance of fertility is probably in favour of the inferior); it results that survival of the fittest, can scarcely at all act in such way as to produce specialities of nature, either bodily or mental. Here the specialities of nature, chiefly mental, which we see produced, and which are so rapidly produced that a few centuries show a considerable change, must be ascribed almost wholly to direct equilibration.

Chapter 16:
Excerpts from "The Principles of Psychology"

The following excerpts are taken from *The Principles of Psychology* by H. Spencer (London: Longman and Green, 1855), Part IV, Chapt. 7: Reason, §197.

As most who have read thus far will have perceived, both the general argument unfolded in the synthetical divisions of this work, and many of the special arguments by which it has been supported, imply a tacit adhesion to the development hypothesis—the hypothesis that Life in its multitudinous and infinitely-varied embodiments, has arisen out of the lowest and simplest beginnings, by steps as gradual as those which evolve a homogeneous microscopic germ into a complex organism.

* * * * * * *

[J]oined with this hypothesis, the simple universal law that the cohesion of psychical states is proportionate to the frequency with which they have followed one another in experience, requires but to be supplemented by the law that habitual psychical successions entail some hereditary tendency to such successions, which, under persistent conditions, will become cumulative in generation after generation, to supply an explanation of all psychological phenomena; and, among others, of the so-called "forms of thought." Just as we saw that the establishment of those compound reflex actions which we call instincts, is comprehensible on the principle that inner relations are, by perpetual repetition, organized into correspondence with outer relations; so, the establishment of those consolidated, those indissoluble, those instinctive mental relations constituting our ideas of Space and Time, is comprehensible on the same principle. If, even to external relations that are frequently experienced in the life of a single organism, answering internal relations are established that become next to automatic—if, in an individual man, a complex combination of psychical changes, as those through which a savage hits a bird with an arrow, become, by constant repetition, so organized as to be performed almost without thought of the various processes of adjustment gone through—and if skill of this kind is so far transmissible, that particular races of men become characterized by particular aptitudes, which are nothing else than incipiently organized psychical connections; then, in virtue of the same law it must follow, that if there are certain relations which are experienced by all organisms whatever—relations which are experienced every instant of their waking lives, relations which are experienced along with every other experience, relations which consist of extremely simple elements, relations which are absolutely constant, absolutely universal—there will be gradually established in the organism, answering relations that are absolutely constant, absolutely universal. Such relations we have in those of Space and Time. Being relations that are experienced in common by all animals, the organization of the answering relations must be cumulative, not in each race of creatures only, but throughout successive races of creatures, and must, therefore, become more consolidated than all others. Being relations experienced in every action of each creature, they must, for this reason too, be responded to by internal relations that are, above all others, indissoluble. And for the yet further reason that they are uniform, invariable, incapable of being absent, or reversed, or abolished, they must be represented by irreversible, indestructible connections of ideas. As the substratum of all other external relations, they must be responded to by conceptions that are the substratum of all other internal relations. Being the constant and infinitely-repeated elements of all thought, they must become the automatic elements of all thought—the elements of thought which it is impossible to get rid of—the "forms of thought."

Such, as it seems to me, is the only possible reconciliation between the experience-hypothesis and the hypothesis of the transcendentalists: neither of which is tenable by itself. Various insurmountable difficulties presented by the Kantian doctrine, have already been pointed out; and the antagonist doctrine, taken alone, presents difficulties that I conceive to be equally insurmountable. To rest with the unqualified assertion that, antecedent to experience, the mind is a blank, is to ignore the all-essential questions—whence comes the power of organizing experiences? whence arise the different degrees of that power possessed by different races of organisms, and different individuals of the same race? If, at birth, there exists nothing but a passive receptivity of impressions, why should not a horse be as educable as a man? or, should it be said that language makes the difference, then why should not the cat and dog, out of the same household experiences, arrive at equal degrees and kinds of intelligence? Understood in its current form, the experience-hypothesis implies that the presence of a definitely organized nervous system is a circumstance of no moment—a fact not needing to be taken into account! Yet it is the all-important fact—the fact to which, in one sense, the criticisms of Lieb-nitz and others pointed—the fact without which an assimilation of experiences is utterly inexplicable. The physiologist very well knows, that throughout the animal kingdom in general, the actions are dependent on the nervous structure. He knows that each reflex movement implies the agency of certain nerves and ganglia; that a development of complicated instincts, is accompanied by a complication of the nervous centres and their commissural connections; that in the same creature in different stages, as larva and imago for example, the instincts change as the ner-vous structure changes; and that as we advance to creatures of high intelligence, a vast increase in the size and complexity of the nervous system takes place. What is the obvious inference? Is it not that the ability to co-ordinate impressions and to perform the appropriate actions, in all cases implies the pre-existence of cer-tain nerves arranged in a certain way? What is the meaning of the human brain? Is it not that its immensely numerous and involved relations of parts, stand for so many *established* relations among the psychical changes? Every one of the countless connections among the fibres of the cerebral masses, answers to some permanent connection of phenomena in the experiences of the race.

* * * * * * *

In the sense, then, that there exist in the nervous system certain pre-established relations answering to relations in the environment, there is truth in the doctrine of "forms of thought"—not the truth for which its advocates contend, but a parallel truth. Corresponding to absolute external relations, there are developed in the ner-vous system absolute internal relations—relations that are developed before birth; that are antecedent to, and independent of, individual experiences; and that are automatically established along with the very first cognitions. And, as here under-stood, it is not only these fundamental relations which are thus pre-determined; but also hosts of other relations of a more or less constant kind, which are con-genitally represented by more or less complete nervous connections. On the other hand, I hold that these pre-established internal relations, though independent of the experiences of the individual, are not independent of experiences in general; but that they have been established by the accumulated experiences of preceding organisms. The corollary from the general argument that has been elaborated, is, that the brain represents an infinitude of experiences received during the evolution of life in general: the most uniform and frequent of which, have been successively bequeathed, principal and interest; and have thus slowly amounted to that high intelligence which lies latent in the brain of the infant—which the infant in the course of its after life exercises and usually strengthens or further complicates—and which, with minute additions, it again bequeaths to future generations. And thus it happens that the European comes to have from twenty to thirty cubic inches more brain than the Papuan. Thus it happens that faculties, as that of music, which scarcely exist in the inferior human races, become congenital in the superior ones. Thus it happens that out of savages unable to count even up to the number of their fingers, and speaking a language containing only nouns and verbs, come at length our Newtons and Shakspeares [sic].

Chapter 17:
William James and the Broader Implications of a Multilevel Selectionism

In his commentary in this book, Peter Godfrey-Smith points out that Spencer missed the opportunity to combine a selectionist adaptationism with a selectionist psychology. He cites Dennett (1978) as an example of someone who did not miss that opportunity, but it is worth noting that Spencer's near-contemporary William James was there first, and that he remains the first and only thinker to fully discuss the implications of the kind of theory being pursued in this book. In so doing, James demonstrates the vast range of topics to which the models now under development could be applied.

James work in these areas is not widely known because it was embedded in a range of disciplines which (until now) were not easily related to each other—physiology, evolutionary theory, psychology, sociology and philosophy of science. For this reason, no single exemplary reading will do. I offer instead a series of extended passages culled from his entire career, along with some commentaries of my own which attempt to elucidate the connections of these passages to each other, and to the contemporary research enterprise.

In modern terms, James's account could be rendered as follows. The biosphere and its components are, quite literally and quite plausibly, brainlike mechanisms mediating mindlike processes, via a *modus operandum* shared by brains and biospheres: selection among variants. Our brains represent a small part of the larger information processing networks with which we are involved, and, over evolutionary time, our actions influence and become part of the evolution of these larger, organismlike systems. Like us, these systems are engaged in an active process of adaptation to their environments. Also like us (James eventually suggests), these systems may be endowed with mentalities similar to our own. And whether one takes that last leap or not (James allows) it must always be considered that individuals can make the crucial difference in determining the course of evolution.

James realized that the essence of the Darwinian hypothesis is the interplay of two processes—variation and selection—and had something important to say about each. In his very first publications (reviews of Darwin's *Variation of Animals and Plants Under Domestication*), James differed with Darwin over the relative importance of variation vs. selection, emphasizing the importance of variation. That he saw this as a blow for individual freedom is best demonstrated in the first article in which he extended the scope of selectionist theory beyond its conventional applications. "A remarkable parallel," wrote James, "obtains between the facts of social evolution on the one hand, and of zoological evolution as expounded by Mr. Darwin on the other" (*Will to Believe*, p. 163).

> If we look at an animal or a human being, distinguished from the rest of his kind by the possession of some extraordinary peculiarity, good or bad, we shall be able to discriminate between the causes which originally produced the peculiarity in him and the causes that maintain it after it is produced. . . .. It was the triumphant originality of Darwin to see this, and to act accordingly. Separating the causes of production under the title of 'tendencies to spontaneous variation,' and relegating them to a physiological cycle which he forthwith agreed to ignore altogether, he confined his attention to the causes of preservation. . .studied them exclusively as functions of the cycle of the environment (*Will to Believe*, p. 167).

But James is explicitly interested in the topic Darwin "agreed to ignore"—and (in a sociological context at first) he gives credit for the origins of adaptations to *plastic individuals*!

In an essay entitled, "Great Men, Great Thoughts, and the Environment," James argued that human history reflects the *interaction* of great men with their environments. Taking issue with those who adopted a strictly deterministic and environmentalist view of history (in which the times make the man, but not vice versa), James argued that great men are social *variations* which interact with, and influence, social *selection pressures*.

> [The social philosopher] must simply accept geniuses as data, just as Darwin accepts his spontaneous variations. For him, as for Darwin, the only problem is. . .How does the environment affect them, and how do they affect the environment. Now, I affirm that the relation of the visible environment to the great man is in the main exactly what it is to the "variation" in the

Darwinian philosophy. It chiefly adopts or rejects, preserves or destroys, in short *selects* him. And whenever it adopts and preserves the great man, it becomes modified by his influence in an entirely original and peculiar way. He acts as a ferment, and changes its constitution, just as the advent of a new zoological species changes the faunal and floral equilibrium of the region in which it appears ("Great Men and Their Environment," in *Will to Believe*, p. 170).

Thus, the relation of variant to the system within which it is selected is interactive, and not linearly causal. Since the great man influences his environment and its subsequent selection pressures, he can therefore make a difference, even if (especially if) an evolutionary interpretation of history is adopted.

Already, James's three critical contributions to variation-selection theory can be recognized. First, variation is a process Darwin's theory does not explain, but whose role in evolution should not be underestimated. This was especially important to James because he believed that spontaneous variation elevated the importance of individual achievements at the same time as it undermined the claims of determinism.

Second, evolution is an interactive process in which organism and environment influence each other.

> . . .evolution is a resultant of the interaction of two wholly distinct factors—the individual, deriving his peculiar gifts from the play of physiological and infra-societal forces, but bearing all the power of initiative and origination in his hands; and, second, the social environment, with its powers of adopting or rejecting both him and his gifts. Both factors are essential to change. The community stagnates without the impulse of the individual. the impulse dies away without the sympathy of the community (*Will to Believe*, p. 174).

And third, variation-selection theory applies to a broader domain than biological evolution *per se*.

In "Great Men, Great Thoughts," James actually made this extension to mental evolution as well as to human history, again rebutting the view that the individual is a passive product of his environment.

> [T]hroughout the whole extent of those mental departments which are highest. . .the new conceptions, emotions, and active tendencies which evolve are originally produced in the shape of random images, fancies, accidental out-births of spontaneous variation in the functional activity of the excessively instable human brain, which the outer environment simply confirms or refutes, adopts or rejects, preserves or destroys,—selects, in short, just as it selects morphological and social variations due to molecular accidents of an analogous sort (*Will to Believe*, p. 184).

This last extension of selectionism into the individual is the most important yet least explicit, even though it is arguably the theoretical cornerstone of the *Principles of Psychology*, and much of his later work.

It is in the *Principles* that James explicitly suggests that variation and selection are processes that occur within individuals, not just in the environment acting upon the individual. This claim is first advanced in the chapter called "The Automaton Theory":

> [T]he study of the phenomena of consciousness which we shall make throughout the rest of this book will show us that consciousness is at all times primarily a *selecting agency*. Whether we take it in the lowest sphere of sense, or in the highest sphere of intellection, we find it always doing one thing, choosing one out of several of the materials so presented to its notice, emphasizing and accentuating that and suppressing as far as possible all the rest. The item emphasized is always in close connection with some *interest* felt by consciousness to be paramount at the time (*Principles of Psychology*, p. 91).

In this passage, James is making so many important moves that it will be well to enumerate them. First, and foremost for our purposes, he is developing an implicit analogy: consciousness is to interests as natural selection is to selection pressures. Second, he is making an explicit biopsychological claim which demonstrates the multileveledness of his view. Consciousness is a biological adaptation evolved through natural selection; its biological function is, in turn, to act as selector of mental variants, selecting those variants which enhance the individual's survival. Third, he is using the biological claim to argue that mind has causal efficacy: if consciousness has evolved through natural selection, it must *have* had positive causal consequences; and to have causal consequences it must have causal efficacy. James acknowledges the integrity of the Huxleyan view that mind is an epiphenomenon, but he insists that the biological facts argue overwhelmingly against that view. In fact, says James, consciousness may well be, as it naïvely seems to be, the agency responsible for the very purposiveness that characterizes intelligent beings.

James' argument that the brain is a structure in which selection is mediated, needed, and exploited, is an easily overlooked thread of cross references in the *Principles*. Traversing a discussion of reflexes in decapitated frogs, and a discussion how recovery from brain damage occurs, James argues that the brain "is essentially a place of currents, which run in organized paths" (*Principles of Psychology*, p. 78).

> As far as the cortex itself goes, since one of the purposes for which it actually exists is the production of new paths, the only question before us is: is the formation of these particular "vicarious" paths too much to expect of its plastic powers? (*Principles of Psychology*, p. 78).

He answers in the affirmative, by invoking the interaction of random neural "accidents" and nonrandom cortical selection.

The normal paths are only paths of least resistance. If they get blocked or cut, paths formerly more resistant become the least resistant paths under the changed conditions. It must never be forgotten that a current that runs in has got to run out *somewhere*; and if it only succeeds by accident in striking into its old place of exit again, the thrill of satisfaction which the consciousness connected with the whole residual brain then receives will reinforce and fix the paths of that moment and make them more likely last to be struck into again. The resultant feeling that the old habitual act is at last successfully back again, becomes itself a new stimulus which stamps all the existing currents in" (*Principles of Psychology*, p. 79).

Modern readers may find this a surprising anticipation of Thorndike's reinforcement theory, but they should not be surprised. Thorndike was one of James's students and conducted his famous experiments on problem solving in cats in James's basement! Indeed a *stimulus-response theory* of reinforcement, complete with diagrams of neural networks, can be found in James's chapter on the Will.

But how is it then, that the behaviorists who carried the day in the mid-twentieth century, and whose arguments and theories were so clearly anticipated by James, ended up drawing such radically different conclusions? How is it that James saw the importance of selection and our ability to explain it physiologically as grounds for affirming the efficacy of consciousness, whereas the behaviorists saw it as the grounds for denying the admissibility of the concept in scientific discussions? I think the answer can be found by examining the lessons the behaviorists did *not* learn from James. First, the early behaviorists (most obviously, Hull) explicitly sought a science of behavior modeled upon Galilean and Newtonian physics. In contrast, James was imbued with field biology and Darwinism from the beginning of his undergraduate days at Harvard. Second, like Huxley, the behaviorists took selectionism as an argument for determinism, and under Thorndike's and Skinner's influence they treated it as *the* prime explanatory principle. In contrast, Darwin was quite clear in his assertions that selection was only *one* relevant explanatory principle, and, as we have seen, James treated the indeterminateness of variation as an antidote to the seeming statistical determinateness of selection. Third, behaviorists recognized only one selective agent—the environment—and emphasized its influence upon the organism while minimizing the organism's influence on the environment; as we have seen, James attacked that position quite explicitly in "Great Men, Great Thoughts." Fourth, neither Spencer nor the early behaviorists shared James's view that the *organism itself* could also act as a selector, whereas James saw the selectionist model operating at so many levels, that the distinction between the generators and selectors of variants could not be mapped simply onto the distinction between organism and environment (c.f., Dennett, 1975, for a modern restatement of this view). Indeed, in the next section of this chapter we will see that, as rendered in the *Principles*, the "character" of a multileveled variation-selection based system (like the brain, or the person) is the character of an intelligent and free-willed agent, not that of a passive and mindless automaton.

A HEIRARCHY OF SELECTORS

The multileveledness of James's notion of selection can hardly be underestimated. We have already seen that biological evolution, social evolution, mental selection, and neural selection all received explicit treatment in James's writings. But this list fails to do justice to the breadth and depth of his vision, for mental selection was itself conceived as multileveled. This claim is spelled out in detail in the conclusion to the chapter in the *Principles* on "The Stream of Thought."

The chapter is concerned with five "characters in thought," the last of which is that

> It is always interested more in one part of its object [thought] than in another, and welcomes and rejects, or chooses, all the while it thinks.
>
> The phenomena of selective attention and of deliberative will are of course patent examples of this choosing activity. But few of us are aware how incessantly it is at work in operations not ordinarily called by these names. Accentuation and Emphasis are present in every perception we have (*Principles of Psychology*, p. 273).

To document this claim, James then takes the reader through a breathless tour of the psyche. He begins with the senses:

> To begin at the bottom, what are our very senses themselves but organs of selection? Out of the infinite chaos of movements, of which physics teaches us that the outer world consists, each sense-organ picks out those which fall within certain limits of velocity. To these it responds but ignores the rest as completely as if they did not exist. It thus accentuates particular movements in a manner for which objectively there seems no valid ground; for as Lange says, there is no reason whatever to think that the gap in Nature between the highest sound-waves, and the lowest heat-waves is an abrupt break like that of our sensations.... Out of what is itself an indistinguishable, swarming *continuum*, devoid of distinction or emphasis, our senses make for us, by attending to this motion and ignoring that, a world full of contrasts of sharp accents, of abrupt changes, of picturesque light and shade (*Principles of Psychology*, p. 273–274).

He then proceeds to show how attention selects among sensations to produce our conceptions of the objects of our attention.

> If the sensations we receive from a given organ have their causes thus picked out for us by the conformation of the organ's termination, Attention, on the other hand, out of all the sensations yielded, picks out certain ones as worthy of its notice and suppresses all the rest.... Helmholtz says that we notice only those sensations which are signs to us of *things*. But what are

things? Nothing...but special groups of sensible qualities, which happen practically or aesthetically to interest us, to which we therefore give substantive names, and which we exalt to this exclusive status of independence and dignity. But in itself, apart from my interest, a particular dust-wreath on a windy day is just as much of an individual thing, and just as much or as little deserves an individual name as my own body does.

And then, among the sensations we get from each separate thing, what happens? The mind selects again. It chooses certain of the sensations to represent the thing most *truly*, and considers the rest as its appearances, modified by the conditions of the moment. Thus my table-top is named *square*, after but one of an infinite number of retinal sensations which it yields.... In like manner, the real form of the circle is deemed to be the sensation it gives when the line of vision is perpendicular to its centre.... The reader knows no object which he does not represent to himself by preference as in some typical attitude, of some normal size, at some characteristic distance, of some standard tint, etc. etc. But all these essential characteristics, which together form for us the genuine objectivity of the thing and are contrasted with what we call the subjective sensations it may yield us at a given moment, are mere sensations like the latter. The mind chooses to suit itself, and decides what particular sensation shall be held more real and valid than all the rest (*Principles of Psychology*, p. 274–275).

At this point, James widens his view, to consider selective processes which occur over longer stretches of the individual's life, arguing that selection further influences the character of a person's experience, ultimately determining his character. At the same time, James demonstrates his commitment to the same kind of selective interactionism he argued for in "Great Men, Great Thoughts": even as selected experiences influence the character of the individual, so does the character of the individual determine what kinds of experiences will be selected.

A man's empirical thought depends on the things he has experienced, but what these shall be is to a large extent determined by his habits of attention. A thing may be present to him a thousand times, but if he persistently fails to notice it, it cannot be said to enter into his experience. We are all seeing flies, moths, and beetles by the thousand, but to whom, save an entomologist, do they say anything distinct? On the other hand, a thing met only once in a lifetime may leave an indelible experience in the memory. Let four men make a tour in Europe. One will bring home only picturesque impressions—costumes and colors, parks and views.... To another, all this will be non-existent; and distances and prices, populations and drainage arrangements, door and window fastenings, and other useful statistics will take their place. A third will give a rich account of the theaters...whilst the fourth will perhaps have been so wrapped in his own subjective broodings as to tell little more than a few names of places through which he passed.

Each has selected, out of the same mass of presented objects, those which suited his private interest and has made his experience thereby (*Principles of Psychology*, 275–276).

James next notes that in rational thought "we find selection again to be omnipotent":

[A]ll reasoning depends on the ability of the mind to break up the totality of the phenomenon reasoned about, into parts, and to pick out from among these the particular one which, in our given emergency, may lead to the proper conclusion.... Reasoning is but another form of the selective activity of the mind (*Principles of Psychology*, p. 276).

And then he turns to esthetics and ethics.

If now we pass to its aesthetic department, our law is still more obvious. The artist notoriously selects his items, rejecting all tones, colors, shapes, which do not harmonize with each other and with the main purpose of his work.... That unity, harmony, "convergence of characters"...which gives to works of art their superiority over works of nature, is wholly due to *elimination*. Any natural subject will do, if the artist has wit enough to pounce upon some one feature of it as characteristic, and suppress all merely accidental items which do not harmonize with this (*Principles of Psychology*, p. 276).

It is not clear whether James intended to liken the artist-as-selector to Mother-Nature-as-artist (whose medium is natural selection). But we shall see that in his summary James reverts again to the image of selection as sculpting, using the image to refer simultaneously to mental and natural selection.

Ascending still higher, we reach the plane of Ethics, where choice reigns notoriously supreme. An act has no ethical quality whatever unless it be chosen out of several all equally possible. But more than [this,]...the ethical energy *par excellence* has to go farther and chooses which *interest* out of several, equally coercive, shall *become* supreme. The issue here is of the utmost pregnancy, for it decides a man's entire career.... What he shall *become* is fixed by the conduct of this moment [of ethical choice]. Schopenhauer, who enforces his determinism by the argument that with a given fixed character only one reaction is possible under given circumstances, forgets that, in these critical ethical moments, what consciously *seems* to be in question is the complexion of the character itself. The problem with the man is less what act he shall now choose to do than what being he shall now resolve to become (*Principles of Psychology*, p. 276–277).

Finally, James summarizes, and in so doing draws an analogy between the process of individual self-creation (just described), and the process of self-creation in the race, i.e., the evolution of the human species of which we are integral parts, and potentially consequential variants.

> Looking back then, over this review, we see that the mind is at every stage a theatre of simultaneous possibilities. Consciousness consists in the comparison of these with each other, the selection of some, and the suppression of the rest by the reinforcing and inhibiting agency of attention. The highest and most elaborated mental products are filtered from the data chosen by the faculty next beneath, out of the mass offered by the faculty below that, which mass in turn was sifted from a still larger amount of yet simpler material, and so on. The mind, in short, works on the data it receives very much as a sculptor works on his block of stone. In a sense the statue stood there from eternity. But there were a thousand different ones beside it, and the sculptor alone is to thank for having extricated this one from the rest. Just so the world of each of us, howsoever different our several views of it may be, all lay embedded in the primordial chaos of sensations, which gave the mere matter to the thought of all of us indifferently. We may, if we like, by our reasonings unwind things back to that black and jointless continuity of space and moving clouds of swarming atoms which science calls the only real world. But all the while the world we feel and live in will be that which our ancestors and we, by slowly cumulative strokes of choice, have extricated out of this, like sculptors, by simply rejecting certain portions of the given stuff. Other sculptors, other statues from the same stone! Other minds, other worlds from the same monotonous and inexpressive chaos! My world is but one in a million alike embedded, alike real to those who may abstract them. How different must be the worlds in the consciousness of ant, cuttle-fish, or crab (*Principles of Psychology*, p. 277).

To the modern reader, this last analogy between the process by which our ancestors defined the reality we experience and the process by which each of us further refines it, may seem to depend upon the assumption of Lamarckian inheritance. But it does not. At the conclusion of the *Principles*, James discusses and accepts Weismann's critique of the notion of Lamarckian inheritance. And indeed, he does not need it, for (as the previous passage should have made abundantly clear, and as was well established in the work of James's contemporary and friend James Mark Baldwin) there is no necessary discontinuity in the hierarchy of selective processes which goes from the individual on into the realm of species evolution (see Schull, 1990).

FOOLS RUSH IN

A scientifically embarassing and still unfashionable question arises now for us, as it did for James: if evolution and the mind work according to similar principles, might evolutionary systems themselves be a locus for purposive mentalities? James (whose father was something of a transcendentalist theologian) insisted that the idea could not be rejected. In his empirical classic, *The Varieties of Religious Experience*, James argued that the phenomenological data support, and scientific theory could not refute, the possibility that religious experiences were in some way reality-based. Later in his life, James took it upon himself to popularize and defend the religious writings of Gustav Fechner. Fechner (who is still revered as the hard-science father of experimental psychophysics) postulated the existence of transcendent minds in biological species and ecosystems. He also postulated cosmic mentalities existing in the domains of astronomy and cosmology, eventuating in an all-encompassing God. But it is interesting to note that in these domains it much less clear that the principles of variation and selection apply. And it it is precisely here that James parted company with Fechner. "...it seems to me that Fechner's God is a lazy postulate of his, rather than a part of his system positively thought out" (*Pluralistic Universe*, p. 132).

> Speculatively, Fechner is thus a monist in his theology; but there is room in his universe for every grade of spiritual being between man and the all-inclusive God; and in suggesting what the positive content of all this super-humanity may be, he hardly lets his imagination fly beyond simple spirits of the planetary order. The earth-soul he passionately believes in; he treats the earth as our special human guardian angel; ...but I think that in his system, as in so many of the actual historic theologies, the supreme God marks only a sort of limit of enclosure of the worlds above man (*Pluralistic Universe*, p. 71–72).

Here, and elsewhere in *A Pluralistic Universe* James's arguments are developed in the context of a general attack on monism; so much so that Perry (1935) found the "subordination" of James's usual selective theorizing to be noteworthy (Perry, p. 591). But the biological theory that provides the real rationale for his pluralistic pantheism is not hard to find—it is precisely the kind of multileveled selectionism which we are only now prepared to study scientifically.

James showed his true colors in the book's introduction. While claiming that the modern intellectual climate had rendered obsolete the traditional notions of God-the-absolute, he averred that

> The vaster vistas which scientific evolutionism has opened, and the rising tide of social democratic ideals, have changed the type of our imagination, and the older monarchical theism is obsolete or obsolescent. The place of the divine in the world must be more organic and intimate (*Pluralistic Universe*, p. 18).

The mixing of the ideas of democracy and evolution captures perfectly the vision James propounds: one in which each individual has a vote in the conduct of the larger biological and social systems of which he is a constituent. And James's true colors are again revealed when he defends Fechner against objections which could apply as well to his own extensions of selectionist theory from biology to psychology.

> Most of us, considering the theory...make the mistake of working the analogy too literally.... If the earth be a sentient organism, we say, where are her brain and nerves? ...All the consciousness we directly know seems tied to brains. Can there be no consciousness, we ask, where there is no brain? (*Pluralistic Universe*, p. 74).

> But if [nerve] fibers are indeed all that is needed to do that trick [in animals], has not the earth pathways, by which you and I are physically continuous, more than enough to do for our two minds what the brain-fibers do for the sounds and sights in a single mind? (*Pluralistic Universe*, p. 75).

Thus, just as James had taken pains to show that the brain was the kind of place that could host a process analogous to natural selection, so is he here at pains to show that the earth (or as we would now say, the biosphere) can host processes analogous to those mediated by the brain. In this case, the relation of the biosphere to each individual human could be exactly the same as the relation of each human to his component brain centres about whose consciousness James had inquired in the *Principles*. And the pooled contributions of human consciousnesses into the larger system could give rise to a larger consciousness.

> ...as we are ourselves a part of the earth, so our organs are her organs.... She brings forth living beings of countless kinds upon her surface, and their multitudinous conscious relations with each other she takes up into her higher and more general conscious life" (*Pluralistic Universe*, p. 74).

Indeed, given the preeminence of Darwin's theory of natural selection in explaining the facts of biology, and James's theories of mental and social selection in explaining the phenomena of mind, the conclusion is nearly unavoidable that

> The self-compounding of mind in its smaller and more accessible portions seems a certain fact, and...the speculative assumption of a similar but wider compounding in remoter regions must be reckoned with as a legitimate hypothesis" (*Pluralistic Universe*, p. 131).

> Every bit of us at every moment is part and parcel of a wider self,.... And just as we are co-conscious with our own momentary margin, may not we ourselves form the margin of some more really central self in things which is co-conscious with the whole of us? May not you and I be confluent in a higher consciousness, and confluently active there, though we know it not? (*Pluralistic Universe*, p. 131).

Thus, James's unconventional interpretations of religious experience, abnormal personality states, and the pluralistic pantheism of his later years were almost inevitable outcomes of a lifetime of theorizing about brain physiology, mind, and evolution by natural selection (broadly conceived). Indeed, substitute the phrase "hierarchically organized adaptive system" for the word "God" in the following passage, and you will see that James's religious contentions are rather hard to set aside:

> We are indeed internal parts of God and not external creations.... Yet because God is not the absolute, but is himself a part when the system is conceived pluralistically, his functions can be taken as not wholly dissimilar to those of the smaller parts—as similar to our functions consequently (*Pluralistic Universe*, p. 143).

Thus "having an environment, being in time, and working out a history just like ourselves" (*Pluralistic Universe*, p. 144), the larger mentalities of which James speculated followed from his psychological and biological theorizing at least as much as it did from his empirical study of religious and psychic experience. Indeed, regardless of whether religious experience is to be explained or validated by the heirarchy of selective processes which constitutes organisms in the the biosphere, the possibility that evolving systems embody "larger intelligences" is one which still deserves scientific consideration today (Schull, 1990).

> Philosophies are intimate parts of the universe, they express something of its own thought of itself. A philosophy may indeed be a most momentous reaction of the universe upon itself. It may, as I said, possess and handle itself differently in consequence of us philosophers, with our theories, being here.... This is the philosophy of humanism in the widest sense (*Pluralistic Universe*, p. 143 of Harvard Edition).

That this applies to the human biosocial universe is undeniable, as is the fact that James himself has had, and may continue to have, an important influence on its ongoing evolution.

CONCLUSIONS

Radical as they were, and unconventional as they still seem today, remarkably few of James ideas on evolution, life, and mind are irrelevant to modern inquiry. Indeed, one of the goals of this paper is to acknowledge James as a seminal and still unexhausted resource for a number of scientific enterprises only now gathering momentum.

Certainly, James's emphasis on Darwinian interpretations of evolution is not outdated, notwithstanding misgivings of many of his now-dated contemporaries and successors who saw greater promise in Lamarckian or mechanistic analyses of evolution. Darwin's theory is *the* core idea in biology, and its preeminence has in no way been reduced by the complementary molecular biological discoveries which have followed.

With regard to cultural evolution and the evolution of ideas, James is recognized as a progenitor of contemporary "evolutionary epistemology" (see Campbell, 1974), but he is rarely credited for one of the first sophisticated Darwinian accounts of cultural and social evolution. Some of the most exciting theoretical work in these areas today is due to the application of modern theory to the ground James "broke" over a hundred years ago. Dawkins's (1976) "selfish meme" account of cultural evolution is a population-genetic rewrite of James's pragmatism, which was itself an application of selectionist theory to the idea of truth. However, while cultural evolutionists and sociobiologists may have relatively little to learn from James about how to model social evolution, they still have a great deal to learn from James about the possible "character" of the systems under study.

This lesson applies even more so, of course, to the understanding of individual organisms, as the recent demise of radical behaviorism attests. Nonetheless, the behaviorist emphasis on the notion of selection (albeit extended to mental selection as well) is something even thoroughly modern mentalists recognize as a permanent theoretical fixture (see Dennett, 1978). Furthermore, with regard to the explanation of individual behavior and intelligence, it is precisely by adopting the idea of within-individual heirarchical organization that the limitations of simplistic reductionism and reinforcement theories are being transcended.

William James was a breathtakingly thorough selectionist theorist. He was also scandalously willing to expose the broader implications of this line of theory. In so doing he makes it clear that we have our work cut out for us. We must not only work through the scientific implications of our models, but we will also have to find a scientifically acceptable way to constrain or acknowledge their social and perhaps religious implications. We need not accept James's interpretations of all this, let alone his religion; but I do think it will be scientifically productive to pay close attention to the issues he raised.

REFERENCES

Campbell, D. T. "Evolutionary Epistomolgy." In *The Philosophy of Karl R. Popper*, edited by P. A. Schillp. La Salle, IL: Open Court, 1974.
Dawkins, R. *The Selfish Gene*. Oxford: Oxford University Press, 1976.

Dennett, D. C., ed. "Why the Law of Effect Will Not Go Away." Reprinted in
  Brainstorms: Philosophical Essays on Mind and Psychology. Cambridge, MA:
  MIT Press, 1978.

James, W. *Principles of Psychology*. Great Books of the Western World. Chicago:
  Encyclopedia Britannica, Inc. (Original work published in 1890.)

James, W. "Great Men, Great Thoughts, and the Environment." In *The Will to
  Believe*, by W. James. New York: Longmans, Green, 1902.

James, W. *A Pluralistic Universe*. Cambridge, MA: Harvard Univerisity Press, 1977.
  (Original work published in 1909.)

Perry, R. B. *The Thought and Character of William James*. Boston: Little Brown
  & Co., 1935.

Schull, J. "Are Species Intelligent?" *Behav. & Brain Sci.* **13** (1990): 63–108.

Chapter 18:
Excerpts from "The Phylogeny and Ontogeny of Behavior"

This chapter originally appeared in *Contingencies of Reinforcement* by B. F. Skinner (New York: Appleton-Century-Crofts, 1969), Chapt. 7: The Phylogeny and Ontogeny of Behavior, 172–217. The referencing style has been changed to be consistent with the rest of this volume; only those references that were cited have been included. Reprinted by permission.

Parts of the behavior of an organism concerned with the internal economy, as in respiration or digestion, have always been accepted as "inherited," and there is no reason why some responses to the external environment should not also come ready-made in the same sense. It is widely believed that many students of behavior disagree. The classical reference is to John B. Watson (1924):

> I should like to go one step further now and say, "Give me a dozen healthy infants, well-formed, and my own specified world to bring them up in and I'll guarantee to take any one at random and train him to become any type of specialist I might select—doctor, lawyer, artist, merchant-chief and, yes, even beggerman and thief, regardless of his talents, penchants, tendencies, abilities, vocations, and race of his ancestors." I am going beyond my facts and I admit it, but so have the advocates of the contrary and they have been doing it for many thousands of years.

Adaptive Individuals in Evolving Populations, Ed. R. K. Belew & M. Mitchell, SFI Studies in the Sciences of Complexity, Vol. XXVI, Addison-Wesley, 1996   **257**

Watson was not denying that a substantial part of behavior is inherited. His challenge appears in the first of four chapters describing "how man is equipped to behave at birth." As an enthusiastic specialist in the psychology of learning he went beyond his facts to emphasize what could be done in spite of genetic limitations. He was actually, as Gray (1963) has pointed out, "one of the earliest and one of the most careful workers in the area of animal ethology." Yet he is probably responsible for the persistent myth of what has been called "behaviorism's counterfactual dogma" (Hirsch, 1963). And it is a myth. No reputable student of animal behavior has ever taken the position "that the animal comes to the laboratory as a virtual *tabula rasa*, that species differences are insignificant, and that all responses are about equally conditionable to all stimuli" (K. Breland and M. Breland, 1961).

But what does it mean to say that behavior is inherited? Lorenz (1965) has noted that ethologists are not agreed on "the concept of 'what we formerly called innate.'" Insofar as the behavior of an organism is simply the physiology of an anatomy, the inheritance of behavior is the inheritance of certain bodily features, and there should be no problem concerning the meaning of "innate" that is not raised by any genetic trait. Perhaps we must qualify the statement that a man inherits a visual reflex, but we must also qualify the statement that he inherits his eye color.

If the anatomical features underlying behavior were as conspicuous as the wings of *Drosophila*, we should describe them directly and deal with their inheritance in the same way, but at the moment we must be content with so-called behavioral manifestations. We describe the behaving organism in terms of its gross anatomy, and we shall no doubt eventually describe the behavior of its finer structures in much the same way, but until then we analyze behavior without referring to fine structures and are constrained to do so even when we wish to make inferences about them.

What features of behavior will eventually yield a satisfactory genetic account? Some kind of inheritance is implied by such concepts as "racial memory" or "death instinct," but a sharper specification is obviously needed. The behavior observed in mazes and similar apparatuses may be "objective," but it is not described in dimensions which yield a meaningful genetic picture. Tropisms and taxes are somewhat more readily quantified, but not all behavior can be thus formulated, and organisms selected for breeding according to tropistic or taxic performances may still differ in other ways (Erlenmeyer-Kimling et al., 1962).

The probability that an organism will behave in a given way is a more promising datum, but very little has been done in studying its genetics. Modes of inheritance are not, however, the only issues.

THE PROVENANCE OF BEHAVIOR

Upon a given occasion we observe that an animal displays a certain kind of behavior—learned or unlearned. We describe its topography and evaluate its probability. We discover variables, genetic or environmental, of which the probability is a function. We then undertake to predict or control the behavior. All this concerns a current state of the organism. We have still to ask where the behavior (or the structures which thus behave) came from. What we may call the ontogeny of behavior can be traced to contingencies of reinforcement, and in a famous passage Pascal suggested that ontogeny and phylogeny have something in common. "Habit," he said, "is a second nature which destroys the first. But what is this nature? Why is habit not natural? I am very much afraid that nature is itself only first habit as habit is second nature."

The provenance of "first habit" has an important place in theories of the evolution of behavior. A given response is in a sense strengthened by consequences which have to do with the survival of the individual and the species. A given form of behavior leads not to reinforcement but to procreation. (Sheer reproductive activity does not, of course, always contribute to the survival of a species, as the problems of overpopulation remind us. A few well-fed breeders presumably enjoy an advantage over a larger but impoverished population. The advantage may also be selective. It has recently been suggested (Wynne-Edwards, 1965) that some forms of behavior such as the defense of a territory have an important effect in restricting breeding.) Several practical problems raised by what may be called contingencies of selection are remarkably similar to problems which have already been approached experimentally with respect to contingencies of reinforcement.

AN IDENTIFIABLE UNIT. A behavioral process, as a change in frequency of response, can be followed only if it is possible to count responses. The topography of an operant need not be completely fixed, but some defining property must be available to identify instances. An emphasis upon the occurrence of a repeatable unit distinguishes an experimental analysis of behavior from historical or anecdotal accounts. A similar requirement is recognized in ethology. As Julian Huxley has said, "This concept...of unit releasers which act as specific key stimuli unlocking genetically determined unit behavior patterns...is probably the most important single contribution of Lorenzian ethology to the science of behavior" (J. Huxley, 1964).

THE ACTION OF STIMULI. Operant reinforcement not only strengthens a given response; it brings the response under the control of a stimulus. But the stimulus does not elicit the response as in a reflex; it merely sets the occasion upon which the response is more likely to occur. The ethologists' "releaser" also simply sets an occasion. Like the discriminative stimulus, it increases the probability of occurrence of a unit of behavior but does not force it. The principal difference between a reflex and an instinct is not in the complexity of the response but in, respectively, the eliciting and releasing actions of the stimulus.

ORIGINS OF VARIATIONS. Ontogenic contingencies remain ineffective until a response has occurred. The rat must press the lever at least once "for other reasons" before it presses it "for food." There is a similar limitation in phylogenic contingencies. An animal must emit a cry at least once for other reasons before the cry can be selected as a warning because of the advantage to the species. It follows that the entire repertoire of an individual or species must exist prior to ontogenic or phylogenic selection, but only in the form of minimal units. Both phylogenic and ontogenic contingencies "shape" complex forms of behavior from relatively undifferentiated material. Both processes are favored if the organism shows an extensive, undifferentiated repertoire.

PROGRAMMED CONTINGENCIES. It is usually not practical to condition a complex operant by waiting for an instance to occur and then reinforcing it. A terminal performance must be reached through intermediate contingencies (programmed instruction). In a demonstration experiment a rat pulled a chain to obtain a marble from a rack, picked up the marble with its forepaws, carried it to a tube projecting two inches above the floor of its cage, lifted it to the top of the tube, and dropped it inside. "Every step in the process had to be worked out through a series of approximations since the component responses were not in the original repertoire of the rat" (Skinner, 1938). The "program" was as follows. The rat was reinforced for any movement which caused a marble to roll over any edge of the floor of its cage, then only over the edge on one side of the cage, then over only a small section of the edge, then over only that section slightly raised, and so on. The raised edge became a tube of gradually diminishing diameter and increasing height. The earlier member of the chain, release of the marble from the rack, was added later. Other kinds of programming have been used to establish subtle stimulus control to sustain behavior in spite of infrequent reinforcement, and so on (Skinner, 1968).

A similar programming of complex phylogenic contingencies is familiar in evolutionary theory. The environment may change, demanding that behavior which contributes to survival for a given reason become more complex. Quite different advantages may be responsible for different stages. To take a familiar example, the electric organ of the eel could have become useful in stunning prey only after developing something like its present power. Must we attribute the completed organ to a single complex mutation, or were intermediate stages developed because of other advantages? Much weaker currents, for example, may have permitted the eel to detect the nature of objects with which it was in contact. The same question may be asked about behavior. Pascal's "first habit" must often have been the product of "programmed instruction." Many of the complex phylogenic contingencies which now seem to sustain behavior must have been reached through intermediate stages in which less complex forms had lesser but still effective consequences.

The need for programming is a special case of a more general principle. We do not explain any system of behavior simply by demonstrating that it works to the advantage of, or has "net utility" for, the individual or species. It is necessary to show that a given advantage is contingent upon behavior in such a way as to alter its probability.

ADVENTITIOUS CONTINGENCIES. It is not true, as Lorenz (1965) has asserted, that "adaptiveness is always the irrefutable proof that this process [of adaptation] has taken place." Behavior may have advantages which have played no role in its selection. The converse is also true. Events which follow behavior but are not necessarily produced by it may have a selective effect. A hungry pigeon placed in an apparatus in which a food dispenser operates every twenty seconds regardless of what the pigeon is doing acquires a stereotyped response which is shaped and sustained by wholly coincidental reinforcement. The behavior is often "ritualistic"; we call it superstitious (Skinner, 1948). There is presumably a phylogenic parallel. All current characteristics of an organism do not necessarily contribute to its survival and procreation, yet they are all nevertheless "selected." Useless structures with associated useless functions are as inevitable as superstitious behavior. Both become more likely as organisms become more sensitive to contingencies. It should occasion no surprise that behavior has not perfectly adjusted to either ontogenic or phylogenic contingencies.

UNSTABLE AND INTERMITTENT CONTINGENCIES. Both phylogenic and ontogenic contingencies are effective even though intermittent. Different schedules of reinforcement generate different patterns of changing probabilities. If there is a phylogenic parallel, it is obscure. A form of behavior generated by intermittent selective contingencies is presumably likely to survive a protracted period in which the contingencies are not in force, because it has already proved powerful enough to survive briefer periods, but this is only roughly parallel with the explanation of the greater resistance to extinction of intermittently reinforced operants.

CHANGING CONTINGENCIES. Contingencies also change, and the behaviors for which they are responsible then change too. When ontogenic contingencies specifying topography of response are relaxed, the topography usually deteriorates; and when reinforcements are no longer forthcoming, the operant undergoes extinction. Darwin discussed phylogenic parallels in *The Expression of Emotions in Man and Animals*. His "serviceable associated habits" were apparently both learned and unlearned, and he seems to have assumed that ontogenic contingencies contribute to the inheritance of behavior, at least in generating responses which may then have phylogenic consequences. The behavior of the domestic dog in turning around before lying down on a smooth surface may have been selected by contingencies under which the behavior made a useful bed in grass or brush. If dogs now show this behavior less frequently, it is presumably because a sort of phylogenic extinction has set in. The domestic cat shows a complex response of covering feces which must once have had survival value with respect to predation or disease. The dog has been more responsive to the relaxed contingencies arising from domestication or some other change in predation or disease, and shows the behavior in vestigial form.

MULTIPLE CONTINGENCIES. An operant may be affected by more than one kind of reinforcement, and a given form of behavior may be traced to more than one advantage to the individual or the species. Two phylogenic or ontogenic consequences may work together or oppose each other in the development of a given response and presumably show "algebraic summation" when opposed.

SOCIAL CONTINGENCIES. The contingencies responsible for social behavior raise special problems in both phylogeny and ontogeny. In the development of a language the behavior of a speaker can become more elaborate only as listeners become sensitive to elaborated speech. A similarly coordinated development must be assumed in the phylogeny of social behavior. The dance of the bee returning from a successful foray can have advantageous effects for the species only when other bees behave appropriately with respect to it, but they cannot develop the behavior until the dance appears. The terminal system must have required a kind of subtle programming in which the behaviors of both "speaker" and "listener" passed through increasingly complex stages. A bee returning from a successful foray may behave in a special way because it is excited or fatigued, and it may show phototropic responses related to recent visual stimulation. If the strength of the behavior varies with the quantity or quality of food the bee has discovered and with the distance and direction it has flown, then the behavior may serve as an important stimulus to other bees, even though its characteristics have not yet been affected by such consequences. If different bees behave in different ways, then more effective versions should be selected. If the behavior of a successful bee evokes behavior on the part of listeners which is reinforcing to the speaker, then the speaker's behavior should be ontogenically intensified. The phylogenic development of responsive behavior in the listener should contribute to the final system by providing for immediate reinforcement of conspicuous forms of the dance.

The speaker's behavior may become less elaborate if the listener continues to respond to less elaborate forms. We stop someone who is approaching us by pressing our palm against his chest, but he eventually learns to stop upon seeing our outstretched palm. The practical response becomes a gesture. A similar shift in phylogenic contingencies may account for the "intentional movements" of the ethologists.

Behavior may be intensified or elaborated under differential reinforcement involving the stimulation either of the behaving organism or of others. The more conspicuous a superstitious response, for example, the more effective the adventitious contingencies. Behavior is especially likely to become more conspicuous when reinforcement is contingent on the response of another organism. Some ontogenic instances, called "ritualization," are easily demonstrated. Many elaborate rituals of primarily phylogenic origin have been described by ethologists.

SOME PROBLEMS RAISED BY PHYLOGENIC CONTINGENCIES

Lorenz has recently argued that "our absolute ignorance of the physiological mechanisms underlying learning makes our knowledge of the causation of phyletic adaptation seem quite considerable by comparison" (Lorenz, 1965). But genetic and behavioral processes are studied and formulated in a rigorous way without reference to the underlying biochemistry. With respect to the provenance of behavior we know much more about ontogenic contingencies than phylogenic. Moreover, phylogenic contingencies raise some very difficult problems which have no ontogenic parallels.

The contingencies responsible for unlearned behavior acted a very long time ago. The natural selection of a given form of behavior, no matter how plausibly argued, remains an inference. We can set up phylogenic contingencies under which a given property of behavior arbitrarily selects individuals for breeding, and thus demonstrate modes of behavioral inheritance, but the experimenter who makes the selection is performing a function of the natural environment which also needs to be studied. Just as the reinforcements arranged in an experimental analysis must be shown to have parallels in "real life" if the results of the analysis are to be significant or useful, so the contingencies which select a given behavioral trait in a genetic experiment must be shown to play a plausible role in natural selection.

Although ontogenic contingencies are easily subjected to an experimental analysis, phylogenic contingencies are not. When the experimenter has shaped a complex response, such as dropping a marble into a tube, the provenance of the behavior raises no problem. The performance may puzzle anyone seeing it for the first time, but it is easily traced to recent, possibly recorded, events. No comparable history can be invoked when a spider is observed to spin a web. We have not seen the phylogenic contingencies at work. All we know is that spiders of a given kind build more or less the same kind of web. Our ignorance often adds a touch of mystery. We are likely to view inherited behavior with a kind of awe not inspired by acquired behavior of similar complexity.

The remoteness of phylogenic contingencies affects our scientific methods, both experimental and conceptual. Until we have identified the variables of which an event is a function, we tend to invent causes. Learned behavior was once commonly attributed to "habit," but an analysis of contingencies of reinforcement has made the term unnecessary. "Instinct," as a hypothetical cause of phylogenic behavior, has had a longer life. We no longer say that our rat possesses a marble-dropping habit, but we are still likely to say that our spider has a web-spinning instinct. The concept of instinct has been severely criticized and is now used with caution or altogether avoided, but explanatory entities serving a similar function still survive in the writings of many ethologists.

A "mental apparatus," for example, no longer finds a useful place in the experimental analysis of behavior, but it survives in discussions of phylogenic contingencies. Here are a few sentences from the writings of prominent ethologists which refer to consciousness or awareness: "The young gosling...gets imprinted upon its mind the image of the first moving object it sees" (W. H. Thorpe, 1951); "the infant expresses the inner state of contentment by smiling" (Julian Huxley, 1964); "[herring gulls show a] lack of insight into the ends served by their activities" (Tinbergen, 1953); "[chimpanzees were unable] to communicate to others the unseen things in their minds" (Frankenberger and Kortlandt, 1965).

In some mental activities awareness may not be critical, but other cognitive activities are invoked. Thorpe (1951) speaks of a disposition "which leads the animal to pay particular attention to objects of a certain kind." What we observe is simply that objects of a certain kind are especially effective stimuli. The ontogenic contingencies which generate the behavior called "paying attention" presumably have phylogenic parallels. Other mental activities frequently mentioned by ethologists include "organizing experience" and "discovering relations." Expressions of all these sorts show that we have not yet accounted for the behavior in terms of contingencies, phylogenic or ontogenic. Unable to show how the organism can behave effectively under complex circumstances, we endow it with a special cognitive ability which permits it to do so.

Other concepts replaced by a more effective analysis include "need" or "drive" and "emotion." In ontogenic behavior we no longer say that a given set of environmental conditions first gives rise to an inner state which the organism then expresses or resolves by behaving in a given way. We no longer represent relations among emotional and motivational variables as relations among such states, as in saying that hunger overcomes fear. We no longer use dynamic analogies or metaphors, as in explaining sudden action as the overflow or bursting out of dammed-up needs or drives. If these are common practices in ethology, it is evidently because the functional relations they attempt to formulate are not clearly understood.

Another kind of innate endowment, particularly likely to appear in explanations of human behavior, takes the form of "traits" or "abilities." Though often measured quantitatively, their dimensions are meaningful only in placing the individual with respect to a population. The behavior measured is almost always obviously learned. To say that intelligence is inherited is not to say that specific forms of behavior are inherited. Phylogenic contingencies conceivably responsible for "the selection of intelligence" do not specify responses. What has been selected appears to be a susceptibility to ontogenic contingencies, leading particularly to a greater speed of conditioning and the capacity to maintain a larger repertoire without confusion.

It is often said that an analysis of behavior in terms of ontogenic contingencies "leaves something out of account," and this is true. It leaves out of account habits, ideas, cognitive processes, needs, drives, traits, and so on. But it does not neglect the facts upon which these concepts are based. It seeks a more effective formulation of the very contingencies to which those who use such concepts must eventually turn to explain their explanations. The strategy has been highly successful at the ontogenic level, where the contingencies are relatively clear. As the nature and mode of operation of phylogenic contingencies come to be better understood, a similar strategy should yield comparable advantages.

IDENTIFYING PHYLOGENIC AND ONTOGENIC VARIABLES

The significance of ontogenic variables may be assessed by holding genetic conditions as constant as possible—for example, by studying "pure" strains or identical twins. The technique has a long history. According to Plutarch *(De Puerorum Educatione)* Licurgus, a Spartan, demonstrated the importance of environment by raising two puppies from the same litter so that one became a good hunter while the other preferred food from a plate. On the other hand, genetic variables may be assessed either by studying organisms upon which the environment has had little opportunity to act (because they are newborn or have been reared in a controlled environment) or by comparing groups subject to extensive, but on the average probably similar, environmental histories. Behavior exhibited by most of the members of a species is often accepted as inherited if it is unlikely that all the members could have been exposed to relevant ontogenic contingencies.

When contingencies are not obvious, it is perhaps unwise to call any behavior either inherited or acquired. Field observations, in particular, will often not permit a distinction. Friedmann (1956) has described the behavior of the African honey guide as follows:

> When the bird is ready to begin guiding, it either comes to a person and starts a repetitive series of churring notes or it stays where it is and begins calling....
>
> As the person comes to within 15 or 20 feet...the bird flies off with an initial conspicuous downward dip, and then goes off to another tree, not necessarily in sight of the follower, in fact more often out of sight than not. Then it waits there, churring loudly until the follower again nears it, when the action is repeated. This goes on until the vicinity of the bees' nest is reached. Here the bird suddenly ceases calling and perches quietly in a tree nearby. It waits there for the follower to open the hive, and it usually remains there until the person has departed with his loot of honey-comb, when it comes down to the plundered bees' nest and begins to feed on the bits of comb left strewn about.

The author is quoted as saying that the behavior is "purely instinctive," but it is possible to explain almost all of it in other ways. If we assume that honey guides eat broken bees' nests and cannot eat unbroken nests, that men (not to mention baboons and ratels) break bees' nests, and that birds more easily discover unbroken nests, then only one other assumption is needed to explain the behavior in ontogenic terms. We must assume that the response which produces the churring note is elicited either (1) by any stimulus which frequently precedes the receipt of food (comparable behavior is shown by a hungry dog jumping about when food is being prepared for it) or (2) when food, ordinarily available, is missing (the dog jumps about when food is not being prepared for it on schedule). An unconditioned honey guide occasionally sees men breaking nests. It waits until they have gone, and then eats the remaining scraps. Later it sees men near but not breaking nests, either because they have not yet found the nests or have not yet reached them. The sight of a man near a nest, or the sight of man when the buzzing of bees around a nest can be heard, begins to function in either of the ways just noted to elicit the churring response. The first step in the construction of the final pattern is thus taken by the honey guide. The second step is taken by the man (or baboon or ratel, as the case may be). The churring sound becomes a conditioned stimulus in the presence of which a search for bees' nests is frequently successful. The buzzing of bees would have the same effect if the man could hear it.

The next change occurs in the honey guide. When a man approaches and breaks up a nest, his behavior begins to function as a conditioned reinforcer which, together with the fragments which he leaves behind, reinforces churring, which then becomes more probable under the circumstances and emerges primarily as an operant rather than as an emotional response. When this has happened, the geographical arrangements work themselves out naturally. Men learn to move toward the churring sound, and they break nests more often after walking toward nests than after walking in other directions. The honey guide is therefore differentially reinforced when it takes a position which induces men to walk toward a nest. The contingencies are subtle, but we should remember that the final topography is often far from perfect.

As we have seen, contingencies which involve two or more organisms raise special problems. The churring of the honey guide is useless until men respond to it, but men will not respond in an appropriate way until the churring is related to the location of bees' nests. The conditions just described compose a sort of program which could lead to the terminal performance. It may be that the conditions will not often arise, but another characteristic of social contingencies quickly takes over. When one honey guide and one man have entered into this symbiotic arrangement, conditions prevail under which other honey guides and other men will be much more rapidly conditioned. A second man will more quickly learn to go in the direction of the churring sound because the sound is already spatially related to bees' nests. A second honey guide will more readily learn to churr in the right places because men respond in a way which reinforces that behavior. When a large number of birds have learned to guide and a large number of men have learned to be guided, conditions are highly favorable for maintaining the system. (It is said that, where men no longer bother to break bees' nests, they no longer comprise an occasion for churring, and the honey guide turns to the ratel or baboon. The change in contingencies has occurred too rapidly to work through natural selection. Possibly an instinctive response has been unlearned, but the effect is more plausibly interpreted as the extinction of an operant.)

Imprinting is another phenomenon which shows how hard it is to detect the nature and effect of phylogenic contingencies. In Thomas More's *Utopia*, eggs were incubated. The chicks "are no sooner out of the shell, and able to stir about, but they seem to consider those that feed them as their mothers, and follow them as other chickens do the hen that hatched them." Later accounts of imprinting have been reviewed by Gray (1963). Various facts suggest phylogenic origins: the response of following an imprinted object appears at a certain age; if it cannot appear then, it may not appear at all; and so on. Some experiments by Peterson (1960), however, suggest that what is inherited is not the behavior of following but a susceptibility to reinforcement by proximity to the mother or mother surrogate. A distress call reduces the distance between mother and chick when the mother responds appropriately, and walking toward the mother has the same effect. Both behaviors may therefore be reinforced (Hoffman et al., 1966), but they appear before these ontogenic contingencies come into play and are, therefore, in part at least phylogenic. In the laboratory, however, other behaviors can be made effective which phylogenic contingencies are not likely to have strengthened. A chick can be conditioned to peck a key, for example, by moving an imprinted object toward it when it pecks or to walk away from the object if, through a mechanical arrangement, this behavior actually brings the object closer. To the extent that chicks follow an imprinted object simply because they thus bring the object closer or prevent it from becoming more distant, the behavior could be said to be "species-specific" in the unusual sense that it is the product of *ontogenic* contingencies which prevail for all members of the species.

Ontogenic and phylogenic behaviors are not distinguished by any essence or character. Form of response seldom if ever yields useful classifications. The verbal response *Fire!* may be a command to a firing squad, a call for help, or an answer to the question, *What do you see?* The topography tells us little, but the controlling variables permit us to distinguish three very different verbal operants (Skinner, 1957). The sheer forms of instinctive and learned behaviors also tell us little. Animals court, mate, fight, hunt, and rear their young, and they use the same effectors in much the same way in all sorts of learned behavior. Behavior is behavior whether learned or unlearned; it is only the controlling variables which make a difference. The difference is not always important. We might show that a honey guide is controlled by the buzzing of bees rather than by the sight of a nest, for example, without prejudice to the question of whether the behavior is innate or acquired.

Nevertheless the distinction is important if we are to undertake to predict or control the behavior. Implications for human affairs have often affected the design of research and the conclusions drawn from it. A classical example concerns the practice of exogamy. Popper (1957) writes:

Mill and his psychologistic school of sociology...would try to explain [rules of exogamy] by an appeal to 'human nature,' for instance to some sort of instinctive aversion against incest (developed perhaps through natural selection...); and something like this would also be naïve or popular explanation. [From Marx's] point of view...however, one could ask whether it is not the other way round, that is to say, whether the apparent instinct is not rather a product of education, the effect rather than the cause of the social rules and traditions demanding exogamy and forbidding incest. It is clear that these two approaches correspond exactly to the very ancient problem whether social laws are "natural" or "conventions."

Much earlier, in his *Supplement to the Voyage of Bougainville*, Diderot (1796) considered the question of whether there is a natural basis for sexual modesty or shame *(pudeur)*. Though he was writing nearly a hundred years before Darwin, he pointed to a possible basis for natural selection. "The pleasures of love are followed by a weakness which puts one at the mercy of one's enemies. That is the only natural thing about modesty; the rest is convention." Those who are preoccupied with sex are exposed to attack (indeed, may be stimulating attack); hence, those who engage in sexual behavior under cover are more likely to breed successfully. Here are phylogenic contingencies which either make sexual behavior under cover stronger than sexual behavior in the open or reinforce the taking of cover when sexual behavior is strong. Ontogenic contingencies through which organisms seek cover to avoid disturbances during sexual activity are also plausible.

The issue has little to do with the character of incestuous or sexual behavior, or with the way people "feel" about it. The basic distinction is between provenances. And provenance is important because it tells us something about how behavior can be supported or changed. Most of the controversy concerning heredity and environment has arisen in connection with the practical control of behavior through the manipulation of relevant variables.

INTERRELATIONS AMONG PHYLOGENIC AND ONTOGENIC VARIABLES

The ways in which animals behave compose a sort of taxonomy of behavior comparable to other taxonomic parts of biology. Only a very small percentage of existing species has as yet been investigated. (A taxonomy of behavior may indeed be losing ground as new species are discovered.) Moreover, only a small part of the repertoire of any species is ever studied (see Note 7.3). Nothing approaching a fair sampling of species-specific behavior is therefore ever likely to be made.

Specialists in phylogenic contingencies often complain that those who study learned behavior neglect the genetic limitations of their subjects, as the comparative anatomist might object to conclusions drawn from the intensive study of a single species. Beach, for example, has written (1950): "Many...appear to believe that in studying the rat they are studying all or nearly all that is important in behavior.... How else are we to interpret...[a] 457-page opus which is based exclusively upon the performance of rats in bar-pressing situations but is entitled simply *The Behavior of Organisms?*" There are many precedents for concentrating on one species (or at most a very few species) in biological investigations. Mendel discovered the basic laws of genetics—in the garden pea. Morgan worked out the theory of the gene—for the fruitfly. Sherrington investigated the integrative action of the nervous system— in the dog and cat. Pavlov studied the physiological activity of the cerebral cortex— in the dog.

In the experimental analysis of behavior many species differences are minimized. Stimuli are chosen to which the species under investigation can respond and which do not elicit or release disrupting responses: visual stimuli are not used if the organism is blind, or very bright lights if they evoke evasive action. A response is chosen which may be emitted at a high rate without fatigue and which will operate recording and controlling equipment: we do not reinforce a monkey when it pecks a disk with its nose or a pigeon when it trips a toggle switch—though we might do so if we wished. Reinforcers are chosen which are indeed reinforcing, either positively or negatively. In this way species differences in sensory equipment, in effector systems, in susceptibility to reinforcement, and in possibly disruptive repertoires are minimized. The data then show an extraordinary uniformity over a wide range of species. For example, the processes of extinction, discrimination, and generalization, and the performances generated by various schedules of reinforcement are reassuringly similar. (Those who are interested in fine structure may interpret these practices as minimizing the importance of sensory and motor areas in the cortex and emotional and motivational areas in the brain stem, leaving for study the processes associated with nerve tissue as such, rather than with gross anatomy.) Although species differences exist and should be studied, an exhaustive analysis of the behavior of a single species is as easily justified as the study of the chemistry or microanatomy of nerve tissue in one species.

A rather similar objection has been lodged against the extensive use of domesticated animals in laboratory research (Kavanau, 1964). Domesticated animals offer many advantages. They are more easily handled, they thrive and breed in captivity, they are resistant to the infections encountered in association with men, and so on. Moreover, we are primarily interested in the most domesticated of all animals—man. Wild animals are, of course, different—possibly as different from domesticated varieties as some species are from others, but both kinds of differences may be treated in the same way in the study of basic processes.

The behavioral taxonomist may also argue that the contrived environment of the laboratory is defective since it does not evoke characteristic phylogenic behavior. A pigeon in a small enclosed space pecking a disk which operates a mechanical food dispenser is behaving very differently from pigeons at large. But in what sense is this behavior not "natural"? If there is a natural phylogenic environment, it must be the environment in which a given kind of behavior evolved. But the phylogenic contingencies responsible for current behavior lie in the distant past. Within a few thousand years—a period much too short for genetic changes of any great magnitude—all current species have been subjected to drastic changes in climate, predation, food supply, shelter, and so on. Certainly no land mammal is now living in the environment which selected its principal genetic features, behavioral or otherwise. Current environments are almost as "unnatural" as a laboratory. In any case, behavior in a natural habitat would have no special claim to genuineness. What an organism does is a fact about that organism regardless of the conditions under which it does it. A behavioral process is none the less real for being exhibited in an arbitrary setting.

The relative importance of phylogenic and ontogenic contingencies cannot be argued from instances in which unlearned or learned behavior intrudes or dominates. Breland and Breland (1961) have used operant conditioning and programming to train performing animals. They conditioned a pig to deposit large wooden coins in a "piggy bank." "The coins were placed several feet from the bank and the pig required to carry them to the bank and deposit them.... At first the pig would eagerly pick up one dollar, carry it to the bank, run back, get another, carry it rapidly and neatly, and so on.... Thereafter, over a period of weeks the behavior would become slower and slower. He might run over eagerly for each dollar, but on the way back, instead of carrying the dollar and depositing it simply and cleanly, he would repeatedly drop it, root it, drop it again, root it along the way, pick it up, toss it up in the air, drop it, root it some more, and so on." They also conditioned a chicken to deliver plastic capsules containing small toys by moving them toward the purchaser with one or two sharp straight pecks. The chickens began to grab at the capsules and "pound them up and down on the floor of the cage," perhaps as if they were breaking seed pods or pieces of food too large to be swallowed. Since other reinforcers were not used, we cannot be sure that these phylogenic forms of food-getting behavior appeared because the objects were manipulated under food-reinforcement. The conclusion is plausible, however, and not disturbing. A shift in controlling variables is often observed. Under reinforcement on a so-called "fixed-interval schedule," competing behavior emerges at predictable points (Morse & Skinner, 1957). The intruding behavior may be learned or unlearned. It may disrupt a performance or, as Kelleher (1962) has shown, it may not. The facts do not show an inherently greater power of phylogenic contingencies in general. Indeed, the intrusions may occur in the other direction. A hungry pigeon which was being trained to guide missiles (Skinner, 1960) was reinforced with food on a schedule which generated a high rate of pecking at a target projected on a plastic disk. It began to peck at the food as rapidly as at the target. The rate was too high to permit it to take grains into its mouth, and it began to starve. A product of ontogenic contingencies had suppressed one of the most powerful phylogenic activities. The behavior of civilized man shows the extent to which environmental variables may mask an inherited endowment.

MISLEADING SIMILARITIES

Since phylogenic and ontogenic contingencies act at different times and shape and maintain behavior in different ways, it is dangerous to try to arrange their products on a single continuum or to describe them with a single set of terms.

An apparent resemblance concerns intention or purpose (see page 107 [of original]). Behavior which is influenced by its consequences seems to be directed toward the future. We say that spiders spin webs in order to catch flies and that men set nets in order to catch fish. The "order" is temporal. No account of either form of behavior would be complete if it did not make some reference to its effects. But flies or fish which have not yet been caught cannot affect behavior. Only past effects are relevant. Spiders which have built effective webs have been more likely to leave offspring, and setting a net in a way that has caught fish has been reinforced. Both forms of behavior are therefore more likely to occur again, but for very different reasons.

The concept of purpose has had an important place in evolutionary theory. It is still sometimes said to be needed to explain the variations upon which natural selection operates. In human behavior a "felt intention" or "sense of purpose" which precedes action is sometimes proposed as a current surrogate for future events. Men who set nets "know why they are doing so," and something of the same sort may have produced the spider's web-spinning behavior which then became subject to natural selection. But men behave because of operant reinforcement even though they cannot "state their purpose"; and, when they can, they may simply be describing their behavior and the contingencies responsible for its strength. Self-knowledge is at best a by-product of contingencies; it is not a cause of the behavior generated by them. Even if we could discover a spider's felt intention or sense of purpose, we could not offer it as a cause of the behavior.

Both phylogenic and ontogenic contingencies may seem to "build purpose into" an organism. It has been said that one of the achievements of cybernetics has been to demonstrate that machines may show purpose. But we must look to the construction of the machine, as we look to the phylogeny and ontogeny of behavior, to account for the fact that an ongoing system acts as if it had a purpose.

Another apparent characteristic in common is "adaptation." Both kinds of contingencies change the organism so that it adjusts to its environment in the sense of behaving in it more effectively. With respect to phylogenic contingencies, this is what is meant by natural selection. With respect to ontogeny, it is what is meant by operant conditioning. Successful responses are selected in both cases, and the result is adaptation. But the processes of selection are very different, and we cannot tell from the mere fact that behavior is adaptive which kind of process has been responsible for it.

More specific characteristics of behavior seem to be common products of phylogenic and ontogenic contingencies. Imitation is an example. If we define imitation as behaving in a way which resembles the observed behavior of another organism, the term will describe both phylogenic and ontogenic behavior. But important distinctions need to be made. Phylogenic contingencies are presumably responsible for well-defined responses released by similar behavior (or its products) on the part of others. A warning cry is taken up and passed along by others; one bird in a flock flies off and the others fly off; one member of a herd starts to run and the others start to run. A stimulus acting upon only one member of a group thus quickly affects other members, with plausible phylogenic advantages.

The parrot displays a different kind of imitative behavior. Its vocal repertoire is not composed of inherited responses, each of which, like a warning cry, is released by the sound of a similar response in others. It acquires its imitative behavior ontogenically, but only through an apparently inherited capacity to be reinforced by hearing itself produce familiar sounds. Its responses need not be released by immediately preceding stimuli (the parrot speaks when not spoken to); but an echoic stimulus is often effective, and the response is then a sort of imitation.

A third type of imitative contingency does not presuppose an inherited tendency to be reinforced by behaving as others behave. When other organisms are behaving in a given way, similar behavior is likely to be reinforced, since they would probably not be behaving in that way if it were not. Quite apart from any instinct of imitation, we learn to do what others are doing because we are then likely to receive the reinforcement they are receiving. We must not overlook distinctions of this sort if we are to use or cope with imitation in a technology of behavior.

Aggression is another term which conceals differences in provenance. Inherited repertoires of aggressive responses are elicited or released by specific stimuli. Azrin, for example, has studied the stereotyped, mutually aggressive behavior evoked when two organisms receive brief electric shocks. But he and his associates have also demonstrated that the opportunity to engage in such behavior functions as a reinforcer and, as such, may be used to shape an indefinite number of "aggressive" operants of arbitrary topographies (Azrin et al., 1965). Evidence of damage to others may be reinforcing for phylogenic reasons because it is associated with competitive survival. Competition in the current environment may make it reinforcing for ontogenic reasons. To deal successfully with any specific aggressive act we must respect its provenance. (Emotional responses, the bodily changes we feel when we are aggressive, like sexual modesty or aversion to incest, may conceivably be the same whether of phylogenic or ontogenic origin; the importance of the distinction is not thereby reduced.) Konrad Lorenz's recent book *On Aggression* (1966) could be seriously misleading if it diverts our attention from relevant manipulable variables in the current environment to phylogenic contingencies which, in their sheer remoteness, encourage a nothing-can-be-done-about-it attitude.

The concept of territoriality also often conceals basic differences. Relatively stereotyped behavior displayed in defending a territory, as a special case of phylogenic aggression, has presumably been generated by contingencies involving food supplies, breeding, population density, and so on. But cleared territory, associated with these and other advantages, becomes a conditioned reinforcer and as such generates behavior much more specifically adapted to clearing a given territory. Territorial behavior may also be primarily ontogenic. Whether the territory defended is as small as a spot on a crowded beach or as large as a sphere of influence in international politics, we shall not get far in analyzing the behavior if we recognize nothing more than "a primary passion for a place of one's own" (Ardrey, 1961) or insist that "animal behavior provides prototypes of the lust for political power" (Dubos, 1965).

Several other concepts involving social structure also neglect important distinctions. A hierarchical pecking order is inevitable if the members of a group differ with respect to aggressive behavior in any of the forms just mentioned. There are therefore several kinds of pecking orders, differing in their provenances. Some dominant and submissive behaviors are presumably phylogenic stereotypes; the underdog turns on its back to escape further attack, but it does not follow that the vassal prostrating himself before king or priest is behaving for the same reasons. The ontogenic contingencies which shape the organization of a large company or governmental administration show little in common with the phylogenic contingencies responsible for the hierarchy in the poultry yard. Some forms of human society may resemble the anthill or beehive, but not because they exemplify the same behavioral processes (Allee, 1938).

Basic differences between phylogenic and ontogenic contingencies are particularly neglected in theories of communication. In the inherited signal systems of animals the behavior of a "speaker" furthers the survival of the species when it affects a "listener." The distress call of a chick evokes appropriate behavior in the hen; mating calls and displays evoke appropriate responses in the opposite sex; and so on. De Laguna (1927) has suggested that animal calls could be classified as declarations, commands, predictions, and so on, and Sebeok (1965) has recently attempted a similar synthesis in modern linguistic terms, arguing for the importance of a science of zoosemiotics.

The phylogenic and ontogenic contingencies leading, respectively, to instinctive signal systems and to verbal behavior are quite different. One is not an early version of the other. Cries, displays, and other forms of communication arising from phylogenic contingencies are particularly insensitive to operant reinforcement. Like phylogenic repertories in general, they are restricted to situations which elicit or release them and hence lack the variety and flexibility which favor operant conditioning. Vocal responses which at least closely resemble instinctive cries have been conditioned, but much less easily than responses using other parts of the skeletal nervous system. The vocal responses in the human child which are so easily shaped by operant reinforcement are not controlled by specific releasers. It was the development of an undifferentiated vocal repertoire which brought a new and important system of behavior within range of operant reinforcement through the mediation of other organisms (Skinner, 1957).

Many efforts have been made to represent the products of both sets of contingencies in a single formulation. An utterance, gesture, or display, whether phylogenic or ontogenic, is said to have a referent which is its meaning, the referent or meaning being inferred by a listener. Information theory offers a more elaborate version: the communicating organism selects a message from the environment, reads out relevant information from storage, encodes the message, and emits it; the receiving organism decodes the message, relates it to other stored information, and acts upon it effectively. All these activities, together with the storage of material, may be either phylogenic or ontogenic. The principal terms in such analyses (input, output, sign, referent, and so on) are objective enough, but they do not adequately describe the actual behavior of the speaker or the behavior of the listener as he responds to the speaker. The important differences between phylogenic and ontogenic contingencies must be taken into account in an adequate analysis. It is not true, as Sebeok contends, that "any viable hypothesis about the origin and nature of language will have to incorporate the findings of zoosemiotics." Just as we can analyze and teach imitative behavior without analyzing the phylogenic contingencies responsible for animal mimicry, or study and construct human social systems without analyzing the phylogenic contingencies which lead to the social life of insects, so we can analyze the verbal behavior of man without taking into account the signal systems of other species.

Purpose, adaptation, imitation, aggression, territoriality, social structure, and communication—concepts of this sort have, at first sight, an engaging generality. They appear to be useful in describing both ontogenic and phylogenic behavior and to identify important common properties. Their very generality limits their usefulness, however. A more specific analysis is needed if we are to deal effectively with the two kinds of contingencies and their products.

NOTE 7.1: NATURE OR NURTURE?

The basic issue is not whether behavior is instinctive or learned, as if these adjectives described essences, but whether we have correctly identified the variables responsible for the provenance of behavior as well as those currently in control. Early behaviorists, impressed by the importance of newly discovered environmental variables, found it particularly reinforcing to explain what appeared to be an instinct by showing that it could have been learned, just as ethologists have found it reinforcing to show that behavior attributed to the environment is still exhibited when environmental variables have been ruled out. The important issue is empirical: what *are* the relevant variables?

Whether we can plausibly extrapolate from one species to another is also a question about controlling variables. The ethologist is likely to emphasize differences among species and to object to arguing from pigeons to men, but the environmentalist may object in the same way to the cross-species generalizations of ethologists. If pigeons are not people, neither are graylag geese or apes. To take an important current problem as an example—the population of the world can presumably be kept within bounds without famine, pestilence, or war only if cultural practices associated with procreation can be changed with the aid of education, medicine, and law. It is also possible that man shows or will show when seriously overcrowded a population-limiting instinct, as certain other species appear to do. The question is not whether human procreative behavior is primarily instinctive or learned, or whether the behavior of other species is relevant, but whether the behavior can be controlled through accessible variables.

Extrapolation from one species to another is often felt to be more secure when the species are closely related, but contingencies of survival do not always respect taxonomic classifications. Recent work by Harlow and others on the behavior of infant monkeys is said to be particularly significant for human behavior because monkeys are primates; but so far as a behavioral repertoire is concerned, the human infant is much closer to a kitten or puppy than to an arboreal monkey. The kinship is not in the line of descent, but in the contingencies of survival. The monkey is more likely to survive if infants cling to their mothers, scream and run if left alone, and run to their mothers when frightened. The human baby cannot do much of this, and if it could, the behavior would have no great survival value in a species in which the mother leaves the young while foraging since highly excitable behavior in the infant would attract predators. Mild activity in hunger or physical distress and clinging and sucking when hungry are no doubt important for the human infant, but they lack the extremity of the responses of the infant monkey.

An emphasis on form or structure obscures the difference between inherited and acquired behavior because it means a neglect of the controlling variables in terms of which a distinction can be made. To define imitation simply as behaving as someone else is behaving is to mention stimuli and responses but to neglect the consequences, and it is the consequences which are either phylogenic or ontogenic. To define aggression as behavior which damages others is to fail to make the distinction for the same reasons.

Our increasing knowledge of controlling variables, both phylogenic and onto-genic, has already resolved some traditional issues. Not so long ago it might have been possible to debate whether a pigeon somehow or other learns to build its nest, but now that we have examined the behavior of pigeons under a fairly wide range of contingencies, we can be sure that it does not. A program which would shape the behavior of building a nest, with no contribution whatsoever from genetic endow-ment, can almost certainly not be arranged. If the pigeon had an inherited capacity to be reinforced by various stages in the construction of a nest, the assignment would be less difficult, but still staggering. It is quite out of the question to suppose that the necessary environmental contingencies arise by accident whenever a pigeon builds a nest. At the same time increasing information about how pigeons do build nests clarifies the phylogenic account.

Behavior which is not characteristic of all members of a species but recurs in more or less the same pattern in a few is likely to be said to show an *underlying* nature characteristic of the species. Thus, de Sade is said to have shown that man's "true instincts were to steal, rape, and murder," even though only a small percent-age of men may do these things, at least in de Sade's culture. Without a culture or under extreme provocation, all men may be capable of doing so, but the extremity of the examples offered by de Sade suggests extreme *environmental* circumstances. As we have seen, a schedule of sexual reinforcement may be naturally "stretched" as the amount of behavior required for reinforcement increases with satiation and, on a different time scale, with age.

NOTE 7.2: SPECIES-SPECIFIC BEHAVIOR

A complete inventory of the genetic behavioral endowment of a given species would cover all aspects of its behavior in all possible environments, including

1. Skeletal and autonomic reflexes to all possible eliciting stimuli, including emotional responses under the most extreme provocation.
2. All instinctive responses evoked by identifiable releasers in all possible settings, all necessary materials being available.
3. All the behaviors which may be shaped and maintained by various contingencies of reinforcement, since a species is characterized in part by the positive and negative reinforcers to which it is sensitive and the kinds of topography which are within reach. For example, it is much harder to bring a pigeon under aversive control than a rat, monkey, or man. It is hard to teach a rat to let go of an object by reinforcing it when it does so. It is difficult to shape vocal behavior in most species below man, even when innate responses are common and imitative repertoires easily set up. The speed, order, and direction in which a repertoire can be modified under operant conditioning is also presumably a characteristic of a species.

   (An interesting example of the availability of an unusual response in a porpoise arose when an effort was made to demonstrate operant conditioning to daily audiences (Pryor). A female porpoise was reinforced for a new response each day, and all previously conditioned responses were allowed to go unreinforced. Standard responses such as "porpoising," "beaching," and "tail-slapping" made their appearance and were reinforced, one in each performance. The standard repertoire was soon exhausted however, and the porpoise then began to execute responses which experienced trainers had never seen before and found hard to name or describe. Certain well-defined responses appeared which had previously been observed *only in other strains of porpoises.* These responses would not have been included in an inventory of the strain under observation had it not been for the unusual contingencies which made it highly probable that all available behavior would appear.)
4. Behavior exhibited under unusual or conflicting sets of contingencies, particularly those involving punishment. (A disposition to neurotic or psychotic behavior and the forms taken by that behavior presumably vary among species.)
5. Behavior characteristic of all levels of deprivation—extreme hunger or thirst as well as the most complete satiation.

The concept of a "natural environment" is appealing in part because it permits us to neglect behavior in other environments as if it were not characteristic of the species. Ethologists tend to show no interest, for example, in behavior under laboratory conditions or after domestication. Yet everything is the product of natural processes. We make a useful distinction between animals and men although we know that men are animals, we distinguish the natural from the social sciences although we know that society is natural, and we distinguish between natural and synthetic fibers although we know that the behavior of the chemist is as natural as that of a silkworm. There is nothing which is essentially human, social, or synthetic.

The "natural" environment in which the behavior of a species is studied by ethologists is usually only one of the environments in which the species is now living. It is significant that different natural environments often generate different behaviors. Kortlandt and his associates (Kortlandt, 1967) are reported to have found that chimpanzees living in a rain forest differ greatly (are much less "advanced" or "humanized") than plains-dwelling chimpanzees. But which is the natural environment? Is a chimpanzee learning binary arithmetic in a laboratory (Ferster, 1964) showing chimpanzee or human behavior? The chimpanzees who "manned" early satellites were conditioned under complex contingencies of reinforcement, and their behavior was promptly described as "almost human," but it was the contingencies which were almost human.

NOTE 7.3: INTERRELATIONS AMONG PHYLOGENIC AND ONTOGENIC VARIABLES

Evolution is not appropriately described as a process of trial and error. A mutation is a trial only to those who insist that evolution has direction or purpose, and unsuccessful or lethal mutations do not disappear because they are errors. These terms are likely to turn up, however, in discussions of the evolution of behavior (rather than, say, of anatomical features) because of the currency of trial-and-error theories of learning. But operant conditioning is not, as we have seen, a matter of trial and error either.

A behavioral mutation is not simply a new form of response; the probability that it will be emitted is as important as its topography. A given topography of sexual behavior may be relevant to survival, but so is the probability that it will be displayed. Any susceptibility to reinforcement, positive or negative, has also presumably evolved by degrees rather than by saltatory changes. If the behavior reinforced by sexual contact has survival value, an increase in the power of the reinforcer should have survival value.

The process of operant conditioning has presumably emerged because of its phylogenic consequences, which must also have favored any increase in its speed. The extent to which a given kind of behavior is susceptible to operant reinforcement must also have been important. The human species took a great step forward when its vocal musculature, previously concerned with the production of responses of phylogenic significance, came under operant control, because the social contingencies responsible for verbal behavior could then begin to operate.

Behavior arising from ontogenic contingencies may make phylogenic contingencies more or less effective. Ontogenic behavior may permit a species to maintain itself in a given environment for a long time and thus make it possible for phylogenic contingencies to operate. There is, however, a more direct contribution. If, through evolutionary selection, a given response becomes easier and easier to condition as an operant, then some phylogenic behavior may have had an ontogenic origin. One of Darwin's "serviceable associated habits" will serve as an example. Let us assume that a dog possesses no instinctive tendency to turn around as it lies down but that lying down in this way is reinforced as an operant by the production of a more comfortable bed. If there are no phylogenic advantages, presumably the readiness with which the response is learned will not be changed by selection. But phylogenic advantages can be imagined: such a bed may be freer of vermin, offer improved visibility with respect to predators or prey, permit quick movement in an emergency, and so on. Dogs in which the response was most readily conditioned must have been most likely to survive and breed. (These and other advantages would increase the dog's susceptibility to operant reinforcement in general, but we are here considering the possibility that a particular response becomes more likely to be conditioned.) Turning around when lying down may have become so readily available as an operant that it eventually appeared without reinforcement. It was then "instinctive." Ontogenic contingencies were responsible for the topography of an inherited response. The argument is rather similar to Waddington's (1953) suggestion that useful callouses on the breast of an ostrich, presumably of ontogenic origin, appear before the egg is hatched because a tendency to form callouses has evolved to the point at which the environmental variable (friction) is no longer needed.[1]

Temporal and intensive properties of behavior can also be traced to both ontogenic and phylogenic sources. For example, contingencies of survival and reinforcement both have effects on the speed with which an organism moves in overtaking prey or escaping from predators. A house cat, like its undomesticated relatives, creeps up on its prey slowly and then springs. Relevant contingencies are both phylogenic and ontogenic: by moving slowly the cat comes within jumping range and can then jump more successfully. The stalking pattern is effective because of the characteristic behavior of the prey. If a species comes fairly suddenly into contact with prey which is disturbed by quick movements, the stalking pattern should emerge first at the ontogenic level; but under such conditions, those members of the species most susceptible to differential reinforcement of slow responding should survive and breed. The stalking pattern should then appear more and more quickly, and eventually in the absence of ontogenic contingencies.

Behavior which is not susceptible to operant reinforcement could not have evolved in this way. If the pilomotor response of an enraged cat frightens away its enemies, the disappearance of the enemy may be reinforcing (it could be used, for example, to shape the behavior of pressing a lever), but it is quite unlikely that

---

[1] I am indebted to Professor Leslie Reid for bringing Waddington's suggestion, and its behavioral implications, to my attention.

the consequence has any reinforcing effect on the pilomotor response. It is therefore unlikely that the instinctive behavior had an ontogenic origin.

There are other kinds of interactions among the two kinds of contingencies. Phylogeny comes first and the priority is often emphasized by ethologists, sometimes with the implication that phylogenic problems must be solved before ontogenic contingencies can be studied. Ontogenic changes in behavior affect phylogenic contingencies. A given species does not, as is often said, choose between instinct and intelligence. As soon as a species becomes subject to ontogenic contingencies, phylogenic contingencies become less cogent, for the species can survive with a less adequate phylogenic repertoire. Man did not "choose intelligence over instinct;" he simply developed a sensitivity to ontogenic contingencies which made phylogenic contingencies and their products less important. The phylogenic contingencies still exist but exert less of an effect. The change may have serious consequences. It has often been pointed out, for example, that the ontogenic cultural practices of medicine and sanitation have overruled phylogenic contingencies which would normally maintain or improve the health of the species. The species may suffer when the culture no longer maintains medical and sanitary practices, or when new diseases arise against which only a natural resistance is a defense.

Some phylogenic contingencies must be effective before ontogenic contingencies can operate. The relatively undifferentiated behavior from which operants are selected is presumably a phylogenic product; a large undifferentiated repertoire may have been selected because it made ontogenic contingencies effective. The power of reinforcers must have arisen for similar reasons. It is tempting to say that food is reinforcing because it reduces hunger (Chapter 3), but food in the mouth is reinforcing when not swallowed or ingested, and man and other species eat when not hungry. The capacity to be reinforced by food must be traced to natural selection. Behavior reinforced with food has survival value mainly when an organism is hungry, and organisms which have developed the capacity to be active in getting food *only* when deprived of food have an advantage in being less often needlessly active. A similar variation in the strength of sexual behavior (in most mammals, though not in man) is more obviously of phylogenic origin. In a great many species the male is active sexually only when the behavior is likely to lead to procreation. The bitch in heat emits odors which greatly strengthen sexual behavior in the male dog, and she then cooperates in copulation. It might be argued that this shows a contemporary purpose, as implied in drive-reduction theories: sexual behavior is strong because it leads to fertilization. A plausible connection, however, is to be found in the phylogenic contingencies: under normal contingencies of survival a constantly active sexual behavior when ovulation is not frequent would displace behavior important for survival in other ways. Man appears to be one of a few species which can afford sexual behavior unrelated to ovulation.

The distinction between the inheritance of behavior of specified topography and the inheritance of the capacity to be reinforced by given consequences is relevant not only to imprinting but to the kind of fact offered in support of the concept of a racial unconscious. If archetypal patterns of behavior seem to recur without transmission *via* the environment, it may be because they are independently shaped by recurring contingencies to which racial sensitivities to reinforcement are relevant. The young boy discovering masturbation by himself may seem to be recalling a rhythmic topography exhibited by his ancestors (contributing perhaps to the topography of music and the dance); but the topography may be shaped simply by the reinforcing effects of certain contacts and movements, the capacity to be thus reinforced being possibly all that is inherited.

COMMON FEELINGS. Inherited behavior may differ from learned in the way we feel about it. What we feel are events in, or states of, our body. When we behave primarily to avoid punishment, we may feel responses conditioned by punishing stimuli. We feel them as shame, guilt, or sin, depending upon the source of the punishment. If a culture punishes incestuous behavior, then any move made toward sexual contact with a close relative will presumably generate conditioned responses which are felt as anxiety. Phylogenic contingencies may induce a man to stay away from incestuous contacts either by providing an innate topography from which such contacts are missing, or by imparting a capacity to be automatically punished by them (when they give rise to an "instinctive abhorrence"). If incestuous contacts are automatically punishing for phylogenic reasons, we may look for a difference in the feelings associated with the avoidance of conditioned and unconditioned aversive stimuli. If the feelings differ, we should be able to decide whether incest is a taboo resulting from an instinctive abhorrence or an abhorrence resulting from a taboo.

Several classical issues which have to do with controlling variables are often stated in terms of feelings. When phylogenic contingencies have generated not only behavior having a specific topography but the capacity to be reinforced by the natural consequences of that behavior, the obvious redundancy may operate as a safety device. It may well be true that mothers "instinctively" nurse their young and are at the same time reinforced when they do so through an inherited sensitivity. The relevance of "pleasurable sensations" in accounting for instinctive behavior is an old theme. Cabanis (1802) argued for the importance of the reinforcement. He also reported a curious practice in which a capon was plucked *le ventre*, rubbed with nettles and vinegar, and set on eggs. The eggs were said to give relief from the irritation so that the capon continued to set on them and hatch them. By creating a strong aversive stimulus, from which the capon could escape by setting on eggs, the farmers who resorted to this practice created synthetic hens. Cabanis says that the capon continued to care for the hatched chicks, although the behavior could scarcely have been shaped or maintained through the aversive control. Perhaps all domestic chickens, male and female, possess the behavior in some strength (compare the example of the porpoise above).

The fact that an operant shaped by virtue of an inherited susceptibility to reinforcement may duplicate an instinct arising from the same phylogenic contingencies figured prominently in Darwinian discussions of purpose. There seemed to be an advantage in replacing remote and nearly inscrutable contingencies of survival with ontogenic contingencies where purpose referred to accessible and identifiable consequences (see page 106). Samuel Butler (1917) argued that a hen felt relief after laying an egg and insisted that a poet felt the same kind of relief after writing a poem. We are still likely to say that a man eats to get relief from hunger pangs, and the English language has the idiom of "relieving oneself" to refer to defecation and urination. The argument is close to a theory of reinforcement as drive reduction. Confusion arises from the fact that food is both reinforcing and satiating. The connection is phylogenic: a nourishing substance becomes a reinforcer, so that any behavior leading to its ingestion is likely to be strengthened.

The sucking responses of a newborn infant are probably the best documented instinctive behavior in man. That the tactual and gustatory stimuli inevitably associated with sucking are also reinforcing is a supplementary fact rather than an explanation. The phylogenic contingencies have generated redundant mechanisms.

NOTE 7.4: AGGRESSION

Aggression is sometimes defined as behavior which expresses feelings of hostility or hate, satisfies a need to hurt, is meant or intended to hurt, or can be traced to aggressive instincts or habits. These definitions remain incomplete until we have defined feelings, needs, meanings, intentions, instincts, and habits. Can aggressive behavior be defined in a better way?

Behavior is not aggressive simply because of its topography. Some forms of response, such as baring the teeth or biting, often turn out to be aggressive (as defined below), but this is not always true. Controlling variables must be specified, among them the variables toward which terms like meaning, need, and instinct point. One variable—the effect of the behavior—is important in traditional usage: behavior is aggressive if it harms others (or threatens to do so). A useful distinction may be drawn between phylogenic and ontogenic effects.

PHYLOGENIC AGGRESSION. Tooth-and-claw competition was once the archetypal pattern of natural selection. What evolved was not only efficient teeth and claws but the reflexes and released behaviors in which they played a part. Classical examples with obvious survival value include the aggression of carnivores toward their prey, sexual competition between male and male (the aggression of male against female—in rape—is said to be confined to the human species), a mother's defense of her young, and the protection of a supply of food (the otherwise friendly dog snaps at anyone who tries to take away his bone). These specific contingencies of survival may have given rise to a more general controlling relation. Painful stimuli are associated with combat quite apart from the specific contingencies under which combat makes for survival, and they have come to release aggressive behavior on a great variety of occasions (Azrin et al., 1965). Physical restraint and the absence of characteristic reinforcement ("frustration") are also effective, presumably for similar reasons.

Aggressive behavior of phylogenic origin is accompanied by autonomic responses which contribute to survival at least to the extent that they support vigorous activity. These responses are a major part of what is felt in aggression. Distinctions among jealousy, anger, rage, hatred, and so on, suggest specific phylogenic contingencies. Whether these are different autonomic patterns, or whether what is felt includes more than autonomic behavior, need not be decided here. (The relation of predator to prey is usually regarded as a special case. It may not give rise to "feelings of aggression" although other phylogenic variables may operate in the pursuit or killing of prey.) Many of the dynamic properties of phylogenic aggression remain to be analyzed: eliciting or releasing stimuli become more effective, either in evoking behavior or in arousing feelings, when repeated or when combined with other stimuli having the same effect; a period of active aggression may be followed by a period of quiescence in a kind of satiation which is not simply fatigue; and so on.

ONTOGENIC AGGRESSION. "Damage to others" may act as a reinforcer giving rise to a kind of aggressive behavior under the control of ontogenic variables. When we hurt someone by insulting him, cursing him, or telling him bad news, the topography of our behavior is determined by contingencies arranged by a verbal community. The contingencies have not prevailed long enough to permit any extensive natural selection of the behavior. When we hurt someone by using recently invented weapons, our behavior is also obviously acquired rather than inherited.

It is not enough to define ontogenic aggressive behavior simply by saying that it damages others. What are the dimensions of "damage"? Presumably the actual stimuli which reinforce aggressive action are to be found in the behavior of the recipient as he weeps, cries out, cringes, flees, or gives other signs that he has been hurt. (Counteraggression may be among these behaviors; an aggressive person is reinforced by "getting a rise" out of his opponent.) Aggressive behavior showing a wide range of topographies may be reinforced by these consequences.

Signs of damage also reinforce behavior which is not itself damaging. Thus, they reinforce the spectator at a wrestling or boxing match or professional football game, and he pays admission and watches the match or game because of them. (They are reinforcing even though he does not "identify himself" with the participants; but identification in such a case is also a form of aggressive behavior, largely imitative in nature [Skinner, 1953].)

Damage to others may be reinforcing for several reasons. It may function as a conditioned reinforcer because signs of damage have preceded or coincided with reinforcers which do not otherwise have anything to do with aggression. Effective damage to a sexual competitor becomes reinforcing (if it has not been made so by phylogenic contingencies) when it is followed by unchallenged sexual reinforcement. Damage inflicted upon a thief becomes reinforcing when it is followed by the retention or return of possessions.

We have also to consider the possibility that a capacity to be reinforced by signs of damage may have evolved under the phylogenic contingencies which lead to phylogenic aggression. Individuals should have been selected when they behaved not only in such a way as to drive off predators or sexual competitors, but in such a way as to produce any stimuli commonly preceding these effects, such as the signs of damage associated with successful combat. Indeed the topography of combative behavior should be more quickly shaped and maintained by immediate signs of damage than by eventual success, as the details of a boxer's style are more effectively shaped by the immediate consequences of particular blows than by the final knockout.

Among the reinforcers which shape ontogenic aggression are any conditions which provide the opportunity to act aggressively, either phylogenically or ontogenically. If we are to define aggression in terms of its consequences, we should have to include the behavior of a pigeon pecking a key when the reinforcement is access to another pigeon which can be attacked. The reinforcing effect varies with the incitement, either phylogenic or ontogenic. The probability that the pigeon will peck the key varies with the probability that it will attack another pigeon when a pigeon is already present.

The feelings associated with ontogenic aggression will depend mainly upon the autonomic behavior elicited by the same contingencies. If damage to others is reinforcing simply because it has commonly been followed by such a reinforcer as food, the aggression to which it gives rise may be as "cold" as other forms of food-getting behavior. An innate capacity to be reinforced by damage to others traceable to phylogenic contingencies may give rise to the autonomic pattern associated with phylogenic aggression. To say that we are aggressive because we "take pleasure in hurting" adds no more to the analysis than to say that we eat because we take pleasure in eating. Both expressions simply indicate kinds of reinforcers.

INTERACTIONS AND COMPARISONS. A given instance of aggression can generally be traced to both phylogenic and ontogenic contingencies, since both kinds of variables are generally operative upon a given occasion. The fact that phylogenic contingencies have contributed to the capacity to be reinforced by ontogenic evidences of damage makes the interrelation particularly confusing. It is still worthwhile to look for the effective variables, particularly when an effort is made either to strengthen or weaken aggressive behavior.

The intensity of instinctive aggressive behavior presumably varies roughly with the incitement, at least according to the contingencies originally involved in its selection. If a mother's defense of her young in some modern environment seems exaggerated, we must turn to the original phylogenic contingencies for an explanation. The frequency and energy of ontogenic aggression may range more widely. An intermittent schedule of reinforcement may build a high probability of aggressive behavior even though the net damage is slight. There are natural programming systems having this effect. A man may spend much of his time in the mild aggression called complaining or nagging even though he only rarely evokes signs of damage, such as a burst of anger. He may be programmed into such a condition as the behavior of his listener slowly adapts or extinguishes. Other schedules of differential reinforcement build up violent forms of aggression. Personal systems of attack and counterattack escalate as readily as international if more and more violent behavior is needed to effect damage (to offset improved defenses or to achieve a net positive damage by exceeding the damage done by others). A set of social contingencies in which aggressive behavior escalates has been described elsewhere (Skinner, 1953, p. 309); when two or more people are exchanging aggressive blows, the aversive stimulation of a blow received may evoke a harder blow in return.

Aggressive behavior which does not seem commensurate with its consequences is often puzzling. Killing is called "senseless" when relevant variables can not be identified. But aggression is never senseless in the sense of uncaused; we have simply overlooked either a current variable or a history of reinforcement.

Aggression might be defined as behavior which affects other organisms either phylogenically as a threat to their survival or ontogenically as a negative reinforcer. Both effects have opposites: behavior may promote the survival of others and positively reinforce them. There seems to be no antonym for aggression which covers behavior of both phylogenic and ontogenic origin. "Affection" is close; but it refers to feelings rather than to behavior or its consequences, as hatred refers to the emotional accompaniments of aggression. The phylogenic opposite of aggression has survival value with respect to a different object: survival is furthered by aggression toward competitors and by affection toward members of the same species. Maternal care, foraging for and protecting a mate or mates, and sexual behavior are examples of the latter. The consequences are reinforcing either because of an innate capacity to be reinforced by caring for others or because behavior which positively reinforces others is followed by other kinds of positive reinforcement. Both aggression and affection show a kind of reciprocity. We tend to act aggressively toward those who act aggressively toward us and to be affectionate toward those who show us affection.

A surprising number of the antonyms of aggression have aversive overtones. "Care," "solicitude," and "concern" all suggest anxiety lest the objects of affection be harmed, possibly coupled with a fear that they will no longer show affection. It has often been pointed out that love is close to hatred and that affection and aggression seem to be combined in certain forms of sadistic behavior. This has nothing to do with the essence of love or hatred or with anything in common in the accompanying feelings. It is the consequences which are close to one another and only then because both kinds of effects may be mediated by one person. Affectionate behavior, particularly when built up by intermittent reinforcement, may have strong aversive consequences which in turn evoke aggressive behavior toward the object of affection.

A tendency to kill members of the same species could promote the survival of the species. There may be advantages in limiting a population, in selecting or training especially good fighters who become valuable to the species when they turn on its enemies, and even in cannibalism, in an extreme emergency, as a way of preserving at least a few members. In general, however, intraspecies aggression is rare.

> The tyger preys not on the tyger brood;
> Man only is the common foe of man (Godwin, 1951).

This is sometimes explained by saying that aggression toward members of one's own species is opposed by an instinctive inhibition, except in men. The concept of inhibition is not needed. We do not say that a carnivore refrains from eating vegetables because of an inhibition; its ingestive behavior is evoked only by certain kinds of stimuli. Even if it were true that tigers kill all animals except tigers, we should not need to hypothesize that tiger-killing is inhibited by a special mechanism. Contingencies of survival will explain a discrimination among kinds of prey.

Ontogenic intraspecies aggression also threatens the species. Cultural practices which minimize aggression against other members of a group, such as taboos against killing members of one's own family, tribe, or nation (note the definition of murder), obviously strengthen the group. The cultural sanctions are usually aversive: intragroup aggression is suppressed by punishment or the threat of punishment. This is inhibition in the original meaning of the word: the aggression is forbidden or interdicted. If we do not kill members of our own group, it is not because of some inner inhibition but because of identifiable variables in our culture.

SUICIDE. It is difficult to see how aggressive action toward oneself could have survival value, particularly in the ultimate form of suicide. If suicidal behavior arose as a mutation, it should quickly have eliminated itself. Phylogenic contingencies in which the death of an individual benefits the species would probably favor the selection of behavior in which other members do the killing. (If intraspecies killing threatens the survival of the species, there is a remote chance that suicide would have survival value in making such behavior less probable.) Some forms of instinctive behavior may be damaging and possibly lead to the death of those who display them when the damage is associated with consequences having strong survival value. A difficult but necessary migration may provide the necessary conditions. So may a change of environment if behavior which once had survival value becomes damaging, or lethal in a new setting.

Ontogenic contingencies are more likely to generate behavior which damages the behaver. Behavior which damages others is often damaging to the behaver in the sense that it exposes him to damage or leads him to accept damage without struggle. We may come to submit to damaging consequences because of ultimate positive reinforcement. We take a cold plunge because of the exhilarating glow which follows, submit to danger because we are reinforced by subsequent escape, and hurt ourselves so that others will feel sorry for us and give us attention. We submit to aversive stimuli in order to escape from stimuli which are even more aversive: we go to the dentist and submit to his drill to escape from a toothache. The religious flagellant whips himself to escape from conditioned aversive stimulation which he feels as guilt or a sense of sin. Animals can be induced to take a shock if, in doing so, they are then reinforced positively or negatively, and with careful programming they will continue to do so even when the shock becomes intense.

The ethical group arranges contingencies on this pattern if it gains when an individual inflicts damage on himself. Thus, the group may support a custom of suicide in the old or infirm. A culture which makes much of personal honor may support the practice of hara-kiri or induce heroes to expose themselves to necessarily fatal circumstances. Contingencies arranged by religious systems support mortification and maceration as well as martyrdom. A philosophy of "acceptance of life" recommends submission to aversive and potentially damaging conditions.

Accidental damaging consequences presumably do not define aggression. Although accidental killing was once punished by death, it is now recognized that such measures have no deterrent effect. Nor is the accidental killing of oneself counted as aggression. The man who runs his motor to keep his parked car warm or smokes a great many cigarettes or the citizens of a city who allow the air to be heavily polluted are not, strictly speaking, committing suicide. Nor is the culture whose practices prove fatal when the environment changes. Sanitation and medicine have emerged from ontogenic contingencies having to do with the avoidance of ill health and death, but it is conceivable that a group which maximizes sanitation and medicine may be most vulnerable to a new virus, such as might arise from a mutation or come from some other part of the universe. Practices which up to now have had survival value, although of ontogenic origin, would then prove to have been lethal. Escalation of military power under ontogenic contingencies which seem to favor survival has frequently led to the destruction of civilizations and in the age of nuclear power may lead to the destruction of life on earth.

DEATH INSTINCT. The fact that so much human behavior leads to death has suggested that man possesses a death instinct. There are many different kinds of phylogenic and ontogenic contingencies having this effect, however, and we are not likely to understand them or be able to do much about them if our attention is diverted from effective variables to a fictional cause. Men do behave in ways which are often damaging and even fatal to themselves and others, but a death instinct implies phylogenic contingencies in which this would have survival value. The ontogenic contingencies are much more plausible and conspicuous, and even there the contingencies involve more than damage or death.

THE ENVIRONMENTAL SOLUTION. The four solutions to the problem of aggression discussed in Chapter 3 deserve further comment. The sybaritic solution is to design relatively harmless ways in which people can be aggressive: a man beats another at tennis or chess rather than with a stick; he reads sadistic literature, sees sadistic movies, and watches sadistic sports. These practices probably reinforce aggression rather than "drain it off," unless the preoccupation with harmless forms leaves no time for harmful. To suppress aggression by punishment in the "puritan" solution is simply to shift the role of the aggressor. A chemical solution, as we have noted, may exist in the form of tranquilizers.

The environmental solution becomes more plausible the more we know about the contingencies. Phylogenic aggression may be minimized by minimizing eliciting and releasing stimuli. Behavior acquired because of an inherited tendency to be reinforced by damage to others can be minimized by breaking up the contingencies—by creating a world in which very little behavior causes the kinds of damage which are reinforcing. We can avoid making damage-to-others a conditioned reinforcer by making sure that other reinforcements are not contingent upon behavior which damages. (To put it roughly, people who get what they want without hurting others are less likely to be reinforced by hurting others.) In short, we can solve the problem of aggression by building a world in which damage to others has no survival value and, for that or other reasons, never functions as a reinforcer. It will necessarily be a world in which non-aggressive behaviors are abundantly reinforced on effective schedules in other ways.

NOTE 7.5: A POSSIBLE EXAMPLE OF PROGRAMMED PHYLOGENIC CONTINGENCIES

The hypothesis of continental drift, which has recently received surprising confirmation, may explain certain cases of complex migratory behavior which are otherwise quite puzzling. Both European and American eels, for example, when ready to breed, leave their freshwater environments and journey to overlapping deep-sea breeding grounds in the middle Atlantic. The adults die there, but the young return to the appropriate continents. It is difficult to imagine that this extremely complex pattern in the behavior of both parents and offspring could have arisen in its present form through random mutations, selected by the survival of individuals possessing appropriate behavior. If we assume, however, that Europe and North America were once contiguous and that they moved only very slowly apart, the first journeys of the eels, or of those earlier forms which evolved as eels, could have been quite short. The present extreme behavior would have been gradually "shaped" through survival as the phylogenic contingencies changed. Each year only a slight extension of behavior would be demanded—possibly only a matter of inches—and the new contingencies could be met by most members of the species. Just as an animal with little or no innate tendency to home can be trained by releasing it at slowly increasing distances, so early forms of eels were "trained" by phylogenic contingencies as the distances to be traversed were extended by continental drift. This would help to explain the fact that the breeding grounds of European and American eels are close together or overlap.

The behavior of salmon in the North Atlantic may be the result of a similar program of phylogenic contingencies.[2]

---

[2]Dr. C.W. McCutchen has called my attention to the fact that Dr. Ronald Fraser in *The Habitable Earth*, published in 1964, points out that the green turtle that now migrates between Brazil and Ascension Island, an annual journey of 1,400 miles each way, may originally have gone at most 100 miles. Dr. Fraser does not discuss the importance of this fact for phylogenic programming.

REFERENCES

Allee, W. C. *Cooperation Among Animals*. New York: Abelard-Schuman, 1938.

Ardrey, R. *African Genesis*. New York: Atheneum, 1961.

Azrin, N. H., R. R. Hutchinson, and R. McLaughlin. "The Opportunity for Aggression as an Operant Reinforcer During Aversive Stimulation." *J. Exp. Anal. Behav.* **8** (1965): 171.

Beach, F. A. "The Snark was a Boojum." *Amer. Psychol.* **5** (1950): 115–124.

Breland, K., and M. Breland. "The Misbehavior of Organisms." *Amer. Psychol.* **16** (1961): 681.

Butler, Samuel. *Notebooks*. New York: Dutton, 1917.

Cabanis, P. J. G. *Rapports du physique et du moral de l'homme*. Paris: Crapart, Caille et Ravier, 1802.

De Laguna, G. *Speech: Its Function and Development*. New Haven, CT: Yale University Press, 1927.

Diderot, D. *Supplement au voyage de Bougainville*. (Written in 1774, published in 1796.)

Dubos, R. "Humanistic Biology." *Amer. Scientist* **53** (1965): 4–19.

Erlenmeyer-Kimling, E., J. Hirsch, and J. M. Weiss. "Studies in Experimental Behavior Genetics: III. Selection and Hybridization Analyses of Individual Differences in the Sign of Geotaxis." *J. Comp. Physiol. Psychol.* **55** (1962): 722–731.

Ferster, C. B. "Arithmetic Behavior in Chimpanzees." *Sci. Am.* (May, 1964).

Friedmann, H. Quoted in article entitled "African Honey-Guides." *Science* **123** (1956): 55.

Godwin, W. Motto on title page of *Caleb Williams*; quoted by Arnold Kettle, *An Introduction to the English Novel*. London: Hutchinson, 1951.

Gray, P. H. "The Descriptive Study of Imprinting in Birds from 1863 to 1953." *J. Gen. Psychol.* **68** (1963): 333–346.

Hirsch, J. "Behavior Genetics and Individuality Understood." *Science* **142** (1963): 1436–1442.

Hoffman, H. S., D. Schiff, J. Adams, and J. L. Serle. "Enhanced Distress Vocalization Through Selective Reinforcement." *Science* **151** (1966): 352–354.

Huxley, J. "Psychometabolism." *Perspectives in Biology and Medicine* **7** (1964): 4.

Kavanau, J. L. "Behavior: Confinement, Adaptation, and Compulsory Regimes in Laboratory Studies." *Science* **143** (1964): 490.

Kelleher, R. T. "Variables and Behavior." *Amer. Psychologist* **17** (1962): 659–660.

Kortlandt, A. (and Z. Frankenberger). *Current Anthropology* **6** (1965): 320.

Kortlandt, A. Reported in *Time* magazine, April 21, 1967.

Lorenz, K. *Evolution and Modification of Behavior*. Chicago: University of Chicago Press, 1965.

Lorenz, K. *On Aggression*. New York: Harcourt, Brace & World, 1966. German edition, 1963.

Morse, W. H., and B. F. Skinner. *J. Comp. Physiol. Psychol.* **50** (1957): 279.

Newman, J. H., Cardinal. *The Idea of a University.* Originally published in 1852. London: Longmans, 1923.

Peterson, N. "Control of Behavior by Presentation of an Imprinted Stimulus." *Science* **132** (1960): 1395–1396.

Popper, K. R. *The Open Society and Its Enemies.* London: Routledge & Kegan Paul, 1957.

Pryor, Karen. Personal communication.

Sebeok, T. A. "Animal Communication." *Science* **147** (1965): 1006–1014.

Skinner, B. F. *The Behavior of Organisms.* New York: Appleton-Century, 1938.

Skinner, B. F. "'Superstition' in the Pigeon." *J. Exp. Psychol.* **38** (1948): 168.

Skinner, B. F. *Science and Human Behavior.* New York: Macmillan, 1953.

Skinner, B. F. *Verbal Behavior.* New York: Appleton-Century-Crofts, 1957.

Skinner, B. F. "Pigeons in a Pelican." *Amer. Psychol.* **15** (1960): 28–37. Reprinted in B. F. Skinner, *Cumulative Record: Revised Edition.* New York: Appleton-Century-Crofts, 1961.

Skinner, B. F. *The Technology of Teaching.* New York: Appleton-Century-Crofts, 1968.

Thorpe, W. H. "The Learning Abilities of Birds. Part I." *Ibis* **93** (1951): 1–52.

Tinbergen, N. *The Herring-Gull's World.* London: Collins, 1953.

Waddington, C. H. "The Evolution of Adaptations." *Endeavor* (July, 1953): 134–139.

Watson, J. B. *Behaviorism.* New York: W. W. Norton, 1924.

Wynne-Edwards, V. C. "Self-Regulating Systems in Populations of Animals." *Science* **147** (1965): 1543–1548.

Orazio Miglino and Richard K. Belew

Preface to Chapter 19

To many readers of this volume Jean Piaget (Neu-chatel, 1896—Geneva, 1980) may be completely unknown, or perhaps known only for his contributions to developmental psychology. In fact, to Piaget "...nature has been as prodigal in her gifts of genius...[as she has been] prodigal in her gift of time" (Brown, 1980, p. 4, foreword), and the scholar had time to make seminal contributions to a wide range of disciplines including not only psychology but also biology and even epistemology. Moreover, the themes that brought these disciplines together for Piaget are very similar to those that cause them to come together in this volume.

In fact the selected passages come from a book whose title could have been our own, *Adaptation and Intelligence: Organic Selection and Phenocopy*. To a modern audience, the only unfamiliar word in Piaget's title is "phenocopy," and it is on this concept that we focus as Piaget's contribution. A phenocopy is "...a product of the convergence between a phenotypic variation and a genotypic variation *which comes to take its place*" [emph. added]. As with many of our authors, the concern is with a correspondence between elements of genetic and phenotypic systems. Asserting that "...the resemblance between the exogenous somatic modification and the corresponding mutation is not purely fortuitous...[since] the high frequency of the phenomenon would seem to preclude this," Piaget chose to challenge the natural causal ordering (genotype forming phenotype) to propose that (in the words of Brown's foreword) "...the genotype has 'copied' the adapted phenotypic form" (Brown, 1980, p. 5).

For Piaget, the genetic system is not simply a passive structure operated upon by chance mutations, but a dynamic system capable of transforming itself so as to assimilate phenotypic characteristics produced through interaction with the environment. We are particularly taken with his attempt to formalize the process (p. 23), which we schematize in Figure 1. The force behind such adaptation is the organism's drive towards a homeostatic equilibrium. Exogenous, environmental changes that disrupt this equilibrium can be met, within a bounded range, by compensating phenotypic variation. In some cases this variation may disrupt the epigenetic (i.e., developmental) process. Piaget would have us view a chain of such "progressive and retroactive repercussions occasioned by a loss of equilibrium" as a kind of feedback, from developing phenotype to the genotype.

It is no wonder that Piaget's proposal was branded the senile Lamarckian musing of a previously insightful scientist. In fact, the portions reprinted here come from a book Piaget intended as a response to just such criticisms. These were first leveled against his earlier volume, *Biology and Knowledge* (Piaget, 1971; original French edition, 1967). As he makes clear here, the traps of direct, "...immediate modification of [the] genome...might as well be called transmission of an acquired

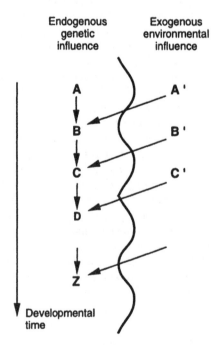

FIGURE 1   Schematic view of Piaget's formalism.

characteristic, in the Lamarckian sense of those words," were well known to Piaget.

But so too were the limitations of an overly dogmatic neo-Darwinian view, "...explaining everything by reference to purely random mutation and selection 'after the event.'" As these same issues played themselves out in a more cognitive guise, especially concerning "nativist" vs. "empiricist" sources of language, Piaget knowingly staked out a middle ground (Piaget, 1980). For his troubles, Piaget notes ironically:

> "A famous pupil of Hull's, D. Berlyne, considered me to be a neo-behaviorist, while another author H. Beilen, refuses to accept this opinion and thinks of me as an innatist, reasoning that I make use of endogenous construction. I am neither one nor the other in that my principle problem is one of the continuous formation of new structures which are not pre-formed either in the environment or in the subject" (Piaget, 1970).

If there is a dated aspect to Piaget's writings, it is due to his ignorance of the enormous strides made by molecular genetics since his passing. Many of his methodological concerns with identifying distinct genotypic and phenotypic forms in the lab and field can now be handled by straightforward genetic assays. However, as our concern moves from morphological variation (plant stem length, shell shape) to behavioral traits whose genetic component is much less clear, Piaget's concerns remain valid. Moreover, for Artificial Life computational modelers, with the luxury of delineating genetic and phenotypic systems exactly, evidence for or against Piaget's phenocopy hypothesis may be realizable.

The portions reprinted here cannot do justice to the life's work from which they are drawn. It is especially worth noting that with *Biology and Knowledge* Piaget took the ambitious *epistemological* step that many in Artificial Life are only now considering: extending a view of adaptation from genetic evolution and individual plasticity to consider social, cultural representations—knowledge. In many respects, then, Piaget's work foreshadows that in Artificial Life. We would do well to listen to our critics and construct our arguments (let alone our sentences!) with as much care.

REFERENCES

Brown, T. A. "Foreword." In *Adaptation and Intelligence: Organic Selection and Phenocopy*, by Jean Piaget. Translated by Stewart Eames. Chicago: University of Chicago Press, 1980.

Piaget, J. *Genetic Epistemology*. Translated by Eleanor Duckworth. New York: Columbia University Press, 1970.

Piaget, J. *Biology and Knowledge: An Essay on the Relations Between Organic Regulations and Cognitive Processes*. Chicago: Chicago University Press, 1971.

Piaget, J. "Opening the Debate: The Psychogeneis of Knowledge and Its Epistemological Significance." In *Language and Learning: The Debate Between Jean Piaget and Noam Chomsky*, edited by M. Piattelli-Palmarini. Cambridge, MA: Harvard University Press, 1980.

Chapter 19:
Excerpts from "Adaptation and Intelligence: Organic Selection and Phenocopy"

The following are excerpts from *Adaptation and Intelligence: Organic Selection and Phenocopy* by Jean Piaget (Chicago: University of Chicago Press, 1980). Translated by Stewart Eames. The original French edition of this work appeared under the title *Adaptation vitale et psychologie de l'intelligence: Sélection organique et phénocopie* (Paris: Hermann, 1974). Reprinted by permission.

PREFACE

The problem of adaptation to the environment is central to the study of evolution. It is particularly relevant to those issues which arise as soon as a connection is sought between organic development and the development of conduct, including the various forms of knowledge. The doctrine on this subject, which was quasi-official among biologists some years ago, consisted in explaining everything by reference to purely random mutation and selection "after the event." Such an explanation now inspires less and less confidence. One reason for this is that the idea of selection has been subjected to serious reconsideration. It was formerly compared with a sifting process, a simple automatic sorting which led only to a broad dichotomy between elimination and survival. Selection has since emerged, however, as a considerably more refined and complex concept, as regards both its results and the mechanism to which they are attributed. Its effect is thought to be the probabilistic modification of the various coefficients and proportions at work in a prevailing state of genetic homeostasis—but it has a further and subsequent effect upon an organism's capacity for modification, the number of its possible responses, and so on. Above all, the operation of selection is increasingly understood to be bound up with factors of choice, in that an organism chooses its environment as well as being dependent upon it. It is bound also to the teleonomic and regulating systems of the organism's internal environment—processes of organic selection as important as those which remain the responsibility of the external environment.

As the concept of selection undergoes this refinement of definition, the role of chance in the production of variants must, to the same extent, be limited. Selection can then readily be imagined, for reasons of symmetry, as tending to operate by means of exploratory "trials" (known also as "scanning"). Such "scanning" will be part subject to chance, but can occur only within zones of possible disequilibrium. In considering the thorough integration of elements of the germ-cell, L. L. Whyte was even led to hypothesize that the multiple conditions which mutations must fulfil [sic] if they are to be compatible with the overall system will have the effect of regulating them. Similarly, R. J. Britten and E. H. Davidson have put forward a model of genetic regulation to explain the properties of a new system starting with the elements of the preceding system.

It should also be remembered that contemporary genetics has concerned itself only with minor (sometimes very minor) changes in organisms' hereditary programming: changes which are therefore easily attributed to chance. The discipline remains, conversely, quite incapable of coping with evolutionary change on a far greater scale, such as that involved in the advent of the vertebrates. It is of course well known, in physics, that a difference in scale can necessitate profound alteration of the structure of explanatory models. In genetics, present theories are, for instance, peculiarly silent about the basic fact that the genetic programming of higher vertebrates is almost a thousand times richer, in terms of units, than that of bacteria. As F. Jacob, speaking in an interview, put it: "This expansion of the genetic programme...represents one of the great difficulties in explaining the mechanism of evolution."[1] It is quite clear, in fact, that it is not a matter of random minor changes, but of what is in the fullest sense a construction. And this structuring raises the possibility that optimization may occur in the processes of equilibration which ensure living creatures their progressive autonomy and capacity for domination of the environment.

The present study arises from a continuing concern with the question of equilibration in the field of cognitive adaptation, and continues, moreover, research

[1][Cet accroissement de programme...représente l'une des grandes difficultés pour expliquer le mécanisme de l'évolution.]

pursued over a number of years into the genotypic and phenotypic variants of a plant species (genus *Sedum*). The question was whether processes giving rise to a "phenocopy," reexamined in the light of these new viewpoints, would not help to simplify certain problems. In broad terms, the phenocopy is a product of convergence between a phenotypic variation and a genotypic mutation which comes to take its place. Its formation is usually ascribed to the genetic processes. P. P. Grassé, however, in a stimulating article which will be quoted again, concludes that "certain phenomena attributed to genetic factors, the phenocopy among others, are perhaps completely different in their origin."[2] A quite central question is raised here which merits close examination, for it raises in turn the problem of relations between environment and hereditary variation. The facts involved are both well-defined in their contexts and far-reaching in their implications.

In broad outline, the solution to be proposed in this study is as follows: (1) There is first the formation of a nonhereditary somatic variation under the influence of external environmental agencies. If it does not produce disequilibrium, this phenotypic modification is simply reconstituted under the influence of the same factors in each succeeding generation. No further effect ensues and there will be no transmission and no phenocopy. (2) Where equilibrium between this exogenous formation and the hereditary epigenetic program is disturbed, the imbalance will have progressive repercussions. There will be no "message" to indicate what is happening or, above all, what measures should be taken, but the disequilibrium will be transmitted by means of local alterations or obstructions in the internal environment. (3) The hierarchic processes of synthesis involved in epigenetic development may not be capable of reestablishing equilibrium at an intermediate level. The repercussions of the original disequilibrium will then extend to the sensitization of those genes which control the syntheses in question. (4) Mutations or genetic variations will then occur in response. These will partly be chance formations, but will then be channelled into areas of disequilibrium. (5) These endogenous variations will effectively be molded within the framework of internal and external environmental influences responsible for the initial somatic form. Now, however, these influences will act selectively until the new variations achieve stability. (6) These new variations will thus eventually converge, by a process of endogenous reconstruction, with the initial semiexogenous phenotypic modifications. Part II of this study will be devoted to corresponding problems in the development of knowledge. It will be seen there that the substitution of endogenous processes for empirical accommodations seems general at every level of development.

The task of assessing this interpretation must be left to the reader, but first I would like to make a remark about how it is related to ideas put forward in an earlier study, *Biology and Knowledge*.[3] The theories expounded in that book seemed, to some colleagues, to be tainted with Lamarckism. Such a suggestion probably

---

[2] [mais certains phénomènes attribués à des facteurs génétiques, les phénocopies, etc., ont peut-être une tout autre origine].

[3] *Biology and Knowledge* (Chicago: The University of Chicago Press, 1971)—Trans.

arose from the principal assertion which was insisted upon throughout: that there was an inherent improbability in ascribing extremely well-organized and extensive powers of synthesis to the genome itself, unless the many regulating forces involved supply it with "feedback" information as to the success or failure of endogenous developments. The earlier study was, however, lacking in precision; the present work is intended to provide this. It seems quite evident that the information feedbacks referred to need comprise no "message," properly so called. (There will therefore be no need to refer, in what follows, to the "reverse transcriptase" passing from RNA to DNA discovered by H. W. Temin and others.) Feedback need consist only in the progressive and retroactive repercussions (by selective obstruction or blocking) occasioned by a loss of equilibrium. In other words, the supposed message may consist of a *noncodified* indication that "something isn't working." When everything is functioning normally, on the other hand, there will be no need for any such indication. The point was made, in *Biology and Knowledge*,[4] that in fact Waddington had gone even further than I had in the direction for which I was subsequently reproached for want of the specifics put forward here.

It is hoped that the present study, which proposes a possible model for the replacement of exogenous by endogenous processes, will find wider acceptance for a writer who has been a tireless opponent of empiricism in the field of epistemology and consequently of pure Lamarckism in biology.

March 1973

J.P.

PRELIMINARIES

The term "phenocopy" is currently applied to an exogenous and nonhereditary (thus phenotypic) somatic modification, which imitates quite precisely the morphological characteristics of a true mutation or hereditary (genotypic) variation. There is, however, general agreement that the formation of this imitative adaptation precedes in time that of the corresponding mutation. The term "copy" of course, in its normal usage, is applied only to the reproduction of a preexisting model: it cannot be said of a precursor, for instance, that he has "copied" the work of a man who comes after him. The major problem posed by the phenocopy, then, is that of the "copying" of the phenotype by the genotype which succeeds it,[5] and not the copying of the phenotype of a mutation not yet produced (unless of course the phenotypic variation

[4]In a footnote to p. 400 of the new French edition (p. 294 in the English edition in 1971)— Trans.
[5]This point is noted by, among others, Konrad Lorenz, writing of behavior patterns which become hereditary in certain species of ducks. Lorenz proposed the term "genocopy" for such a phenomenon. See *L'Agression*, Flammarion, 1969, pp. 72–73 (English edition, *On Aggression*, 1966).

is considered a forerunner, already partly influenced by a genetic mechanism which develops subsequently).

Whatever the terminological difficulties, the problem of the phenocopy is of great theoretical interest. If the resemblance between the exogenous somatic modification and the corresponding mutation is not purely fortuitous (and the high frequency of the phenomenon would seem to preclude this), we have then to understand why it is that, as we believe rightly or wrongly, purely endogenous variations are able to converge with phenotypic modifications, the occurrence of which implies diverse degrees of environmental intervention.

Yet the first problem, of course, is that of knowing which has come first: the phenotypic variation or the new genotype? The phenocopy itself is only rarely observed under laboratory conditions, and it is difficult under natural conditions to establish the chronological sequence of events with any certainty. All that can usually be ascertained is that the exogenous somatic modification concerned is more widespread and frequent, whereas the hereditary modification appears only at some more or less localized point. This suggests the hypothesis that the hereditary modification is formed later. This point can be illustrated by the example of *Sedum album* L., a well-known member of the Crassulaceae. When growing at an altitude of 2000 m or above, this plant always has diminutive stems, leaves, and flowers. As the present writer has often verified, however, it regains its normal size of growth when transplanted into a lowland habitat, beginning at about 1600 m. Such a change can only represent a nonhereditary adaptation to altitude. On the other hand, descendants of a population of the same dwarf form found growing on the summit of Môle (1900 m) in Haute-Savoie maintained their characteristic form, after removal of habitats in Geneva and the plain of Faucigny. Consequently, it must belong to a neighbouring genotypic variety (*micranthum Bast.*). In this case it would seem very likely that the mutation has succeeded the phenotypic modification, and not the reverse. To justify such a conclusion, however, as in all cases of exogenous modification, an extensive and detailed study under natural conditions would be essential.

The development of the phenocopy is thus a potentially enormous subject, perhaps even more so than is often thought. For this reason, preparation for this study has included observations of organisms in many different locations, over a period of some years. These observations have concentrated upon the variations shown by a Mediterranean species of *Sedum*, the *S. sediforme* (also known as *Nicaense* or *altissimum*). In the process, a new hereditary variety was fortunately discovered, on the northern margins of the species' range of distribution, and this new type (variety *parvulum*) provides a good example of a phenocopy. Other observations had previously been made of the races and phenotypes of the aquatic snail *Limnaea stagnalis*, which inhabits lakes and marshes. Detailed analyses of these investigations will be combined as a basis of reference in the present enquiry. The study will be organized, briefly, as follows:

It seems best to begin (chapter 1) by showing clearly how a genotype is recognized under natural conditions and what the respective roles of the genome and the environment are in the course of growth or epigenesis. In fact, genotypes are always embodied in phenotypes (it is the same in the laboratory). The two kinds of phenotype possible (which can also be seen as poles of a continuum) will then be examined (chapter 2). These are distinguished by the extent of direct environmental influence on individual morphology.[6] In one this influence is indeed direct (as in the modification of an organism's size by variation in the availability of nutrients); in the other, environmental effects operate through the medium of innate and specialized regulating mechanisms (to modify, for instance, the skin-pigmentation of a white man or the level of chlorophyll in a plant). Findings from the long-term observations of *Sedum* and *Limnaea* will then be summarized (chapter 3) from the standpoint of relevance to the phenocopy. Hypotheses which have already been put forward to account for the phenocopy and analogous processes will then be reviewed (chapter 4), and the explanatory model which this study proposes will be developed (chapter 5). This has the double aim of countering unfounded suppositions of Lamarckism and of transcending the rather simplistic schemata implicit in orthodox neo-Darwinism. the hypothesis presented being founded upon processes of regulation and equilibration, we shall conclude (in part II) with a comparison between these endogenous processes and those involved in elementary forms of behavior. This will bring us back, through new arguments, to the ideas developed earlier in *Biology and Knowledge*.

This scheme leads us, of course, towards problems of psychology and epistemology, and away from biology. It is comforting to note, however, that the same direction was taken by the discoverer of organic selection and the phenocopy: J. M. Baldwin, after his article of 1896 on the effect which now bears his name, went on to become a great psychologist.

1. GENOTYPE AND EPIGENETIC SYSTEM

It should be recalled at the outset that there is neither simple opposition nor simple parallelism between the concepts of genotype and phenotype. This would be the case only if, after describing the observable characteristics of various phenotypes, one could achieve the same precision with the genotype in its pure state by direct measurement of individuals from a carefully selected strain. It is clear, on the one hand, that a phenotype is dependent at all times upon its genotype, and not only upon the environment. Yet a genotype, on the other hand, is always embodied in phenotypes. Even when it constitutes a pure strain (which is in any case never certain), its representatives inhabit a particular environment, natural or artifical [sic], and are to varying degrees dependent upon it. A genotype must therefore be

---

[6]Some regulating factors, of course, are always operative, but in cases of direct influence their function is only very general.

characterized, in the first place, by the features common to all its phenotypes in all their environments. Second, it must be descriptively differentiated from other genotypes developed contemporaneously in an identical environment.

It will be seen at once how this dual requirement must complicate the diagnosis of hereditary variations and simple phenotypic adaptations—and their distinction is, of course, fundamental to study of the phenocopy. A double experimental approach is consequently advisable. First, in order to distinguish between an indefinite number of genotypes, it is necessary to cultivate those varieties which one believes to be hereditary over several years (or a sufficient number of generations) under the same environmental conditions. In the case of *Sedum sediforme*, our study will be confined to three distinct varieties: *altissimum* (very tall and generally blue-green in color), *medium* (of moderate size and similar coloration), and *parvulum* (a green dwarf form). The characters of these varieties are effectively maintained even when they are grown in the same cultivation bed or the same plant pot. Second, however, it was necessary to repeat these experiments under varying environmental conditions.[7] The plants were therefore grown at various altitudes (400, 900, or 1600 m), in various soils, under various sunlight and shade conditions, and so on—and parts of these cultures were transplanted, from time to time, from one environment to another. The formation of phenotypic modifications could therefore be ascertained, since they would not be stable under altered environmental conditions. For instance, a colony of *medium* may adopt the *parvulum* form under unfavorable conditions, but revert to *medium* type if a further change occurs. Similarly, *parvulum* plants may grow taller and present a *medium* adaptation if grown on well-fertilized garden compost. They will, however, return to the *parvulum* form in the same undisturbed plot when, after a year or so, rain has washed the fertilizer of the soil—unless, of course, more fertilizer is added.

These complicated procedures demand a degree of patience, until the existence of distinct genotypes can be ascertained and their various phenotypic variations recognized. Patience is rewarded, however, from the theoretical viewpoint, in that we are shown the complexity of interactions between the environment and biochemical syntheses of growth controlled by the genetic program. These interactions raise the whole question of epigenesis, of what Waddington has called the "epigenetic system" and Mayr the "epigenotype."

In the first place, it should be remembered that we hardly ever—perhaps even never—find isolated genotypes under natural conditions. We find only populations: systems of interconnected genotypes subject to the possibility of panmixis. The fundamental unit is then no longer the genome characteristic of a genotype more or less isolable in the laboratory, but rather the genetic pool of the population as a system in the sense that it has its own regulating mechanisms, as was shown in the

---

[7]Carefully avoiding hybridization, however, which is very common in *Sedum*.

classic experiments of Dobzhansky and Spassky,[8] and consequently its own "genetic homeostasis." When we speak of the three hereditary varieties *altissimum, medium,* and *parvulum* as genotypes therefore, we make a strictly unwarranted extension of the term, legitimate only as convenient abbreviation. We should in fact almost certainly speak in terms of genetic pools and populations. These populations are thus only represented by individuals with apparently stable hereditary characteristics. Each has its own complex genome which is only a partial image of the genetic pool of the population concerned.

Having said that, we should consider a question fundamental to any interpretation of the phenocopy: that of the part played by environmental influences during epigenesis. It is widely known, of course, that every genome, simple or complex, performs two very basic functions. On the one hand it ensures the hereditary transmission of the information contained in its DNA, and thus determines the characteristics of subsequent generations. On the other hand, it imposes a constant structure on the individual organism during its growth: through the medium of RNA, etc., it ensures protein-synthesis according to an innate program which is ultimately that of its DNA. This process is directed centrifugally, but minute adjustments must take place at every developmental level (germ-cell, intercellular connections, tissues, and organs). These adjustments require the operation of a great number of very delicate regulating mechanisms, particularly those of allosteric nature[9]—and they imply the constant collaboration of all genes in a concerted operation.

What, then, is the part played by the environment during these successive syntheses and their complex regulation? Some writers merely insist upon the importance of its role in nutrition: the growing organism must, after all, be capable of integrating the substances and energies necessary to its development. This integrative process might therefore be understood as merely assimilation of external contributions to endogenous structures imposed by the genetic program, without any concomitant structural alteration. If that were so, then the system's equilibrium could be interpreted as nothing more than the result of complete predetermination. Certainly there would be continuous construction. There could, therefore, be no return to the tenets of static preformism, held by partisans of sperm and egg before P. J. Wolff's discovery rendered the doctrine obsolete. But this construction would still conform to a genetic program which remains, in itself, preformed. This would mean development that could not be modified by external events.

Simple observation should be sufficient to show us that the opposite is true. The inevitable environmental disturbances which occur during epigenesis are in fact quite capable of being translated into significant morphological modifications. One

[8]These writers raised a mixed assembly of fourteen distinct races in a "population cage." After a phase of disequilibrium, they found that equilibrium was progressively restored, and that factors formerly operative were largely reestablished. Populations come to possess, in addition, their own adaptive norms, genetic recombinations being more prevalent than simple mutations; and it will be evident from this that an essential part is played by the multiple heterozygotes.

[9]Through mediators ensuring spatial adjustment.

instance of this is provided by a variation in the shell of *Limnaea stagnalis*. This is a good example in that the operative factors are essentially mechanical, but the tissues are also involved, since the shell is secreted by the epidermis according to a design which is generally constant. This *Limnaea* develops an elongated form in pond habitats (fig. 1). In the more turbulent conditions of larger lakes, however, a variety *lacustris* is known to develop, the form of which is considerably more compact (fig. 2). The mechanism responsible for this variation, insofar as it is a simple phenotypic adaptation, will be described in chapter 2. During the growth of this *lacustris* variant, which is our only concern for the moment, it may happen that the creature's environment is significantly changed. Individuals originating in turbulent lake habitats may be removed to complete their development in the aquarium—or, if the level of the lake were to fall, an individual might complete the construction of its shell in the very different conditions of a pool left isolated above the new waterline. In these two instances there would follow a quite remarkable change in the form of the shell. Half of it would develop in the shorter, compact form, this being a phenotypic modification in imitation of the hereditary variety *lacustris* (fig. 3). The other half would show a reversion to the normal elongated shape of still water varieties.

Such a phenomenon, while confirming the constraining role of the hereditary programming of the elongated varieties, shows at the same time that, during one phase of epigenesis, the environment has in itself exerted a very important morphogenetic influence by imposing the "shorter shell" phenotypic variation. For the present, this environmental influence during epigenesis should simply be noted, without prejudging in any way the explanation of how the hereditary varieties of *lacustris* are formed. The environment has here induced an effect which, while to an extent remaining subject to the hereditary program, was not foreseen by it. We can thus speak here, without exaggeration, of interaction. On the one hand the effect of the genetic program has certainly been modified by the changed environment. On the other, the environmental effect has itself been conditioned by the limits imposed by the hereditary programming upon the range of possible variations. Observations of *Sedum* provide another example. Some whole plants were noted whose branches had apparently begun their growth in the *medium* form (leaves averaging 8–10 mm in length and only slightly convex). Their development had been completed, however, in the *parvulum* form (leaves 5–6 mm long, with a pronounced underside convexity). The resulting branches (5–8 cm in length) showed a clear discontinuity between these growth forms. This modification might arise in response to climatic variation (drought after a period of wet weather, or the reverse), or to variation in light of terrain, etc. Whatever the ultimate cause, transplantation from one location to another has in this case produced a discontinuous alteration of growth, in the shape as well as the size of leaves. These modifications, again, were not foreseen in the innate genetic program, and must therefore be due to environmental factors.

There remains a most important question concerning the nature of the genotype and the part it plays in the course of epigenesis. On the one hand, are these

FIGURE 1 *Limnaea subula.*

FIGURE 2 *Limnaea lacustris.*

FIGURE 3 Result of change in environment during growth of *lacustris* phenotype.

environmental effects upon epigenetic development to be taken as simply unfortunate or inconsequential accidents? Is the power of producing these phenotypic adaptations, on the other hand, indicative of a certain dynamism? In the latter case, such a capability would be extremely useful, in that a greater flexibility would undoubtedly be advantageous to the survival and continued propagation of the genotype concerned. To this second question, however, J. Monod would certainly give a negative answer. Evolution itself, in his opinion, insofar as it involves the production of new genotypes, is "in no way a property of living creatures since it is rooted in the very imperfections of that preservation mechanism which is uniquely their advantage."[10] With good reason, then, the necessity of undergoing all kinds of phenotypic modifications under environmental influence might seem ascribable to "imperfections" even more unfortunate. Dobzhansky, Waddington, and others take the contrary view, seeing these phenotypes as the genotype's responses to the stresses or hostility of the environment. This term "response," however, with its suggestion of activity must be considered in greater detail.

To allow that environmental agencies, exerting their visible effects on the production of phenotypic variations, may have had repercussions extending to the genome—and that the response referred to may arise directly from this—would surely be out of the question. Acceptance of such a hypothesis would involve a virtual return to Lamarckism. It would imply that the environment had produced immediate modification of this genome—not only instigating an accommodative response, but also allowing the possibility that the response might become hereditarily fixed. And this, of course, might as well be called the transmission of an acquired characteristic, in the Lamarckian sense of those words. Since there has never been experimental verification of such a phenomenon, the only remaining possibility is to interpret both the environmental action and the "response" of the genome as occurring at a particuler [sic] stage of epigenetic development. Genetic information passes from DNA to RNA to control the initial selection of amino acids from which proteins are synthesized. Between the starting point of this genetic message and its outcome will be the vast process of epigenetic construction by which the organism attains its adult state. Between origin and culmination there exists, of course, a number of hierarchical levels comparable to relays of one sort or another. Each step of the construction will be characterized by successive processes of synthesis, all evidently directed by genetic information. Since growth, however, involves the continual assimilation of contributions from outside the system, the part played by the environment will be all the more important as the higher stages of development are approached. This will lead to one of two possible results. External factors of nutrition may have no modifying effect at all upon these progressive developments, in which case they will be completed, without deviation, according to the hereditary program. Alternatively, there may be divergence or opposition between the exogenous factors in the environment and the endogenous processes of development. It is then that these endogenous processes will react. Since they are directed by the genome, the reaction they produce at the level of epigenesis concerned can then fittingly be termed a "response of the genome."

But of what does this genomic response consist? Any of the stages of development just outlined will be characterized by a system of regulations or of causal

---

[10][nullement une propriété des êtres vivants, puisqu'elle a sa racine dans les imperfections mêmes du mécanisme conservateur qui, lui, constitue bien leur unique privilège.] *Le hasard et la nécessité* (1970), p. 130.

loops in which every stage is dependent on its predecessors and controls its successors. It is necessary, however, that on the level considered the reactions involved form a system which is both open and at the same time closed in a cycle.[11] In an extremely generalized and simplified form, this might be expressed as follows: $(A \times A' \rightarrow C) \ldots (Y \times Y' \rightarrow B), (B \times B' \rightarrow Z), (Z \times Z' \rightarrow A)$, where $A', B' \ldots Y', Z'$ are external aliments, and $A, B, C \ldots Z, A$ are elements of the cyclic system itself.

Still generalizing excessively but necessarily, the response of the genome, or more accurately of the synthetic apparatus it directs, will be restricted to two possibilities. Environmental change (symbolized by alteration of A' to A'' or B' to B'', etc.) may end by destroying the cycle, in which case growth is obstructed and the organism will consequently die. The environmental action may, alternatively, prove acceptable, in which case the closing of the cyclic system remains possible, on condition that another local alteration takes place. (For instance, the exogenous alteration of A' to A'' would lead to the modification of A into $A2$; but $A2 \times A''$ would continue to give B, other parts of the system remaining unchanged.) The endogenous response, in other words, if it is positive, certainly constitutes an accomodation [sic] of the system to the new environmental situation—but one which maximally conserves its cyclic coherence. There is one sense in which this response is conservative or even restrictive: it excludes unacceptable variations, and preserves (by a kind of optimization) all that can be retained from "normal" processes of synthesis. Yet in another sense the response is dynamic and even innovative, since it ensures adjustement [sic] of the old system to new and unforeseen circumstances.

It should be emphasized, however, that this tentative interpretation of the genome's accommodatory responses in no way concerns the problem of heredity to which we will return in chapter 2. Reactions taking place at the higher stages of epigenetic synthesis do not, in fact, necessarily modify what happens at lower levels of development. These genomic responses are only phenotypic, which means, on the contrary, that once the prevailing environmental influence (resulting from the alteration of A' to A'') is suppressed, the accommodatory response (changing A into $A2$) also comes to an end. The former cycle ($A \times A' \rightarrow B'$, etc.) will then immediately reestablish itself, and regain its former equilibrium. There is thus no question, in this hypothesis, of acquired characteristics being hereditarily fixed or transmitted.

If the ideas broadly outlined here are acceptable, it follows that the greater the number of phenotypic responses available to any genotype, the greater its advantage. Indeed, this plasticity will be a measure of its vitality. A very relevant suggestion is made by P. P. Grassé in a stimulating article[12] concerning the lowland plant species subjected to alpine conditions by Bonnier. These plants developed somatic modifications characteristic of plants found on alpine facies, but these did not prove to be hereditary when they were returned to normal habitats. Grassé argues that this flexibility was more useful to these plants than the formation of new genotypes

[11] Cf. the *intégrons* invoked by F. Jacob.

[12] In *Savoir et action* (November, 1972).

which might have been overspecialized. This is perhaps to take the idea too far, but it does seem certain that, from the point of view of adaptation and preadaptation, a genotype with multiple phenotypes would be at a definite advantage. For example in a study undertaken years ago of several species of terrestrial mollusks widespread in Valais, we established a positive correlation between the size variation of these species in lowland, and the altitude at which they were able to live in their smaller forms.[13] In this case, therefore, the number of distinct phenotypes available initially constitutes a factor in possible adaptation to mountainous terrain.

This leads us to the problem of norms of reaction. It is known that, in terms of the values $V1 \ldots Vn$ of a particular environmental variable, a genotype can react by adopting the characteristic forms $F1 \ldots Fn$, these being the given genotype's possible phenotypic variations in relation to this variable. Such a "norm of reaction" is especially interesting from the viewpoint of the limits tolerated by the genotype concerned. These limits can only be transgressed when new variations or genetic combinations are initiated, which will in turn set up new norms, distinct from the former ones. In this connection there are, however, two points which should be noted. One is that the norm of reaction, though helping to explain interactions between epigenetic processes and environment as has just been considered, in no way permits us to distinguish the precise border between the effects of exogenous and endogenous factors in the phenotypes concerned. We only know that both kinds of factor are at work: the former because nonhereditary modifications occur, the latter because they are limited. Yet the precise boundary between the two spheres of influence remains indeterminate. The second point is that though norms of reaction with one or two environmental variables are easily constructed, the total combination of possible conditions is never known. Nor, as a consequence, can one know all the phenotypes which might potentially arise, even though they have yet to be observed. Whence arises, of course, the great difficulty of identifying or characterizing genotypes under natural conditions.

The foregoing remarks would tend to indicate that two antagonistic forces or tendencies are at work in the heart of the epigenetic system. These were pointed out as significant from the outset, since their equilibrium or lack of it plays a most important part in the formation, or lack of formation, of phenocopies. The first of these two tendencies is naturally conservative. In the course of the successive syntheses ensured by the genome, this tendency compels development to abide by the genetic program, to follow the "necessary paths" or "chreods" of which Waddington speaks. It also compels the synthetic processes to cancel or compensate any deviation, by means of a stabilizing mechanism bearing on the trajectories, and resulting, therefore, in "homeorhesis" rather than simply homeostasis. The second force at work in the epigenetic system is one which also demands very careful attention. In no way is it simply an inverse tendency towards variation—that would

---

[13] J. Piaget, Corrélation entre la répartition verticale des mollusques du Valais et les indices de variations spécifiques [Correlation between the vertical distribution of the mollusks of Valais and indices of species variation], *Revue suisse de zoologie* **28** (1920): 125–153.

be absurd. Rather, it encourages a kind of selective flexibility, making modifications imposed by the environment more or less acceptable according to circumstances. As a result, the processes by which the norm of reaction is reestablished are very complex. For instance, there will obviously be a great difference in the degree of acceptability to the epigenetic program between somatic modifications due to the etiolation of a plant and simple variations in its size. In the former instance, a *Sedum* half-choked and deprived of light by surrounding weeds of greater height can nevertheless live on for weeks or months, and regain its normal form once the ground is cleared. In such a case, modification takes the form of very long, thin stems: leaves are widely spaced on the stem, and the whole plant is very pale in color. Accommodations have clearly occurred which are necessary to current conditions, but which are in no sense desirable. On the other hand, more commonly occurring variations in its size are of much more lasting use to the plant. They allow adjustments between its surface area and volume, which are of value in relating parts of the plant engaged in its nutrition with the mass requiring sustenance. Such modifications are thus responsive both to environmental conditions and to the requirements of synthesis, a hereditary program allowing considerable margins in its realization, since it must take account of many very variable factors.

To summarize: A genotype's norm of reaction would certainly seem to depend upon, among other factors, a compromise between two tendencies. These are quite distinct and, in a sense, mutually opposed. One is conservative, acting against potentially harmful environmental modifications which might endanger equilibrium. The other is a tendency towards flexibility. It aims to utilize exogenous modifications within those limits where their contribution will be advantageous, even though the effect of their multiple combinations cannot be foreseen. This also, however, raises a great problem: that of the kinds of equilibration achieved by phenotypes. This problem is, moreover, quite liable to reappear in different forms where the formation of hereditary modifications is concerned. Phenotypic equilibration might, on the one hand, constitute a simple return after accommodation to the preceding form of equilibrium or another analogous form. It might, on the other hand, be accompanied by some kind of optimizing effect. In other words, the form of equilibrium pursued would then be the best that could be attained under the given conditions.

Peter M. Todd

Preface to Chapter 20

As humans, we tend to view learning as a behavior of paramount importance: it shapes who we are, enables our cultural advancement, and allows us to partake of the uniquely defining human characteristic of language. Because learning is so crucial to our identity, we tend to classify other species by how much or how little of this magical ability they possess, how near or how far from our "pinnacle of plasticity" they lie. But in our rush to judge the tutorability of other animals, we can all too easily forget that learning is not a necessity for life—many creatures get along fine with no learning at all, or as little as possible. Nor is it a free lunch— all of its benefits come at a price. As Aeschylus said nearly 2500 years ago, "He who learns must suffer." In this paper, Timothy Johnston lays out the necessary dimensions of that suffering.

Learning has evolved, as an adaptation shaped by countless generations of selection, because it conferred certain fitness benefits on the individuals bearing this ability. These benefits necessarily outweighed the costs associated with learning, but the costs also exerted pressures that influenced the ultimate form learning took. On the other hand, in species lacking a particular type of learning, the potential costs may have outweighed any potential benefits. The benefits of learning have long been discussed, and Johnston reiterates them here, but the major contribution of his paper is a discussion of the varieties of costs that may be associated with learning.

Adaptive Individuals in Evolving Populations, Ed. R. K. Belew & M. Mitchell, SFI Studies in the Sciences of Complexity, Vol. XXVI, Addison-Wesley, 1996

Much of what Johnston says is speculative and provisional, owing to the paucity of data needed to substantiate some of these ideas. Most studies have not considered learning as an adaptive process that answers the challenges of an organism's environment and is thus subject to the cost/benefit analysis of evolution. For too long, behaviorist approaches to the study of learning have focused on the animal's reactions in a sterile laboratory setting, and as a result the accumulated results of learning theory have little to say about the actual adaptive functions learning might have evolved to play in the creature's normal life. Johnston's paper is thus also a call to reorient approaches to the study of learning: by making experiments more ecologically and environmentally valid, further insights into the evolutionary roles of learning may be gained. But there are also many places where theoretical approaches to the study of the evolution of learning, including simulation approaches, can add to this new, functionally minded empirical work.

Considering the benefits of learning is in many ways an easier task than contemplating the costs, because the potential benefits seem fewer, and therefore clearer. The main advantage Johnston cites for within-lifetime plasticity is the commonly accepted ability to adjust to important environmental changes that occur faster than evolution can track. Johnston discusses the importance of such an ability for adapting to the distribution of food, the locations of nests and territories, the identities of conspecifics, and the like—all features of the external environment that can change rapidly. But an organism's *internal* environment can also change throughout life (especially during early growth and development)—skeletons and muscles grow, eyes develop, abilities alter—and these changes must be adapted to as well. Important instances of this kind of learning can happen automatically, as for example when the visual system wires itself up through exposure to stimuli (see Singer, 1988); or they can be the result of self-initiated behavior, as for example when monkeys engage in play to get practice using their growing bodies in various adult activities (Smith, 1982).

The second benefit that Johnston covers is the specific ability of birds to learn songs for attracting mates. Johnston states that there are "rather few kinds of learning skills that might evolve by sexual selection, namely, those that involve the development of courtship behavior." But because mating and courtship are central to an individual's behavior, and indeed to the course of evolution itself, it seems likely that Johnston has rather underestimated the importance of learning in this domain. In addition to the song learning Johnston focuses on, sexually selected forms of learning can include the acquisition of mate preferences, adapting one's courtship behaviors to the preferences of others, learning the proper mechanical sequence for successful fertilization, and so on. Todd (Chapter 21) discusses the nature and consequences of such forms of learning, all applied to the task of reproduction in the varying genetic environment rather than survival in the varying energetic environment.

Johnston writes that "One way in which the search for relevant data [about the environment-tracking functions of learning] might be stimulated is by the development of more refined theoretical models that predict the evolution of learning in particular environments." This is clearly an opening for simulation studies in this area, provided that we can model different types of environments closely enough. Efforts at developing a theory of environments and characterizing their evolutionary effects are now starting to appear in the simulation literature (see Todd and Wilson, 1993; Todd, Wilson, Somayaji, and Yanco, 1994; Menczer and Belew, Chapter 13). As Johnston states, it is important in this work to set up situations of limited resources, so that competition can allow the benefits of learning to emerge. But it is also important to establish the conditions in which simple reactivity, rather than adaptivity, will suffice. In some cases, storing knowledge in the world (such as performing a simple search for a home nest site each time one returns from foraging) could be more advantageous than storing it in the head (learning the nest site location). Here is where the costs of learning become important, and it is in considering these costs that Johnston's paper is most unique and valuable.

It is difficult to give a clear account of the costs that can go into the evolution of learning (or absence thereof). In part, this is because the possible costs of learning are more numerous than the potential benefits, and they are more variable across species. But equally vexing is the fact that even less empirical data has been collected concerning costs than benefits. Correspondingly, it is very easy to disagree with some of the specific costs that Johnston proposes. For instance, one could argue that delayed reproduction is not a cost of learning, but rather is first caused by something else (another evolutionary pressure, such as the need to amass enough energy to create offspring), and then, once in place, the delay provides an opportunity for learning to evolve *without* incurring this cost. As another instance, increased juvenile vulnerability seems like a less prevalent cost than Johnston allows. Johnston assumes that learning juveniles will start without the knowledge that nonlearning juveniles have, thus making them more vulnerable. But this certainly need not be the case—this is a false dichotomy between having innate knowledge and learning from an initial complete absence of knowledge. Learning juveniles can logically start at the same level of knowledge as nonlearners and thereafter gain more, making them at all times *less* vulnerable than their rival nonlearners. And finally, increased learning does not logically necessitate increased parental investment. One can imagine baby fish that can learn where to hide to avoid being eaten by their father, and that therefore do at least as well as their nonlearning siblings that cannot pick up this protective maneuver. Here the learners will outsurvive the nonlearners, even with equal lack of parental solicitude in both cases.

But despite some disagreement about specific cases, it is essential to consider these and other possible costs nonetheless. Costs create selection pressures that operate even when the cost itself is not exacted—that is, to speak loosely, evolution will seek alternative routes where possible to *avoid* paying a cost lurking down a particular phenotypic path. For example, as Johnston discusses, there are definite costs to bigger brains, but learning species may not always pay these costs, instead discovering ways to accomplish learning in a small brain. Similarly, the cost of delayed reproduction may be avoided by evolving learning that occurs at the earliest possible stages of life (see Hurford [1991] for a model of this phenomenon in the case of language acquisition). Simulations of the evolution of learning tend to lack carefully considered costs, instead focusing solely on the benefits of plasticity. (For exceptions, see Hurford [1991] for a model that includes costs corresponding to delayed reproduction and juvenile vulnerability, and Todd and Miller [1991] for a model with genome/nervous system complexity costs. Hinton and Nowlan [Chapter 25] also incorporate a form of delayed reproduction cost in their model—the more learnable "synapses" an individual has in their simulation, the longer it will take to learn, and the fewer offspring it will be able to have.) But because costs and their attendant selective pressures can so readily shape the kinds of learning that evolve, we must include them in our models to even hope to be looking at realistic learning abilities. On a more positive note, simulations that incorporate costs can have the benefit of helping us understand how those costs interact throughout the lifetime of the organism. In this case, our models can generate hypotheses about life histories and learning that we might otherwise never imagine.

Costs not only shape existing learning abilities in any given species, but will also tend to minimize those abilities: in the words of Johnston, "...selection acts to favor those individuals that possess the least costly learning abilities permitting successful reproduction in the environment in which the population is evolving." This minimizing of learning abilities is another important reason to include costs in our simulation models. Here the costs help us to avoid evolving and misanalyzing the kind of surplus, environmentally irrelevant learning abilities that Johnston warns us against throughout his paper. Without this cost-induced minimizing effect, our evolved learning models could perform as oddly, and as unenlighteningly, as ducklings forced to imprint on moving paint cans in the laboratory. Johnston makes a final analogy that hits very close to home for researchers in this area: "Analysis of an animal's learning abilities may be likened to the process of analyzing the operation of a complex system such as a computer program"—that is, in both cases it is essential to know what the system was designed to do (the functional benefits), and what constraints operated to affect how those functions are achieved (the attendant costs). Both sides of the story, benefits and costs, must be taken into account to ensure that our "complex computer program" simulations actually model the kinds of learning evolved by real organisms adapting to the real world.

REFERENCES

Hinton, G. E., and S. J. Nowlan. "How Learning Can Guide Evolution." This volume.

Hurford, J. R. "The Evolution of the Critical Period for Language Acquisition." *Cognition* **40** (1991): 159–201.

Menczer, F., and R. K. Belew. "Latent Energy Environments." This volume.

Singer, W. "Ontogenetic Self-Organization and Learning." In *Brain Organization and Memory: Cells, Systems, and Circuits*, edited by J. L. McGaugh. New York: Oxford University Press, 1988.

Smith, P. K. "Does Play Matter? Functional and Evolutionary Aspects of Animal and Human Play." *Behav. & Brain Sci.* **5** (1982): 139–184.

Todd, P. M. "Sexual Selection and the Evolution of Learning." This volume.

Todd, P. M., and G. F. Miller. "Exploring Adaptive Agency II: Simulating the Evolution of Associative Learning." In *From Animals to Animats: Proceedings of the First International Conference on Simulation of Adaptive Behavior*, edited by J.-A. Meyer and S. W. Wilson, 306–315. Cambridge, MA: MIT Press/Bradford Books, 1991.

Todd, P. M., and S. W. Wilson. "Environment Structure and Adaptive Behavior From the Ground Up." In *From Animals to Animats 2: Proceedings of the Second International Conference on Simulation of Adaptive Behavior*, edited by J.-A. Meyer, H. L. Roitblat, and S. W. Wilson, 11–20. Cambridge, MA: MIT Press/Bradford Books, 1993.

Todd, P. M., S. W. Wilson, A. B. Somayaji, and H. A. Yanco. "The Blind Breeding the Blind: Adaptive Behavior Without Looking." In *From Animals to Animats 3: Proceedings of the Third International Conference on Simulation of Adaptive Behavior*, edited by D. Cliff, P. Husbands, J.-A. Meyer, and S. W. Wilson, 228–237. Cambridge, MA: MIT Press/Bradford Books, 1994.

Timothy D. Johnston

Chapter 20:
Selective Costs and Benefits in the Evolution of Learning

This paper originally appeared in *Advances in the Study of Behavior* **12** (1982): 65–106. Reprinted by permission. The table of contents from the original has been omitted.

I. INTRODUCTION

To a large extent, the study of animal learning owes its existence to the theory of evolution, for it was the concern of Darwin (1871, 1872), Romanes (1884), and C. L. Morgan (1896) with demonstrating intellectual and, hence, evolutionary continuity among animal species that led to the rapid rise of comparative psychology after 1900 (see Gottlieb, 1979). The comparative study of learning received from its founders an emphasis on uncovering similarities in learning ability among different species, and has retained that focus in large measure for almost 100 years. While that emphasis may be appropriate enough for an emerging comparative science, it should be noted that the theory of evolution, which provides the theoretical justification for comparative study, is equally receptive to an alternative emphasis; namely, that different animals possess adaptive specializations that equip them for survival in the particular environments in which they live. These two issues, continuity of descent and specialization of adaptation, are familiar sides of the evolutionary coin. The psychology of learning has, throughout its history, tended to emphasize the former to the virtual exclusion of the latter.

In this article I wish to consider one issue that arises when learning is considered from the viewpoint of evolutionary adaptation: the problem of the selective costs and benefits attendant on its evolution. Looked at from a teleological point of view, learning is manifestly a very useful ability. Almost any individual animal would appear to gain some advantage from the possession of more varied or more efficient learning abilities. However, from an evolutionary point of view, what is important is not "usefulness" in the colloquial sense, but "benefit" in the technical, reproductive sense (Williams, 1966; see Section IV). I shall argue that, although learning might well be considered quite generally useful, it does not confer a universal selective benefit. Indeed, it will be seen that learning incurs a number of important selective costs that tend to oppose its evolution. An appreciation of the relative costs and benefits of learning is essential to a proper perspective on the problem of its evolution.

II. LEARNING AND EVOLUTION—HISTORICAL BACKGROUND AND CURRENT CONCERNS

The comparative study of learning has drawn repeatedly on evolutionary thinking throughout history, albeit with different emphases and different degrees of success. At first, the primary concern was to establish the *fact* of mental evolution by demonstrating intellectual continuity among animal species, including humans. The largely anecdotal accounts compiled by Romanes (1884) were too uncritically accepted by him as evidence of feats of learning and insight rivaling those of humans, but they provided the basis for Lloyd Morgan's (1896) more critical assessment and paved the way for later experimental studies of learning. Thorndike (1911, p. 22) declared that the main purpose of the experimental study of animal learning was "to learn the development of mental life down through the phylum." By that he meant that its aim should be to establish continuities in learning mechanisms among species, showing how the capabilities of more advanced forms might have been derived in evolution from those of more primitive animals. Thorndike realized that controlled experimentation offered the only hope of any real insight into the mechanisms of learning, and his work established the field of animal learning as a major branch of experimental psychology.

The elucidation of phylogenetic relationships among different forms of learning has been a major concern for comparative psychology, as represented by modern workers such as Harlow (1958), Warren (1965), and Bitterman (1965, 1975). This brand of comparative study has been criticized for evolutionary naiveté (Hodos and

Campbell, 1969; Capitanio and Leger, 1979), and it is less prominent now than it once was. Although phylogenetic questions of continuity and discontinuity among learning mechanisms have certainly been preeminent in the comparative study of learning, in the early history of the field there was at least passing interest in the question of the adaptive significance of learning. That question was the primary motivation for the functionalist school of psychology. (James, 1890; Angell, 1907; see Boring, 1950; Cravens and Burnham, 1971), which, drawing on the pragmatic philosophy of Charles Peirce and William James, attempted to relate the operations of the mind (such as learning) to the everyday requirements of the environment. The major learning theorist most clearly influenced by functionalism was Clark Hull, whose early writings include analyses of the adaptive significance of conditioning mechanisms (Hull, 1929, 1937). As late as 1943, Hull wrote, "Since the publication by Charles Darwin of the *Origin of Species*, it has been necessary to think of organisms against a background of organic evolution and to consider both organismic structure and function in terms of *survival*" (Hull, 1943, p. 17). Despite his early evolutionary interests, the main body of Hull's theory of learning lacks any evolutionary content and the same is true of other learning theorists who have, on occasion, expressed their sympathy with evolutionary thinking (see, e.g., J. B. Watson, 1924; Skinner, 1974).

The emphasis on phylogenetic continuity among the learning abilities of different species was bolstered by the widespread acceptance of a clear-cut dichotomy between learning and instinct, particularly during the early development of psychological learning theory (see Marquis, 1930; McGraw, 1946; Anastasi, 1954). Instinct was viewed as an endogenous provision of "nature" (evolution) that permits an animal to deal with the particular set of problems posed by its environment. Learning, on the other hand, was a process whereby "nurture" (experience) could supplement the provisions of instinct, permitting an animal to deal with unusual or unexpected circumstances by acquiring the necessary behavioral skills.

Despite the cautions of some contemporary writers (e.g., Kuo, 1922; Carmichael, 1925; Gesell, 1933; McGraw, 1946), the tacit acceptance of a dichotomy between learning and instinct had profound effects, both methodological and conceptual, on the subsequent development of psychological learning theory. Methodologically, it sanctioned and even demanded the use of biologically arbitrary tasks for the study of learning, for only in that way could the psychologist study "the association process, free from the helping hand of instinct" (Thorndike, 1911, p. 30). The most cursory glance at the methodological approaches currently used for the study of animal learning will reveal the continuing effects of that bias. Conceptually, the learning/instinct dichotomy cleared the way for the acceptance of learning as a general process (Seligman, 1970), free from any species-specific "contaminations." Since learning, in this view, is a "supraspecific" characteristic, the adaptive demands placed on particular species by their environments cannot be relevant to its understanding.

Comparative study soon revealed that species differ in their performance on learning tasks. Such findings prompted a distinction between performance on a task and the learning processes that underlie performance. That distinction allowed the idea of a general process to be retained, by arguing that species differences reflect differences in performance variables such as perceptual and motor capabilities: Species-specific adaptations related to performance are thus overlaid on a general learning process, but are not the primary concern of learning theory (see LoLordo, 1979). Variations in the learning process itself have also been admitted, but were limited to quantitative differences in rates of conditioning, and to the presence or absence of a limited number of kinds of learning, such as habituation, operant conditioning, or discrimination learning (Tolman, 1949; Thorpe, 1963; Bitterman, 1965, 1975; Lorenz, 1969; Razran, 1971). In the latter case, each kind of learning has been seen as a general process in that its characteristics transcend species boundaries.

The psychology of learning thus stands in a curiously ambiguous relation to evolutionary theory. Although, as a comparative science, it draws its main justification from the theory of evolution, many important evolutionary implications, in particular those regarding adaptive specialization, have been overlooked. Ethologists have frequently taken psychology to task for this omission (e.g., N. Tinbergen, 1951; Lorenz, 1965, 1969), but, since ethology has not shown the same broad concern with problems of learning, it has not produced theories of learning that might challenge those proposed by psychologists. As a result, its criticisms have been somewhat blunted.

In recent years there has been a number of criticisms of the study of animal learning from within the field itself, criticisms that indicate a resurgence of functionalism in psychology (Shettleworth, in press; see also Petrinovich, 1979). Those criticisms originated in the findings of Garcia and Koelling (1966) that rats will readily learn to avoid a sweet-tasting solution if its ingestion is paired with toxicosis but not if ingestion is paired with foot-shock. Subsequent experiments demonstrated that such learning occurs even when the delay between ingestion and toxicosis is as long as 2 hr (Garcia et al., 1966). These findings appear to conflict with certain central assumptions of general process learning theory (Seligman, 1970) and the implications of this conflict have been explored in detail by a number of theorists (Rozin and Kalat, 1971; Revusky, 1971, 1977; Shettleworth, 1972, in press; Hinde, 1973; Logue, 1979; LoLordo, 1979; Johnston, 1981). The similar implications of data from other areas of animal learning have been discussed by Bolles (1970, 1971), Seligman and Hager (1972), and Hinde and Stevenson-Hinde (1973).

One implication that has been drawn from the research just cited is that there are "biological constraints" on an animal's learning abilities (cf. N. Tinbergen, 1951, p. 145), constraints that are drawn in large part by the evolutionary history of the species concerned. It has been proposed, notably by Rozin and Kalat (1971) and Hinde (1973), that a proper understanding of an animal's learning abilities requires that they be analyzed in terms of the animal's overall biological adaptation to its environment. There are two related senses in which we might seek to understand learning in relation to the concept of adaptation: first, as a *product* of evolutionary adaptation, that is, as the outcome of a history of selection pressures acting on the gene pool of the population to which the individual belongs; second, as a *process* of ontogenetic adaptation by which the individual adjusts to certain characteristics of the environment over the course of its own lifetime. It is in regard to learning as a product of evolution that I shall discuss the question of its selective costs and benefits. Elsewhere (Johnston and Turvey, 1980; Johnston, 1981, in press) a theoretical approach to the study of learning as a process of ontogenetic adaptation has been discussed in some detail.

III. AN ECOLOGICAL CONCEPTION OF LEARNING

If we are to be able to discuss the selective costs of benefits of learning, it is important to begin with an understanding of what phenomena are to count as instances of learning. Providing a detailed account of a conception of learning that is an adequate basis for evolutionary analysis would be a substantial theoretical endeavor in its own right and I do not propose to attempt it within the scope of this article (see Johnston and Turvey, 1980). It is clear, however, that the conception must be an ecological one; that is, it must be given in terms of the relationship between the animal and its environment, rather than in terms of the animal alone. The selective history of a population is a history of interactions between an evolving gene pool and a particular changing environment, and if learning is to be understood in the context of that history it must be understood in terms of both the animal and its environment.

The requirement for an ecological conception of learning means that traditional conceptions of learning as conditioning (e.g., Kimble, 1967), which have been articulated within the nonecological, general process tradition, are likely to be of very limited usefulness. The general process tradition (Seligman, 1970) is one in which demonstrating the relevance of an experimental learning task to the ecological demands that an animal normally faces in its environment has been of no concern whatsoever. It is entirely possible that the kinds of learning abilities revealed by those tasks (and embodied in the conceptions of learning to which they have given rise) are never employed by the animal under normal circumstances. If this is so, then selection has never acted on them and any inquiry into their selective history is largely meaningless.

Defining learning in terms of conditioning (Bolles, 1975; Rescorla and Holland, 1976) is therefore likely to be quite inappropriate and, in particular, too restrictive for the purposes of this article. Some forms of conditioning may well be involved in an animal's ontogenetic adaptation to its environment, but there are other relevant developmental phenomena that are excluded by definitions of conditioning (see Gottlieb, 1976a; Johnston and Turvey, 1980). For my present purposes, I shall use the term "learning" to refer to any process in which, during normal, species-typical ontogeny, the organization of an animal's behavior is in part determined by some specific prior experience. In the absence of the requisite experience, either some behavioral ability will be altogether lacking, or its organization will be different from that of similar individuals for whom the experience was available.

The requirement that the requisite experience be specific is designed to exclude such phenomena as the effects of inadequate nutrition on behavioral development (Leathwood, 1978). Such effects are highly nonspecific and I do not see that anything is gained by admitting them as instances of learning. Specificity is, of course, a matter of degree (Bateson, 1976), so that the preceding definition permits the identification of developmental phenomena as more or less typical instances of learning, rather than as either learning or not learning (see Johnston and Turvey, 1980, and Johnston, in press, for arguments in support of such a conception of learning).

IV. COST-BENEFIT ANALYSIS AND THE EVOLUTION OF ADAPTATIONS

The purpose of this section is to review, very briefly, some of the evolutionary theory that motivates the analysis of learning in terms of selective costs and benefits. Nothing in this account is either original or controversial, but it may help to avoid misinterpretation of subsequent arguments if their premises are stated explicitly here.

The aim of a cost-benefit analysis is to allow us to make statements about the conditions under which particular kinds of adaptations are likely to evolve and, hence, explain why these adaptations are observed in some populations and not in others. Such statements may be either qualitative or quantitative and they may account to a greater or lesser degree for the details of an adaptive trait. The present analysis is limited to qualitative considerations and will not attempt to account for any of the details of particular learning abilities. My aim will be to elucidate the selective costs and benefits associated with the evolution of *any* developmental system in which the organization of a behavioral characteristic is in part determined by prior experience.

The term "evolution" in the above formulation refers to a change in the population-typical phenotype over some period of time, and I shall not attempt to provide an account in genetic terms. The genetic basis of any learning ability is likely to be very complex, since learning is a function of entire epigenetic systems, not of single genes. Selection for the evolution of a learning ability cannot be between competing alleles within a population (the usual model of population genetics) but must be between competing genetic systems, either whole genotypes or parts of genotypes.[1] The study of such evolutionary problems is at present in a very early

[1]This is not to say that single allele substitutions do not affect learning; there is abundant evidence that they may do so (e.g., Aceves-Piña and Quinn, 1979; Dudai, 1979). However, models that assume selection between the hypothetical alleles *learning* and *no-learning* are unlikely to provide an adequate account of the evolution of learning.

stage (DeBenedictis, 1978; Lloyd, 1977) and the application of rigorous genetic analyses to the evolution of learning is likely to remain a very difficult task for some time to come (see Frazzetta, 1975, for further comments on the problems encountered in analyzing the evolution of developmental adaptations).

Phenotypic models of evolution assume that individuals in the population vary with regard to the phenotypic trait of interest, in this case in the manner in which development of a behavioral ability is influenced by prior experience, and that there is a correlation between phenotypic and genetic variation in this regard. If individuals with a high degree of developmental sensitivity to certain prior experiences have a greater reproductive success than those with a lower degree of sensitivity, then this phenotypic trait is said to confer a *net selective benefit* on its possessor. If this state of affairs persists, then possession of the trait will gradually come to be typical of individuals in the population. If the reverse is true, then the trait incurs a *net selective cost* and the possession of a low degree of developmental sensitivity will become, or remain, typical.

The possession of any phenotypic trait may be expected to have a number of consequences, each of which, taken separately, may be associated with a certain selective cost or benefit. Only when the total benefits outweigh the total costs is there a net benefit to the individual. By considering various costs and benefits of learning in isolation, a cost-benefit analysis can indicate the situations under which learning may carry a net selective benefit and so is expected to evolve. The discussion of individual selective costs and benefits in this article will assume that "other things are equal." Consider the proposition that a given selective cost is incurred by an individual who has a greater dependence on learning in the development of some behavioral skill than do other individuals in the population. That does not mean that the learning ability confers no selective advantage. Rather, it means that if all individuals develop equivalent behavioral skills, regardless of the extent to which those skills are learned, then the cost in question will oppose the evolution of that learning ability. Obviously, learning must carry a net selective benefit under some conditions, for many animals do in fact learn. However, in order to understand what those conditions are, it is necessary first to consider the various costs and benefits in isolation, "other things being equal." Some of the complex interactions among individual costs and benefits that are certain to occur as learning evolves are discussed later in the article (Section VII,A).

It is important to emphasize that the costs and benefits associated with learning are to be understood in purely reproductive terms. Although a particular learning ability may enable an animal to perform certain tasks more efficiently or skillfully, and to that extent may be seen as beneficial to the individual, a change in the population-typical phenotype can only occur if there is also a reproductive benefit (Darwin, 1859; Williams, 1966; Ghiselin, 1974; Lewontin, 1974, 1978).

There are a number of problems of interpretation involved in the analysis of the selective costs and benefits of learning. Some of these are common to all attempts to infer the selective history of an adaptive trait, since direct evidence is usually unavailable and we must rely on indirect evidence from present-day populations.

We therefore search among extant species for correlations between the existence of learning abilities, for example, and the existence of putative selective costs and benefits. Assuming that the same principles of natural selection hold today as in the past, we may then infer that these costs and benefits for which the expected correlations hold were involved in the evolution of learning. This approach has some well-known drawbacks and limitations (see J. L. Brown, 1975; Hinde, 1975), but it is usually the only course open to us. In some cases it may be possible to make direct tests of hypotheses concerning the cost or benefit of a trait, by making experimental alterations and observing their reproductive effects (e.g., N. Tinbergen et al., 1963). Unfortunately, where learning is the trait in question, we cannot make the necessary modifications to the phenotype. However, the method of adaptive correlation has proved very useful in this study of behavioral evolution (see Brown, 1975; Alcock, 1979), not least as a means of stimulating the investigation of questions whose importance was not previously recognized, and I make no apology for employing it here.

There is a second problem, peculiar to the evolutionary study of learning; namely, that only a few studies of learning have investigated the development of ecologically relevant behavioral skills. Thus, the data we possess on the learning abilities of different species frequently do not allow us to make any statements about the role that learning may play in the species-typical development of behavior under normal circumstances. Such data are of very limited relevance to any evolutionary inquiry: A learning ability that is never manifest under natural conditions cannot reveal anything about the selection pressures that may be involved in the evolution of learning, since it cannot affect the reproductive success of individuals that possess it. While it is possible that such "laboratory" learning abilities may once have been of adaptive significance to their possessors, the complete lack of information about the nature of past environments in which this may have been the case prevents us from using such abilities as the basis for evolutionary analysis. The question of why those abilities should have evolved at all will be considered later in this article (Section VII,B).

The evidence that we do possess about ecologically relevant learning abilities is therefore most often indirect. Where it is more direct, complementary data on possible selective pressures in extant populations are usually lacking. I hope that by drawing attention to these lacunae in our knowledge, and by pointing out their importance to the evolutionary study of learning, this article can help to stimulate further study of ecologically relevant learning abilities and, thus, permit the questions that are raised here to be posed in a more rigorous manner, and eventually answered.

V. THE SELECTIVE BENEFITS OF LEARNING

It might seem that the selective benefits of learning are so obvious as hardly to merit extended discussion. Surely an organism that can learn can obtain far more information about its environment, and, hence, can adapt more successfully to it, than one that is unable to learn. On this view, selection ought always to reinforce and extend current learning abilities and to favor the evolution of new ones. The idea that evolutionary insights can be obtained by ordering organisms on a scale of learning ability (e.g., Bitterman, 1965, 1975; Yarezower and Hazlett, 1977; see Hodos and Campbell, 1969) appears to reflect allegiance to some such view. In a recent paper, Mayr provides a clear statement of this point of view:

> The great selective advantage of a capacity for learning is, of course, that it permits storing far more experiences, far more detailed information about the environment, than can be transmitted in the DNA of the fertilized zygote. Considering this great advantage of learning, it is rather curious in how relatively few phyletic lines genetically fixed behavior patterns have been replaced by the capacity for the storage of individual acquired information (Mayr, 1974, p. 652).

This is a curious position for an evolutionary theorist to adopt, because it draws no distinction between the colloquial and technical senses of the "advantage" of a characteristic to its possessor. The potential *usefulness* of a learning ability to some organism is not the same thing as its *selective benefit* and it has nothing to do with the likelihood that the ability will evolve (Williams, 1966). In addition, Mayr's formulation implies that the selective advantage of a learning ability is independent of the environment in which evolution is occurring. As many authors have pointed out, adaptation specifies a relation that may or may not exist between an organism and some environment (Sommerhoff, 1950; Williams, 1966; Ghiselin, 1966, 1974; Slobodkin and Rapoport, 1974). Adaptation is not an environment-neutral property of either an organism or a particular characteristic of an organism, such as its learning abilities. The adaptive significance and selective benefits of learning can only be assessed with respect to particular sets of environmental factors, not in an ecological vacuum.

A. ADAPTATION TO ENVIRONMENTAL VARIABILITY

1. NATURE OF THE BENEFIT. The environment of an organism may be characterized by a list of ecological factors (including abiotic, biotic, and social ones) that affect its well-being and survival (Mason and Langenheim, 1957). Insofar as relatively invariant (i.e., constant or only slowly changing) ecological factors are important to the relationship between an organism and its environment, adaptation to them may be effected by natural selection, regardless of the existence of developmental sensitivity on the part of individuals within the population. Such factors produce sustained selection pressures that constrain the range of genetic variation to just those genotypes that give rise to well-adapted phenotypes (Lewontin, 1974). Other factors that are more variable may still permit natural selection to effect the necessary adaptation if some constant pattern of variation is manifest within individual life spans. Light intensity, for example, varies on both a diurnal and an annual cycle, but since the patterns of variation remain invariant over long periods of time, natural selection can effect adaptation to those patterns in animals with life spans of more than 1 year.

Many organisms interact with their environments in ways that require them to adapt to ecological factors that do not exhibit such invariant properties. The nature and distribution of food sources, location of a burrow or nest, routes of travel, individual identities of parents or mate, and social relationships with conspecific neighbors are all ecological factors that may vary importantly between the life spans of successive generations and/or within that of a single generation. Clearly, natural selection cannot effect adaptation to such factors in the absence of some degree of developmental sensitivity to the environment, since the direction of selection varies stochastically from one generation to the next and there can be no net adaptive change in the gene pool of the population.

Under such circumstances, a selective benefit will accrue to those individuals that possess the requisite developmental sensitivity. The ability to learn, as one form of developmental sensitivity to the environment, has as its primary selective benefit that it permits adaptation to ecological factors that vary over periods of time that are short in comparison with the lifetime of an individual (G. Bateson, 1963; Williams, 1966; Slobodkin, 1968; Slobodkin and Rapoport, 1974; Plotkin and Odling-Smee, 1979; Johnston and Turvey, 1980).

2. AVAILABLE EVIDENCE. Because of the paucity of ecologically motivated studies of learning, and the extreme difficulty of testing adaptive hypotheses about learning skills, evidence that adaptation to environmental variability confers a selective advantage on individuals able to learn is hard to come by. Generally speaking, the most compelling evidence would be to find closely related species, or populations of the same species, that exhibit differences in learning ability clearly related to differences in environmental variability. For example, Sasvári (1979) demonstrated that great tits (*Parus major*) are more adept at learning feeding techniques by observation than either blue tits (*P. caeruleus*) or marsh tits (*P. palustris*). The great

tit occupies a wider range of habitats than either of the other two species (Perrins, 1979), and so individuals may be expected to encounter different kinds of prey, depending on the location in which they are reared. If different feeding techniques are required for efficient exploitation of different prey types, as suggested by the work of Kear (1962) and Partridge (1976), then observational learning may permit great tits to exploit the wider range of prey that they typically encounter.

Gray (1979, 1981; Gray and Tardif, 1979) has studied the feeding diversity of several species of deermice (*Peromyscus* spp.). The results from two species, *P. maniculatus sonoriensis* and *P. leucopus*, are of special interest. *P. m. sonoriensis* inhabits southwestern desert environments and *P. leucopus* inhabits more northerly woodlands. Gray and Tardif (1979, Experiment 2) found that laboratory-reared *P. leucopus* had lower adult feeding diversity than wild-caught adults, whereas laboratory rearing had no effect on adult feeding diversity in *P. m. sonoriensis*. These results imply that experiences may affect the development of feeding diversity to a greater extent in *P. leucopus* than in *P. m. sonoriensis*, an implication that was borne out by further experiments (Gray and Tardif, 1979, Experiment 3). Gray (1981) suggests that the different susceptibilities of the two species to the effects of laboratory rearing is related to the fact that *P. leucopus* inhabits an environment that may be more variable than that inhabited by *P. m. sonoriensis*, so that selection has favored the evolution of developmental sensitivity in the former species but not in the latter. More detailed study of the two species' environments is needed, however, before this hypothesis can be properly evaluated.

Emlen (1969, 1970) has shown that early experience is important for the development of stellar navigation skills in the indigo bunting (*Passerina cyanes*), a nocturnal migrant. Birds that are exposed to the night sky prior to their first migration season subsequently select an appropriate migratory orientation by reference to star patterns, whereas birds lacking this experience do not. The precision of the earth's axis changes the relation between stellar and geographical directions at the rate of about 3° every 1000 years. Emlen (1975) suggests that this rate of change may be too rapid to effect adaptive changes in the gene pool of the population, and that individual learning has been selected to permit adaptation to this aspect of the environment.

3. EVALUATION. Only as more investigators turn to the study of learning from an ecological perspective will it become possible to accumulate data to address the hypothesis that learning confers the selective benefit of adaptation to environmental variation. Although it is undeniably true that learning can in principle confer such a benefit, further data are needed before it can be concluded that it *does* do so in evolving populations. Studies such as Gray's, which relate differences in learning to differences in environmental variation among related species, are likely to offer the most valuable insights in this regard.

One way in which the search for relevant data might be stimulated is by the development of more refined theoretical models that predict the evolution of particular kinds of learning in particular environments. Treatment of this issue lies outside the scope of the present article, but the reader is referred to the work of Estabrook and Jesperson (1974), Bobisud and Potratz (1976), and Arnold (1978) for examples of such models.

The ability to learn is likely to be of greatest selective advantage in resource-limited populations in which individuals must compete for food, nest sites, and other resources. In such relatively *K*-selected species (MacArthur and Wilson, 1967), the most successful individuals tend to be those that utilize the available resources most efficiently. Learning is likely to contribute substantially to such efficiency when the nature and distribution of limited resources is variable. Under those conditions, an important selective benefit will be realized by individuals that possess the requisite learning skills. In more *r*-selected species (MacArthur and Wilson, 1967), by contrast, reproductive output rather than efficiency of resource exploitation is the primary determinant of reproductive success, and learning is unlikely to be of great selective advantage in such species (see also Section VI, C and D).

B. SEXUAL SELECTION

1. NATURE OF THE BENEFIT. Sexual selection has been proposed by Nottebohm (1972) as the agent responsible for the evolution of vocal learning in many species of songbirds. According to this argument, females select males with the most elaborate vocal repertoires; therefore, developmental mechanisms that increase the number of song types in the repertoire will tend to evolve. Proponents of the argument claim that the genome is limited in the amount of information it can encode for elaborate species-typical song, and that sexual selection favors the evolution of learning as a mechanism for increasing the vocal repertoire.

2. AVAILABLE EVIDENCE. Nottebohm presented no evidence in support of his hypothesis and, so far as I know, none has yet been forthcoming. J. L. Brown (1975) criticized Nottebohn's suggestion on the grounds that large repertoires occur primarily in monogamous species, rather than in polygamous ones in which sexual selection would presumably be stronger. He suggests that a more important function of vocal learning may be to maintain pair bonds. There is some evidence that selection may favor large repertoires, but not on the basis of female choice. Kroodsma (1976) found that female canaries (*Serinus canarius*) exposed to recordings of large repertoires (35 syllable types) built nests faster and laid more eggs than did females exposed to small repertoires (5 syllable types). Whether or not this is an example of sexual selection may be open to question.

3. EVALUATION. Even if repertoire size does influence female choice, we would only expect selection for large repertoires to result in the evolution of learning if there are limits on the capacity of the genome to encode the relevant information. The finding that male song sparrows (*Melospiza melodia*) develop larger repertoires when given the opportunity to learn from adult song models than when reared in isolation (Kroodsma, 1977) lends some indirect support to that suggestion, but arguments based on the information encoding capability of the genome tread on very insecure ground. Our present state of knowledge about the way in which genetic information is translated into the structure of behavior in the course of development is almost nonexistent, and in Section VI,E, I shall present some arguments that learning may actually involve more, rather than less, genetic information. At present, the issue probably cannot be decided. There are in any case only rather few kinds of learning skills that might evolve by sexual selection, namely, those that involve the development of courtship behavior.

C. LACK OF VARIATION FOR OTHER ADAPTIVE SOLUTIONS

1. NATURE OF THE BENEFIT. Natural selection is an opportunistic process—it must work with whatever variation is present in the population and can only effect adaptation to the extent that appropriate variation is available (see further Section VIII,B). Even where a style of development that is wholly or largely independent of experience could well effect adaptation to some aspect of the environment, the ability to learn the requisite skills may still evolve, if variation in regard to learning is the only appropriate kind of variability available in the population. In such a situation, the learning ability confers a selective benefit on its possessor, even though learning is not more adaptive than an alternative style of development (Lewontin, 1978, 1979; Gould and Lewontin, 1979). Styles of development that are either more or less dependent on experience might occupy equivalent adaptive peaks (Wright, 1932) and the style that evolves would then depend upon the available variation in the population.

2. AVAILABLE EVIDENCE. Direct evidence in favor of this selective benefit is unattainable, since it depends upon knowing the nature of phenotypic and genetic variation in extinct populations and such information is forever lost to us. Indirect evidence may be obtained by comparing species in which similar kinds of behavioral skills develop with greatly differing degrees of dependence on the nature of experience. If the skill in each species effect adaptation to equally invariant aspects of the environment, then we may argue that the different styles of development represent equivalent adaptive peaks.

Consider the contrasts that are found among a number of species of songbirds, in which normal song development shows quite different degrees of dependence on auditory experience (reviews by Marler and Mundinger, 1971; Nottebohm, 1970; Marler, 1975). For example, song sparrows (*Melospiza melodia*) reared without exposure to conspecific adult song models develop song that is indistinguishable from, although less variable than, that of normally raised birds (Mulligan, 1966; Kroodsma, 1977). In another species, the white-crowned sparrow (*Zonotrichia leucophrys*; Marler, 1970), exposure to adult song is an absolute requirement of normal song development. Both species develop elaborate song repertoires under normal circumstances and it is not clear what, if any, adaptive significance is to be assigned to the developmental differences between them (Marler, 1967; Nottebohm, 1972).

A second line of indirect evidence comes from studies of the developmental effects of self-stimulation. To give but one example, Gottlieb (1978) has shown that surgically devocalizing Peking ducklings (*Anas platyrhynchos*) just prior to hatching prevents the development of a highly specific approach response to the maternal assembly call, which is shown by normal (embryonically vocal) ducklings after hatching. Other aspects of the prenatal auditory environment, such as exposure to maternal calls, are not required for normal development of this behavior (Gottlieb, 1971). The self-produced auditory environment of the embryo is quite invariant between generations and, indeed, artificially varying that environment, by playing altered recordings of the embryonic call to devocalized embryos, does not produce a corresponding change in the postnatal approach preference (Gottlieb, 1980). It is difficult to see what the adaptive significance of an experientially dependent style of development might be in this situation, since complete independence from the nature of individual experience would seem able, in principle, to produce the same behavioral end-point.

3. EVALUATION. Since the preceding lines of evidence rely on the lack of any apparent adaptive interpretation for experience-dependent development in the examples discussed, they are obviously somewhat unsatisfactory. The lack of adaptive interpretations may simply reflect our incomplete understanding of the ecological context surrounding these instances of development and such explanations may be forthcoming in the future. That learning may confer a selective advantage when other adaptive solutions are unavailable cannot be seriously doubted in principle, but demonstrating the involvement of this benefit in particular instances of the evolution of learning is likely to remain extremely problematic. One value of pointing out the probable existence of this selective benefit is to stress that adaptive explanations for the possession of learning abilities may not always be available (Lewontin, 1979; Gould and Lewontin, 1979; see Section VIII,B).

VI. THE SELECTIVE COSTS OF LEARNING

The selective benefits of learning have been discussed before, by a number of authors, but the selective costs involved in the evolution of learning have received much less attention. No complete understanding of learning as a product of evolutionary adaptation can be achieved until we gain an appreciation of those costs and the extent to which they are incurred in evolving populations. In this section I shall identify six potential selective costs of learning and review evidence that these costs are in fact incurred in the course of evolution.

A. DELAYED REPRODUCTIVE EFFORT AND/OR SUCCESS

1. NATURE OF THE COST. In general, the earlier an organism can begin reproduction, the more offspring it can produce at one time, and the more frequently it can reproduce, the greater will be its selective advantage. However, any reproductive effort is a heavy drain on an organism's resources of time and energy (Ricklefs, 1974) and so there will tend to be selection against investment in reproduction at those times during an individual's life when such investment would unduly compromise the survival of either the parent(s) or offspring. As the development of those behavioral skills that are involved in individual survival and parental care becomes more dependent on individual experience, there is a period in early life during which the developing organism's behavioral competence is relatively limited in certain respects,[2] namely, in those that involve the skills being learned. During this time, diversion of resources into reproductive effort is likely to be a poor investment, since the individual is in a particularly precarious adaptive position. Its lack of competence not only compromises its own survival as a result of reproductive effort but also makes it less likely that adequate preparation can be made for the birth and early development of its offspring.

There are two selective consequences of this situation. If an organism's dependence on learning is high, the number of offspring that it can expect to rear during this period of relative behavioral incompetence will be relatively low. In comparison with other individuals in the population whose behavioral skills develop in a manner that, while equally adaptive, is less dependent on learning, it is therefore at a selective disadvantage. In addition, there may be selection in favor of delaying the age of first reproduction until full behavioral competence is reached. This is likely to occur if early reproductive effort places such a strain on the individual's resources that either its own survival is threatened or the success of subsequent reproduction is compromised.

---

[2]I do not wish to imply that the young animal is merely an inadequate adult; there are certainly many characteristics of juvenile behavior and morphology that are adaptations to the specific ecological niche occupied by the young animal (see Oppenheim, 1980, 1981; Galef, 1981). My attention in this discussion is limited to those skills involved in reproductive activities, rather than to the entire behavioral repertoire.

2. AVAILABLE EVIDENCE. Evidence that the selective cost of delayed reproduction is incurred by the evolution of learning requires two types of studies: demonstrations of reduced breeding success in young animals in comparison with conspecific adults; and demonstrations of a role for experience in the development of certain crucial behavioral skills.

a. JUVENILE BREEDING SUCCESS. Lack (1954) proposed that the delayed repro-duction found in many animals, especially birds and mammals, might be an adap-tation to the difficulties to be expected by an unskilled juvenile in supporting both itself and its offspring. Later (Lack, 1966) he supported this argument with data from a number of studies that demonstrate lower breeding success in young birds of several species. Particularly complete data concerning the relationship between age and breeding success are available for the great tit (*Parus major*; Perrins, 1965), the kittiwake (*Rissa tridactyla*; Coulson and White, 1961; Wooler and Coulson, 1977), the yellow-eyed penguin (*Megadytes antipodes*; Richdale, 1957), and the European blackbird (*Turdus merula*; Snow, 1958). In these species, clutch size, hatching rate, and fledging rate are all lower for birds breeding in their first year of life than for those breeding in their second or subsequent years. In the penguin and kittiwake this juvenile disadvantage extends into the second and perhaps third years.

b. DEVELOPMENT OF ADULT BEHAVIORAL COMPETENCE. Among the behavioral skills that would be expected to contribute most directly to an individual's ability to raise offspring are foraging and nest-site selection. A number of studies have shown that the foraging efficiency of the young birds is lower than that of adults, but, unfortunately, relevant data are not available for the species mentioned above, whose breeding biology has been studied particularly thoroughly. However, it has been noted, e.g., by Ashmole (1963), that breeding tends to be especially long de-layed (up to 5 or 6 years in some cases) in species of predatory birds that appear to require a considerable degree of skill in catching their prey. Marked dispari-ties between juvenile and adult foraging efficiencies have been reported in several such species, including the little blue heron (*Florida caerula*; Recher and Recher, 1969), the royal tern (*Thalasseus maximus*; Buckley and Buckley, 1974), the sharp-skinned hawk (*Accipiter striatus*; Mueller and Berger, 1970), the herring gull (*Larus argentatus*; Verbeek, 1977), the sandwich tern (*Sterna sandivicensis*; Dunn, 1972), and the brown pelican (*Pelecanus occidentalis*; Orians, 1969). Bené (1945) observed that the young black-chinned hummingbirds (*Archilochus alexandri*) often attempt to feed from leaves and twigs as well as flowers but later concentrate their feeding attempts more appropriately. Although he made no direct tests of the involvement of learning in this increase in foraging efficiency, he did find that adults could learn to adjust their foraging behavior in response to changes in the distribution and concentration of nectar sources, suggesting that learning may also be involved in the earlier development of efficient foraging.

    In the case of the brown pelican, Blus and Keahey (1978) have found that immature birds (1–3 years old) lay smaller clutches, hatch fewer eggs per clutch, and fledge fewer young than do adult birds (>3 years old). This study also found that immature birds construct more nests in low-lying areas and, as a result, lose many more clutches to flooding than do adult birds. Nest placement does not seem to be a result of competition for nest sites, since there is frequently unoccupied high ground near a flooded nest. Blus and Keahey suggest that experience may play a role in nest-site selection in this species. No data are available on the breeding biology of the other species whose age-related foraging efficiency has been studied, but in the arctic tern (*S. paradisaea*), a close relative to the sandwich tern, Coulson and Horobin (1976) found that hatching and fledging success increase between 3 and 8 years of age and that most birds breed for the first time at 3 or 4 years.

3. EVALUATION. These data on foraging efficiency and breeding success suggest that, especially where foraging involves the capture of mobile, elusive prey, young birds may be unable to catch enough food to breed as successfully as older birds. The question arises whether their lower foraging efficiency reflects the learning of this skill or is due to some other factor. So far as I know, in none of the species mentioned has the development of foraging been studied to elucidate the role of experience. It might be, as Groves (1978) has suggested, that the difference between adults and juveniles simply reflects the death of the least successful juveniles. This is an unconvincing explanation, because juvenile mortality is certainly high enough in almost all species of birds (see Lack, 1966) to provide strong selection pressure for high foraging efficiency early in life. That such efficiency has not evolved in many species suggests quite strongly that selection among alternative behavioral phenotypes is not sufficient to ensure high efficiency and that individual experience is required for its development. An alternative explanation for these data is that adult efficiency must await the maturation of some physiological system independently of experience, such as the growth of a sufficiently powerful musculature to permit certain necessary manipulations of the prey. Deciding between these alternatives in particular cases must await further experimental analysis of the factors involved in the development of these skills.

It is of interest that in Groves' (1978) study of foraging in the ruddy turnstone (*Arenaria interpes*), the difference between adult and juvenile success lay mainly in prey handling and search times. One of the few studies in which the role of experience in the development of species-typical foraging skills has been studied directly is that of Davies and Green (1976) on the reed warbler (*Acrocephalus scirpaceus*). They found that whereas success in catching flies (i.e., captures per attempt) was not affected by prior experience with flies, handling time per fly caught was significantly lower in experienced than in nonexperienced birds. Thus, in this one instance, experience does appear to affect an aspect of foraging behavior that has been shown to improve with age. Further research along the lines of Davies and Green's study is needed to establish the generality of such experimental effects in the development of foraging behavior.

In general, there is some circumstantial evidence in favor of the proposition that learning some behavioral skills involves a reduction in early reproductive success and/or a postponement of the onset of breeding. This selective cost will not be incurred to an equal extent by the evolution of all learning abilities. Evolving the ability to learn those skills, such as foraging and nest-site selection, that contribute most directly to breeding success will incur this selective cost most heavily. Skills that can be learned completely before the individual is physiologically or anatomically capable of breeding, or before environmental conditions are suitable for breeding, will in general be more or less exempt from the selective cost of delayed reproduction.

The crucial data that are lacking concern the role of experience in the development of skills such as foraging and nest-site selection. There is considerable scope for research on these problems, which would make a substantial contribution to our understanding of the factors involved in the evolution of learning. The work of behavioral ecologists has provided sophisticated descriptive accounts of the strategies employed by adult animals in behavior such as foraging (e.g., Pyke et al., 1977; Krebs, 1978; Pyke, 1978), and some students of learning have begun to turn their attention to the development of such behavior (e.g., Staddon, 1980). If it can be shown that the role of learning is greatest in those species in which breeding is longest delayed, or that have the most reduced breeding success as juveniles, then the argument for a relationship between learning and this selective cost would be further strengthened.

B. INCREASED JUVENILE VULNERABILITY

1. NATURE OF THE COST. The more an individual must rely on learning for the development of adaptive behavioral skills, the more vulnerable it is during the period before the necessary experience has been obtained, or while it is being obtained. If one individual's recognition of its food supply, for example, depends more on learning than does that of another individual, then the former is more likely to select an occasional nonnutritious or even poisonous item. An individual whose social behavior is less genetically constrained than another's runs the risk of responding inappropriately as a juvenile in situations where such a response may be dangerous. Other things being equal then, individuals whose development is more dependent on learning are at higher risk as juveniles than those whose development is less so and, therefore, they will tend to be selected against.

Although a good prima facie case can thus be made for the existence of juvenile vulnerability as a selective cost of learning, evidence in its favor is hard to come by. In fact, at first sight it would seem that the evidence tends to support the contrary hypothesis, since juvenile mortality is typically very much higher in animals such as fish, in which behavior appears to owe rather little to individual experience, than in mammals, in which learning is an important component of behavioral development.

The reason for this may be that prenatal care is much more prevalent in the latter (see Section VI,C) and this tends to offset juvenile mortality.

2. AVAILABLE EVIDENCE. The studies of foraging efficiency discussed previously suggest that in species with complex foraging skills an inexperienced juvenile animal expends more time and energy to obtain each gram of food than does a more experienced adult. Juveniles will thus be more liable to nutritional deficiencies, especially at times when food is scarce, which increases the risk of disease, predation, and accidental death. Bolles (1970) has suggested that the costs in terms of juvenile and even adult mortality incurred by learning to avoid predators are so high that such learning is highly unlikely to evolve in any population. He argues that instead animals have evolved species-specific defense reactions (SSDRs) commensurate with their ability to hide, escape, or fight and that these are elicited by any dangerous situation, independently of any specific learning.

One way to demonstrate the cost of juvenile vulnerability is to compare juvenile mortality in species having different degrees of dependence on learning a particular skill, when parental care is removed. For example, if two related species differ in the extent to which foraging skills are learned, then that species with the greater dependence on learning should have higher mortality from nutritional deficiencies (and related causes) when parental care is removed. Although I know of no such comparative studies, Stirling and Latour (1978) found that polar bear (*Thalarctos maritimus*) cubs less than 2.5 years old were less successful hunters than adults and concluded, on the basis of an admittedly small number of tagging returns, that their mortality is very high if they are deprived of parental care. They suggest that hunting skill in this species depends on experience but concede that there is no direct evidence that this is the case.

3. EVALUATION. Direct evidence that dependence on learning incurs the cost of increased juvenile mortality is virtually nonexistent, partly because it is hard to design the appropriate experiments. In principle, examining the increase in juvenile mortality following removal of parental care in two species that differ in their dependence on learning would provide the relevant data, but some important difficulties of interpretation with such study should be noted. If, as will be argued in Section VI,C, parental care evolves to offset juvenile mortality due to learning, it may also offset mortality from other causes that have nothing to do with learning, such as inadequate thermoregulation or disease transmission by ectoparasites that are removed by grooming. It is important, therefore, that the causes of mortality be determined in any such study and this may be very difficult, especially under field conditions.

C. INCREASED PARENTAL INVESTMENT IN EACH OFFSPRING

1. NATURE OF THE COST. In all natural populations, some mortality between birth and maturity is to be expected regardless of the presence of any learning ability. Two parental strategies are available, in principle, to counter the effects of this selection: An individual may produce large numbers of offspring of which at least a few will survive, and invest very little in each one, or it may produce small numbers of offspring and invest heavily in each one. In general, organisms are expected to adopt some compromise between these two alternatives, and the question is, in which direction from the median will the strategy lie? Each end of the continuum has its own costs associated with it. As I have argued, organisms whose behavioral development is heavily dependent on learning may tend to be at higher risk as juveniles than organisms whose development is less dependent on learning and so the requirements to offset expected mortality tend to be higher. In general, increasing the dependence of behavioral development on learning also requires greater structural complexity (see Section VI,D) and so the energetic investment in the production of each offspring tends to be relatively high. A parental strategy in the direction of overproduction of offspring is thus unlikely to be selected for; instead, it is to be expected that an increase in learning capacity will carry with it the requirement for an increased parental investment in each offspring. This strategy reduced the number of offspring that any individual can rear at one time and, if the period of dependency extends beyond the next breeding season, may reduce the number of times an individual can breed during its lifetime, and, hence, reduce its reproductive success in comparison with other members of the population.

2. AVAILABLE EVIDENCE. The low reproductive rates of many seabirds and large predatory birds such as condors or eagles have been attributed to the necessity for a long period of postfledging care (Amadon, 1964; Ashmole, 1963). These birds must forage over wide areas to obtain food, expending a large amount of energy to capture each food item, which places a premium on efficient flight and optimum techniques of prey capture. The study of Mueller and Berger (1970) already referred to showed that juvenile sharp-shinned hawks tended to select inappropriate prey more often than did adults. Many seabirds that range over wide ocean areas in search of prey have very long periods of parental care that extend well beyond the time when the young are fully capable of sustained flight (Ashmole and Tovar, 1968). W. Y. Brown (1976) found that parental care in the brown noddy (*Anous stolidus*) extends for over 100 days postfledging and Simmons (1970) noted a similar period of dependency in the brown booby (*Sula leucogaster*). Both authors comment on the skilled fishing techniques of these species and suggest that long experience is required before an individual becomes sufficiently skilled to feed itself.

Houston (1976) made determinations of the energy budget of Rüppell's griffon vulture (*Gyps rueppellii*) in East Africa and concluded that during at least part of the rearing period the adults are unable to obtain enough food to meet both their own and their young's energy requirements. A similar conclusion was reached by Hainsworth (1977) in his analysis of the energy budget of breeding sparkling violetears (*Colibri coruscans*, a species of hummingbird). Since both species breed when food is maximally abundant, these findings suggest that foraging efficiency may be an important limiting factor in their respective energy budgets and that experience may well be necessary for adequate foraging efficiency to develop. Houston (1976) notes that young vultures spend a considerable amount of time after fledging practicing gliding. These birds must cover large areas by gliding flight in search of food and Houston's observations suggest the involvement of learning in at least this component of foraging behavior.

3. EVALUATION. The finding that seabirds with complex foraging skills show very long periods of parental care suggests that parental care may be a selective cost of learning but, once again, the crucial data that are lacking concern the role of experience in the development of these skills. It seems hard to question the view that increased dependence on learning requires increased parental care but evidence in its support remains somewhat scanty. A corollary of this selective cost is that where mortality factors that cannot be offset by any conceivable behavioral skill (such as disease or extreme environmental fluctuations) exert strong selection on the population, a high degree of parental investment in each offspring is unlikely to evolve. In such populations, characteristic of r-selected species (MacArthur and Wilson, 1967), the best reproductive strategy is to produce large numbers of offspring, since even an individual that can attain a high level of behavioral competence as a result of learning has no greater chance of survival, and demands a greater investment in its production, than an individual that cannot learn as much.

D. GREATER COMPLEXITY OF THE CENTRAL NERVOUS SYSTEM

1. NATURE OF THE COST. As the development of behavior becomes more dependent on the nature of experience, the individual must process increasing amounts of information during its development. The nature of this processing depends on the particular conception of learning that is being entertained. It may involve the detection of complex relationships among objects and events in the environment (Humphrey, 1933; Johnston, 1978), the formation of associative connections between st r li and responses (Rescorla and Holland, 1976), or the storage and retrieval of information in memory (Estes, 1973). In any event, as the information detection or processing requirements associated with learning increase, the complexity of the physiological basis for learning in the central nervous system (CNS) also increases. Since the capacity of a single neuron to encode information is largely fixed, increasing the capacity of the CNS as a whole requires more nerve cells and a greater degree of connectivity among them.

There are two components to the selective cost of an increase in the size and/or complexity of the nervous system: the high energetic cost of maintaining nerve tissue, and the requirement of shielding the functioning brain from even minor physiological fluctuations. In adult humans, the brain, comprising about 2% of the total body weight, accounts for 20% of the total basal oxygen metabolism (Sokoloff, 1960). In the first decade of life, the consumption of the brain may exceed 50% of the total basal metabolism (Kennedy and Sokoloff, 1957). This high respiratory rate (3.5 mg $O_2/100$ g/min in adults) is maintained even during severe metabolic disorders such as hypoglycemia and diabetic acidosis, indicating that during metabolic stress, the cost of maintaining the brain may be proportionately much higher. The brain has an obligatory dependence on glucose metabolism and, since glucose cannot be stored, the brain requires a continuous supply, either by food intake or from other metabolic processes.

Recent work (e.g., Oleson, 1971; Raichle, 1975; Lassen et al., 1978) has demonstrated that the local metabolism of individual brain structures is dependent upon their functional activity. A large brain not only incurs a high biological maintenance cost, but also a high operating cost. Such costs are incurred by the diversion of available food resources away from direct investment in reproduction, putting their bearer at a selective disadvantage with respect to other, smaller-brained individuals in the population.[3]

In addition to this direct energetic cost, there are indirect costs to be borne by the possessor of a large, complex nervous system. Nerve impulse transmission, the significant event of CNS functioning, involves small changes in ion balance, weak bioelectric phenomena, and the diffusion of small amounts of transmitter substances across synaptic gaps. All of these events are sensitive to even minor changes in the chemical milieu that invests the CNS (Tschirgi, 1960). Biochemical noise will be introduced into CNS functioning by changes in the organism's physiological state unless it is shielded from such fluctuations by mechanisms capable of precise homeostatic control. Such mechanisms, of which the blood-brain barrier is the best known, have indeed evolved and their development and maintenance is a further cost incurred by the large brain that an increased learning capacity requires. Several experiments have shown that CNS functioning is readily disturbed by minor fluctuations in body temperature (Roots and Prosser, 1962; Tebēcis and Philips, 1968; Peterson and Prosser, 1972) and diversion of resources, whether of time (Heath, 1965) or materials (Wheeler, 1978), into thermoregulatory strategies to control brain temperature represent another indirect cost to the possessor of a large, complex nervous system.

Finally, it may be pointed out that the very complexity of the nervous system makes it increasingly vulnerable to accidental mechanical and biochemical disturbances that may render its possessor incapable of successful or prolific reproduction.

---

[3] It may be worth repeating here that this selective cost is incurred presuming "other things equal." That is, a large-brained individual will be at a disadvantage with respect to a small-brained one if they are both equally competent in those abilities that affect reproductive success, whether directly or indirectly. Of course there will be situations in which larger-brained individuals have a reproductive advantage (*vide* the brain size of *Homo sapiens*), but only when the selective costs discussed here are outweighed by the benefits of possessing such a larger brain.

2. AVAILABLE EVIDENCE. Granted that central nervous tissue is a "biologically expensive" commodity, this expense will only be incurred as learning evolves if an increase in learning ability entails a more elaborate CNS. Evidence concerning the relationship between learning and CNS complexity comes from two sources: comparative studies of learning ability in animals that differ in brain size, and artificial selection experiments involving learning and brain indices.

a. COMPARATIVE STUDIES. Rensch (1956, 1959) presented data from a variety of species that, he argues, indicate a positive correlation between absolute brain size and learning ability. Unfortunately, his conclusion is hard to evaluate properly because almost none of the comparisons he makes are based on data from the same study and may well reflect differences in task difficulty and/or training procedures. More recent studies of this issue have been equivocal. Ridell and Corl (1977) reviewed data from a number of primate studies supporting Rensch's conclusions, whereas Miller and Tallarico (1974) found no relation between brain size and detour-learning ability in two species of birds.

None of the studies cited above used learning tasks that are natural ones for the species involved and so their relevance to the question of selection for larger brains may be questioned. Eisenberg and Wilson (1978), on the other hand, found that among bats, those species that normally feed on patchily distributed food that is both spatially and temporally unpredictable tend to have larger relative brain sizes than those that feed on more uniformly distributed food. They suggest that the foraging strategies of the former group of species require more learning than those of the latter and offer this as a possible selection pressure for larger relative brain size, but direct experimental data on the role of learning in the development of feeding behavior in bats are not available.

Jerison (1970, 1973) presents evidence from fossil studies that the relative brain size (i.e., brain/body ratio) of both carnivores and herbivores showed a progressive increase until recent times, with carnivores having relatively larger brains than the herbivores on which they preyed. This he attributes to reciprocal selection pressures for more elaborate behavioral adaptations for offense and defense by carnivores and herbivores respectively. If such adaptations included the ability to modify behavior on the basis of prior experience, then these data might constitute evidence for the requirement of a larger brain to support learning. Recently, however, Radinsky (1978) has questioned the basis of Jerison's conclusions and the proper interpretation of these fossil data remains in doubt.

b. SELECTION EXPERIMENTS. Since the early experiments of Tryon (1940) it has been known that selection can produce both increases and decreases in learning abilities of different sorts. More recent studies have examined the correlation between learning ability and various brain indices in differently selected lines, mainly of mice. The data from these studies have been reviewed by Wahlsten (1972), who finds no compelling evidence for a correlation between learning and brain size as a result of selection (see also Jensen, 1977).

3. EVALUATION. Both comparative studies and selection experiments pose serious difficulties of interpretation. As noted above, the majority of comparative studies have used highly artificial learning tasks, e.g., reversals of discrimination and delayed alternation (Riddell and Corl, 1977). No evidence is presented that proficiency in such tasks is of any adaptive significance in the species tested and, if it is not, then the relevance of such tasks to the evolution of learning is very limited. The comparative approach is also limited by the impossibility of controlling for differences other than learning ability among the species tested. Jerison (1973) discusses a number of other selection pressures besides learning ability that might be responsible for differences in brain indices. He suggests that increasing demands for sophisticated perceptuomotor skills are likely to have been of particular importance in the evolution of the brain, and such skills will usually be very hard to dissociate from learning ability in comparative studies (see also Welker, 1976).

Intraspecies selection experiments offer a more direct test of relationships between learning ability and brain size in evolution. However, most experiments have involved selecting for smaller or larger brains and examining the effects of this selection on learning ability (Wahlsten, 1972). A more appropriate design would be to select for learning ability (preferably of some ecologically relevant kind) and examine the effects of this selection on brain size or other indices, since under natural conditions it is learning ability, not brain size, that is the determinant of reproductive success. While such an experiment is certainly possible it does not appear to have been done.

In summary, the evidence supporting the popular notion of a relation between brain size and learning ability in the course of evolution is very scanty and mostly equivocal. Perceptual requirements may have played a more important role than learning in brain evolution (Jerison, 1973; Welker, 1976).

E. GREATER COMPLEXITY OF THE GENOME

1. NATURE OF THE COST. Williams (1966) has suggested that a plastic, or facultative, developmental system, such as is implied by the possession of an enhanced learning capacity, requires a more elaborate genetic specification than one that is less plastic. If this suggestion is correct, then it identifies a further selective cost of learning, since a greater amount of genetic material is more susceptible to the randomizing effects of mutation. In order to offset these effects, either selection must supply mechanisms for the recognition and repair of damaged sequences of DNA (Bessman et al., 1974; Richardson et al., 1964; see J. D. Watson, 1970, p. 292, for a discussion of repair mechanisms and their evolutionary significance) or reproductive effort must be increased to compensate for losses due to deleterious mutations (for some examples of neurological mutations and their effects see Sidman et al., 1965).

2. AVAILABLE EVIDENCE. It is hard to establish the existence of such a selective cost since our present knowledge of the involvement of the genome in developmental processes such as learning is virtually nonexistent. While it seems reasonable to suppose that developmental plasticity requires more elaborate genetic specification, it seems equally plausible, in principle, to argue that if the structure of behavior is more closely specified by the animal's experience (as a result of learning) less of the specification need be supplied by the genome. Indeed, limits on the capacity of the genome to encode information have been offered as one source of selection pressure in favor of learning (Nottebohm, 1972; Mayr, 1974) and, on this view, learning would require less rather than more genetic material.

Two lines of evidence suggest that the first of these two arguments may be the correct one, although both lines are indirect. In the first place, any facultative developmental process should, at the molecular level, require more genetic material than a nonfacultative one, since more regulatory genes must exist to control the underlying change in protein synthesis. Consider the *lac* operon in the bacterium *Escherichia coli*, which is concerned with the production of enzymes for lactose metabolism. Here, three structural genes are responsible for the production of the enzyme β-galactosidase and two other enzymes involved in lactose metabolism. In the absence of lactose in the environment, these genes are repressed by a repressor substance produced by a neighboring regulatory gene. In the presence of lactose, the repressor is inactivated and transcription of the operon is induced. The structural gene in this case comprises some 5100 DNA base pairs; the regulatory gene adds another 1020 base pairs (Watson, 1970). If the *lac* operon were not facultative but instead produced β-galactosidase continuously, only 83% of the total DNA involved in its functioning would be necessary.

This example is not intended to suggest that the behavioral events of learning correspond in any simple manner to the regulated activity of single genes. Ultimately, however, all developmental events are determined by an interaction between the coordinated activity of the genome and the environment and the preceding example suggests the need for a more complex genetic base for such determination when it is facultative rather than obligatory.

The second line of evidence is provided by studies of the proportion of transcribable DNA that is present in different body tissues. Grouse et al. (1972), using RNA hybridization techniques, found that 11% of the DNA of brain neurons is transcribable (i.e., actively involved in protein synthesis and regulation), compared with 4–5% in liver, kidney, and spleen. That means that a greater proportion of the total DNA is involved in CNS function than in the function of other tissues, suggesting that if the evolution of a more complex CNS is required to support an increased learning capacity, then this may involve the acquisition of more DNA sequences.

A set of related findings of interest is that rats reared in an enriched environment have larger brains with higher connectivity (Rosenzweig, 1966), show an enhancement of learning ability on at least some learning tasks (Krech et al., 1962), and have brains with a higher proportion of transcribable DNA (Grouse et al., 1978), than do rats reared under impoverished conditions. The proper interpretation of these findings is not entirely clear, but they point to some interesting relationships among CNS complexity, genome complexity, and learning ability that invite further investigation, preferably involving ecologically relevant learning tasks.

3. EVALUATION. Any claims that learning requires a more complex genetic basis are, at best, highly speculative. The indirect evidence cited above is intriguing but it is offered more in the hope that it will stimulate further thought on this issue than as evidence in favor of the proposition.

F. DEVELOPMENTAL FALLIBILITY

1. NATURE OF THE COST. The very flexibility of behavior that is the primary selective benefit of learning (Section V, A) implies that by not being exposed to, or not attending to, the appropriate kinds of environmental influences, an animal may develop maladaptive behavior patterns. Thus there will tend to be selection against forms of learning that do not incorporate some form of "protection" against such maladaptive developmental responses. We therefore expect to find extant examples of learning associated with some form of protection.

2. EVIDENCE. Protection against maladaptive development may be of two kinds: social or developmental. Social protection includes parental care which both ameliorates the adverse consequences of behavioral errors during early learning and helps to increase the uniformity of the early environment, ensuring that appropriate experiences for learning are available. The possibility of an association between increased parental care and learning has already been discussed (Section IV,C).

Developmental protection includes the provision of constraints on development that specifically preclude maladaptive developmental responses to the environment. There is extensive evidence that such constraints exist (e.g., Shettleworth, 1972) and their possible role as protection against developmental error has recently been discussed by Marler (1977) in regard to song learning. Many species of songbirds require exposure to an adult song model for normal vocal development (Marler and Mundinger, 1971). Under natural conditions, the young bird is surrounded by a complex acoustic environment, yet it selectively learns the conspecific song rather than others that it may hear. This selectively reflects developmental constraints on the stimuli that are accepted as song models (Nottebohm, 1970; Marler and Peters, 1977; Marler, 1977), a phenomenon sometimes referred to as an "auditory template" (e.g., Marler, 1976). In some cases, social bonds may also play a role in ensuring normal song development (Immelmann, 1969; Dietrich, 1980).

Maternally naive ducklings, and other precocial birds, show a selective approach response to the maternal call of their species shortly after hatching (Gottlieb, 1971). This response may serve to prevent the formation of inappropriate social attachments (Gottlieb, 1965; Johnston and Gottlieb, 1981), despite the very wide range of artificial objects to which attachments may be formed under laboratory conditions (Sluckin, 1973).

3. EVALUATION. The existence of adaptive constraints in development that prevent learning from going astray is probably very widespread. Learning can only be a successful component of an animal's adaptive strategy to the extent that such constraints are provided and, in their absence, learning may be expected to incur a heavy selective cost for its possessor.

VII. LEARNING AND THE ADAPTIVE COMPLEX

In many cases, the existence of selective costs and benefits associated with the evolution of learning rests largely on prima facie arguments that are supported by only limited amounts of empirical evidence. It is clear that more evidence is needed before the selective pressures associated with the evolution of learning can be identified with any confidence and I hope that by indicating the kinds of evidence that are required, the preceding discussion may contribute to the eventual solution of this problem. In order to simplify the discussion, I have presented each of the selective costs and benefits independently of one another. However, it would be a mistake to leave the impression that in any real instance of evolution, the selective picture will be so simple. Interactions among selective pressures are likely to be the rule rather than the exception. It is, furthermore, a biological truism that organisms evolve as adapted wholes rather than as assemblages of adaptive traits. In order to gain some insight into the evolution of particular traits, such as the various learning abilities that an animal may possess, it is necessary to abstract the trait(s) of interest from this adaptive complex and consider them in isolation, but it is important not to lose sight of the numerous interactions that occur in evolution, or of the consequences of those interactions. In this section, I wish to consider some of the more obvious problems that arise when one attempts to consider the evolution of learning within the context of the overall adaptive complex.

A. INTERACTIONS OF SELECTIVE PRESSURES AND ADAPTIVE TRAITS

Interactions among the various costs and benefits of learning are unlikely to be linearly additive, since there will most likely be reciprocal interactions among the various selection pressures that act on an evolving population. The evolution of different learning abilities will not involve the same selective costs and benefits, nor will the same cost or benefit always be involved to the same extent. Different stages of the life cycle will be exposed to different selection pressures and will involve different kinds and degrees of learning so that the selective context vis-à-vis learning will change, probably in complex ways, as the organism develops.

Given the current lack of data on ecologically relevant learning abilities that might permit an evaluation of selective costs in the context of life history traits, breeding biology, and other relevant characteristics of particular organisms, it is probably futile to attempt a detailed integration of the overall selective picture attendant on the evolution of learning. To avoid leaving an overly simplified impression, however, it is worthwhile to indicate, albeit briefly and in general terms, some of the interactions that probably do exist. At the very least, such a discussion may help to stimulate further thought on the problems involved.

Those selective costs most directly concerned with breeding success (i.e., delayed reproduction, juvenile vulnerability, and parental care) are perhaps most likely to show strong nonadditive interactions during the evolution of learning. Juvenile mortality and parental care are in many respects complementary costs: Animals that incur the latter are less likely to have to bear the former as well. If the evolution of learning incurs the cost of added CNS complexity, on the other hand, then that of parental care may become obligatory, because of the energetic investment represented by each offspring. As more parental care is provided, the cost implied by the fallibility of learning may become less important, although this will depend on the nature of the care provided as well as on its extent (in terms of days postpartum).

The costs of learning interact not only with each other but also with other elements of the organism's overall adaptation to the environment. Numerous examples of such interactions may easily be thought of. If selection favors the evolution of complex perceptuomotor skills for reasons unrelated to learning, requiring an increase in CNS complexity, then the subsequent or concurrent evolution of a learning ability may be partially exempt from that cost. The evolution of a complex social system, especially one that involves a relatively close-knit family unit, may relieve some of the costs of parental care, perhaps by kin selection acting through non-reproducing relatives (e.g., the "aunt" role in some primate species: Hrdy, 1976; McKenna, 1979). On the other hand, the evolution of such a social system will probably require the ability to identify individual group members (among other requirements) and such identities must of course be learned. The evolution of a learning ability may not incur the cost of delayed reproduction if selection acts to favor such a delay because of other physiological or ecological requirements of breeding, such as attaining a certain body size or obtaining a breeding territory.

Not all selective pressures act equally throughout the life cycle. Delayed reproduction is a cost that is incurred only early in life and may be offset if an increase in learning ability enables the reproductive period to be extended into later life, or permits a greater reproductive success when reproduction does begin. Parental care, on the other hand, cannot be a selective factor until after the onset of reproduction. The interaction of these costs with other elements of the selective context will therefore change during the life cycle and they may tend to be offset by appropriate scheduling of other developmental events. These considerations suggest that theories of life-history evolution (Stearns, 1977), which are primarily concerned with reproductive parameters such as age at maturity and clutch size, might include the evolution of learning as part of their analyses.

Finally, it should be noted that the evolution of one kind of learning ability may make the evolution of others less expensive. If selection favors, say, a delay in reproduction to permit the evolution of one learning ability, then the evolution of others will not incur this selective cost and there may be a "snowball" effect, permitting the relatively rapid evolution of several learning abilities, even if their selective benefits, taken individually, are relatively slight. Of course, if the selective picture changes, so that the first ability becomes less advantageous, then the other abilities may rapidly incur a heavy selective cost and individuals in which are less well developed may be at a sudden selective disadvantage.

As in any discussion of evolutionary phenomena, such complexities might be multiplied almost indefinitely and no useful purpose would be served by extending this discussion further. Frazzetta (1975) provides a most illuminating discussion of the problems inherent in explaining the evolution of complex adaptations, of which learning is certainly one. A mature evolutionary account of learning will have to cope with these problems and in this preliminary discussion I can do no more than draw attention to their existence.

B. LIMITS ON ADAPTIVE PRECISION

It has sometimes been argued, e.g., by Cody (1974), that natural selection produces an optimally adapted phenotype, that is, a phenotype that is precisely tailored to the requirements of the organism in its environment. This view has been justly criticized by Ghiselin (1966, 1974), Darlington (1977), Lewontin (1978, 1979), and Gould and Lewontin (1979)—indeed, Darwin (1859) was well aware of the limits of natural selection in effecting adaptation. Several factors limit the possibilities for adaptive precision, including insufficiently strong selection pressure, competing adaptive demands on the organism, pleiotropy, genetic linkage, correlated growth effects, and lack of appropriate genetic or phenotypic variation. An evolutionary account of learning, to which this article is a contribution, must avoid what Lewontin (1979) calls the "adaptationist fallacy" of assuming that all characteristics of an organism are tailored to subserve some adaptive function.

Let us suppose that in some population, selection favors the evolution of a particular learning ability, L, say the ability to adopt new foraging patterns on the basis of changes in prey distribution or abundance (e.g., Bené, 1945; L. Tinbergen, 1960). Does this mean that the outcome of selection (assuming selection to be successful) will be *only* those characteristics necessary to effect L, and that individuals in the selected population will exhibit no other new learning abilities? Natural selection theory strongly suggests that this will not be the case and to argue otherwise is to commit the adaptationist fallacy. While selection will produce *at least* those characteristics sufficient to effect L, there is no reason to suppose that this will be the only outcome of selection, a fact that is clearly shown by studies of artificial selection on other traits. For example, Pyle (1976, 1978) has shown that selection for positive or negative geotaxis in *Drosophila melanogaster* can have correlated effects on other characters that were not directly selected. These characters include female oviposition site preferences (Pyle, 1976), mating speed, courtship duration, locomotor activity, and aristal morphology (Pyle, 1978). Similar sorts of effects are very likely to occur in the course of selection for learning abilities.

There are two kinds of account that may be given of this outcome (discussed by Pyle, 1978). If the genes controlling geotaxis are pleiotropic or are closely linked to other genes with different functions, then some correlated characters may be incidental by-products of selection. Natural selection can establish such characters in a population if they are adaptively neutral, as is particularly the case for what may be called "hidden phenotypic characters." Characters may be hidden either physically (e.g., the color of internal organs; Lewontin, 1978) or ecologically, if they are not revealed under the conditions in which the animal lives and in which selection is operating. I shall return to the latter possibility. On the other hand, some correlated characters may, unknown to the investigator, be essential to the performance of the task being selected—Pyle (1978) suggests that the aristae may be involved in geotactic behavior and that changes in aristal morphology may subserve the changes in geotactic behavior that were the direct outcome of selection in his experiment.

The probable widespread existence of such correlated effects of selection has some important consequences for the evolution of learning. As one learning ability is selected for in a population, others may evolve as correlated effects of selection. This may be the result of pleiotropism or linkage, or it may be that the physiological mechanisms involved in the selected learning ability incidentally support other abilities of no adaptive relevance. It is likely that such abilities will remain hidden in individuals that possess them, because the experiences necessary for them to be revealed in development will not be provided by the environment in which the population lives. We might call such abilities "ecologically surplus" learning abilities because they make no contribution to the animal's ontogenetic adaptation to its environment (see Boice, 1977). If circumstances change, such surplus abilities may be available as preadaptations (Bock, 1959; Bock and von Wahlert, 1965; Gans, 1979). Surplus abilities may also be revealed in experiments on animal learning, in which a nonnatural environment is provided for the animal, and in the concluding section of this article I shall consider some of the implications of this point for the study of learning.

VIII. IMPLICATIONS FOR THE STUDY OF LEARNING

In this article I have provided a preliminary account of some of the selective costs and benefits that are involved in the evolution of learning by natural selection. Although in some cases rather little evidence for the existence of a particular cost or benefit is available, overall I feel that we are justified in concluding that despite its selective benefits, the evolution of learning almost always incurs some selective costs and that these costs may sometimes be very heavy. On this analysis, it does not seem particularly surprising that learning is not more widespread in the animal kingdom (cf. Mayr, 1974). The situations in which learning confers a net selective benefit on its possessors may indeed be few and in attributing "obvious" advantages to learning, we may merely be reflecting an unwarranted anthropocentric bias (Nottebohm, 1972).

If this is an accurate assessment of the situation attendant on the evolution of learning, then we are led to a position that may have important implications for the way we should go about the study of learning. It seems rather unlikely that a "general learning ability" exists that has been increased and refined by natural selection in the course of evolution, since this view implies a more or less universal advantage to the ability to learn (Rozin, 1976). A more appropriate view might be that selection acts to favor those individuals that possess the least costly learning abilities permitting successful reproduction in the environment in which the population is evolving. Nothing is gained, and much may be lost, by an individual that possesses ecologically surplus learning abilities. Animals may thus be expected to possess close to the minimum effective learning ability that is adaptive for them, although, based on the arguments in Section VII,B, we expect to find at least some ecologically surplus abilities in most animals that are able to learn.

The distinction between adaptive and surplus abilities is not a trivial one. The former abilities reflect a particular ecological relationship between an animal and its environment that is the result of many generations of selection. The later [sic], by contrast, reflect what we might call the "operating characteristics" of the underlying physiological and/or genetic support for learning. The possible interpretations that may be given to these two kinds of ability, i.e., the roles that they play in theories of learning, are clearly quite different and it is very important that our theories offer some way to distinguish between them (see Johnston, 1981). Current theories of learning provide no basis for such a distinction because they assign no theoretical significance to descriptions of the animal's natural environment; and it is description of the environment that is precisely what is required to distinguish adaptive from surplus learning abilities.

An ecological theory of learning, incorporating a description of the' learner's natural environment as an integral part, would provide the basis for distinguishing between adaptive and surplus abilities and would stand in sharp contrast to current theories of learning, which are thoroughly nonecological. A possible approach to such a theory has been discussed in detail elsewhere (Johnston and Turvey, 1980; Johnston, 1981); at present, let me simply remark on some implications of accepting the surplus/adaptive distinction for the conduct of research on animal learning.

Many, perhaps most, of the tasks used to investigate the learning abilities of animals have no discernible ecological relevance. To the extent that animals are able to perform them, they presumably either mimic tasks that are ecologically relevant or tap ecologically surplus learning abilities. In either case, the use of such tasks may make a contribution to ecological theories of learning, but only if we know which of these two possibilities is the case. If the former, then the task can serve as a simplified experimental situation for the study of an ecologically relevant learning ability. In the case of a surplus ability, the contribution is of a different kind, since the nature of such an ability cannot be understood in terms of its contribution to ecological adaptation. Its analysis may however contribute to an understanding of the mechanisms underlying ecologically relevant learning abilities, in the same way that an analysis of developmental pathologies (i.e., responses to stimuli for which the developing system is evolutionarily unprepared) can contribute to our understanding of normal development. It is obviously important that we be able to distinguish in our theories between normal and pathological developmental processes and this is just as true of theories of learning as it is of theories of other developmental phenomena.

Analysis of an animal's learning abilities may be likened to the process of analyzing the operation of a complex system such as a computer program. When a human designer produces such a system, he or she will usually try to anticipate the kinds of inappropriate inputs it may receive when in use and include some error-catching facilities in the design. If a program is intended to operate on numerical data, alphabetic input may produce a warning message, rather than a program crash or, worse, uninterpretable output. Since natural selection cannot anticipate, the use of biologically inappropriate stimuli in learning experiments may well produce behavior that is uninterpretable from an ecological point of view. Unfortunately, human ingenuity is such that some interpretation of the behavior will usually be found, but such interpretations only contribute to our understanding of contrived and possibly artifactual phenomena.

These various considerations suggest a change in emphasis in the kinds of problems that are studied by investigators of animal learning. Very few of the problems currently under investigation (see Bitterman et al., 1979) have any clear relevance to the problems faced by animals outside the artificial environment of the laboratory[4]; at least, if they do, their relevance remains to be demonstrated by those engaged in such investigation. In an ecological approach to learning, the primary focus of experimental analysis would be on the role that experience plays in the development of behavioral skills that the animal normally employs in its dealings with the environment (Marler, 1975; Gottlieb, 1976b). Ethologists have been making this suggestion for many years (e.g., N. Tinbergen, 1951; Lorenz, 1965), and the arguments in this article lend it additional support. The literature of ethology and behavioral ecology abounds with detailed descriptions of such skills and there is enormous scope for the application of laboratory techniques of developmental analysis to the study of their acquisition (for examples of such application, see Marler, 1970; Emlen, 1972; Hall et al., 1977; Freeman and Rosenblatt, 1978; Gottlieb, 1979; Gray and Tardif, 1979). Studies of this kind will lay the basis for any future ecological theory of learning, which will be necessary to an understanding of the evolution of this important mode of ontogenetic adaptation.

ACKNOWLEDGMENTS

I am grateful to Gilbert Gottlieb, David B. Miller, Ronald W. Oppenheim, and the editors of [the original] volume for many helpful comments on earlier versions of the manuscript. The paper was completed while I was a Visiting Scientist at the Institute of Animal Behavior, Rutgers University, and I am grateful to Jay S. Rosenblatt and Colin Beer for providing that opportunity. Discussions with Gregory Ball and Sarah Lenington clarified my thinking on a number of points. Preparation of this paper was supported in part by Grant No. HD-00878 from the National Institute of Child Health and Human Development.

---

[4]Possible exceptions to this general statement are studies of taste aversion learning (Rozin and Kalat, 1971; Kalat, 1977) and of avoidance learning (Bolles, 1970, 1971; see Shettleworth, in press).

REFERENCES

Aceves-Piña, E. O., and W. G. Quinn. "Learning in Normal and Mutant *Drosophila* Larvae." *Science* **206** (1979): 93–96.

Alcock, J. *Animal Behavior: An Evolutionary Approach*, 2nd ed. Sunderland, MA: Sinauer, 1979.

Amadon, D. "The Evolution of Low Reproductive Rates in Birds." *Evolution* **18** (1964): 105–110.

Anastasi, A. "The Inherited and Acquired Components of Behavior." In *Genetics and the Inheritance of Integrated Neurological and Psychiatric Patterns*, edited by D. Hooker and C. C. Hare, 67–75. Baltimore, MD: Williams & Wilkins, 1954.

Angell, J. R. "The Province of Functional Psychology." *Psychol. Rev.* **14** (1907): 61–91.

Arnold, S. J. "The Evolution of a Special Class of Modifiable Behaviors in Relation to Environmental Pattern." *Am. Nat.* **112** (1978): 415–427.

Ashmole, N. P. "The Regulation of Numbers of Tropical Oceanic Birds." *Ibis* **103b** (1963): 458–473.

Ashmole, N. P. and S. H. Tovar. "Prolonged Parental Care in Royal Terns and Other Birds." *Auk.* **85** (1968): 90–100.

Bateson, G. "The Role of Somatic Change in Evolution." *Evolution* **17** (1963): 529–539.

Bateson, P. P. G. "Specificity and the Origins of Behavior." In *Advances in the Study of Behavior*, edited by J. S. Rosenblatt, R. A. Hinde, E. Shaw, and C. Beer, Vol. 6, 1–20. New York: Academic Press, 1976.

Bené, F. "The Role of Learning in the Feeding Behavior of Black-chinned Hummingbirds." *Condor* **47** (1945): 3–22.

Bessman, M. J., N. Mazycka, M. F. Goldman, and R. L. Schnaar. "Studies on the Biochemical Basis of Spontaneous Mutation. II. The Incorporation of a Base and Its Analogue into DNA by Wild-type, Mutator, and Anti-mutator DNA Polymerases." *J. Mol. Biol.* **88** (1974): 409–421.

Bitterman, M. E. "Phyletic Differences in Learning." *Am. Psychol.* **20** (1965): 396–410.

Bitterman, M. E. "The Comparative Analysis of Learning." *Science* **188** (1975): 699-709.

Bitterman, M. E., V. M. LoLordo, J. B. Overmier, and M. E. Rashotte, eds. *Animal Learning: Survey and Analysis.* New York: Plenum, 1979.

Blus, L. J. and J. A. Keahey. "Variation in Reproductivity with Age in the Brown Pelican." *Auk.* **95** (1978): 128–134.

Bobisud, L. E., and C. J. Potratz. "One-Trial Learning Versus Multi-Trial Learning for a Predator Encountering a Model-Mimic System." *Am. Nat.* **110** (1976): 121–128.

Bock, W. J. "Preadaptation and Multiple Evolutionary Pathways." *Evolution* **13** (1959): 194–211.

Bock, W. J., and G. von Wahlert. "Adaptation and the Form-Function Complex." *Evolution* **19** (1965): 269–299.

Boice, R. "Surplusage." *Bull. Psychon. Sci.* **9** (1977): 452–454.

Bolles, R. C. "Species-Specific Defense Reactions and Avoidance Learning." *Psychol. Rev.* **77** (1970): 32–48.

Bolles, R. C. "Species-Specific Defense Reactions." In *Aversive Conditioning and Learning*, edited by F. R. Brush, 183–233. New York: Academic Press, 1971.

Bolles, R. C. *Learning Theory.* New York: Holt, 1975.

Boring, E. G. "The Influence of Evolutionary Theory upon American Psychological Thought." In *Evolutionary Thought in America*, edited by S. Persons, 268–298. New Haven, CT: Yale University Press, 1950.

Brown, J. L. *The Evolution of Behavior.* New York: Norton, 1975.

Brown, W. Y. "Prolonged Parental Care in the Sooty Tern and Brown Noddy." *Condor* **78** (1976): 128–129.

Buckley, F. G., and P. A. Buckley. "Comparative Feeding Ecology of Wintering Adult and Juvenile Royal Terns." *Ecology* **55** (1974): 1053–1063.

Capitanio, S. J., and D. W. Leger. "Evolutionary Scales Lack Utility: A Reply to Yarczower and Hazlett." *Psychol. Bull.* **86** (1979): 876–879.

Carmichael, L. "Heredity and Environment: Are They Antithetical?" *J. Abnorm. Soc. Psychol.* **20** (1925): 245–260.

Cody, M. L. "Optimization in Ecology." *Science* **183** (1974): 1156–1164.

Coulson, J. C., and J. Horobin. "The Influence of Age on the Breeding Biology and Survival of the Arctic Tern, *Sterna paradisaea*." *J. Zool.* **178** (1976): 247–260.

Coulson, J. C., and E. White. "An Analysis of the Factors Influencing the Clutch Size of the Kittiwake." *Proc. Zool. Soc. Lond.* **136** (1961): 207–217.

Cravens, H., and J. C. Burnham. "Psychology and Evolutionary Naturalism in American Thought, 1890–1940." *Am. Quart.* **23** (1971): 635–657.

Darlington, P. J. "The Cost of Evolution and the Imprecision of Adaptation." *Proc. Natl. Acad. Sci. U.S.A.* **74** (1977): 1647–1651.

Darwin, C. R. *On the Origin of Species by Means of Natural Selection, or the Preservation of Favoured Races in the Struggle for Life.* London: John Murray, 1859.

Darwin, C. R. *The Descent of Man, and Selection in Relation to Sex.* London: John Murray, 1871.

Darwin, C. R. *The Expression of the Emotions in Man and Animals.* London: John Murray, 1872.

Davies, N. B., and R. E. Green. "The Development and Ecological Significance of Feeding Techniques in the Reed Warbler (*Acrocephalus scirpaceus*)." *Anim. Behav.* **24** (1976): 213–229.

DeBenedictis, P. A. "Are Populations Characterized by Their Genes or by Their Genotypes?" *Am. Nat.* **112** (1978): 155–175.

Dietrich, K. "Vorbildwahl in der Gesangsentwicklung beim Japanischen movchen (*Lonchura striata* var. *domestica*, Estrildidae)." *Z. Tierpsychol.* **52** (1980): 57–76.

Dudai, Y. "Behavioral Plasticity in a *Drosophila* Mutant, *dunce*^{DR276}." *J. Comp. Physiol.*, *A* **130** (1979): 271–275.

Dunn, E. K. "Effect of Age on the Fishing Ability of Sandwich Terns, *Sierna sandvicensis*." *Ibis* **114** (1972): 360–366.

Eisenberg, J. F., and D. E. Wilson. "Relative Brain Size and Feeding Strategies in the Chiroptera." *Evolution* **32** (1978): 740–751.

Emlen, S. T. "The Development of Migratory Orientation in Young Indigo Buntings." *Living Bird* **8** (1969): 113–126.

Emlen, S. T. "Celestial Rotation: Its Importance in the Development of Migratory Orientation." *Science* **170** (1970): 1198–1201.

Emlen, S. T. "The Ontogenetic Development of Orientation Capabilities." In *Animal Orientation and Navigation*, edited by S. R. Galler, K. Schmidt-Koenig, G. J. Jacobs, and R. F. Belleville, 191–210. Washington, DC: NASA (SP-262), 1972.

Emlen, S. T. "The Stellar-Orientation System of a Migratory Bird." *Sci. Am.* **233(2)** (1975): 102–111.

Estabrook, G. A., and D. C. Jesperson. "Strategy for a Predator Encountering a Model-Mimic System." *Am. Nat.* **108** (1974): 443–457.

Estes, W. K. "Memory and Conditioning." In *Contemporary Approaches to Conditioning and Learning* edited by F. J. McGuigan and D. B. Lumsden, 265–286. New York: Wiley, 1973.

Frazzetta, T. H. *Complex Adaptations in Evolving Populations.* Sunderland, MA: Sinauer, 1975.

Freeman, N. C. G., and J. S. Rosenblatt. "The Interrelationship Between Thermal and Olfactory Stimulation in the Development of Home Orientation in Newborn Kittens." *Dev. Psychobiol.* **11** (1978): 437–457.

Galef, B. G. "The Ecology of Weaning: Parasitism and the Achievement of Independence by Altricial Mammals." In *Parental Care*, edited by D. J. Gubernick and P.H. Klopfer, [no page numbers given]. New York: Plenum, 1981.

Gans, C. "Momentarily Excessive Construction as the Basis for Protoadaptation." *Evolution* **33** (1979): 227–233.

Garcia, J., and R. A. Koelling. "Relation of Cue to Consequence in Avoidance Learning." *Psychon. Sci.* **4** (1966): 123–124.

Garcia, J., F. R. Ervin, and R. A. Koelling. "Learning with Prolonged Delay of Reinforcement." *Psychon. Sci.* **5** (1966): 121–122.

Gesell, A. "Maturation and the Patterning of Behavior." In *A Handbook of Child Psychology*, edited by C. Murchison, 2nd ed., 209–235. Worcester, MA: Clark University Press, 1933.

Ghiselin, M. T. "On Semanatic Pitfalls of Biological Adaptation." *Philos. Sci.* **33** (1966): 147–153.

Ghiselin, M. T. *The Economy of Nature and the Evolution of Sex.* Berkeley, CA: University of California Press, 1974.

Gottlieb, G. "Imprinting in Relation to Parental and Species Identification by Avian Neonates." *J. Comp. Physiol. Psychol.* **59** (1965): 345–356.

Gottlieb, G. *Development of Species Identification in Birds.* Chicago: University of Chicago Press, 1971.

Gottlieb, G. "Conceptions of Prenatal Development: Behavioral Embryology." *Psychol. Rev.* **83** (1976a): 215–234.

Gottlieb, G. "The Roles of Experience in the Development of Behavior and the Nervous System." In *Studies on the Development of Behavior and the Nervous System, Vol. 3, Neural and Behavioral Specificity,* edited by G. Gottlieb, 25–54. New York: Academic Press, 1976b.

Gottlieb, G. "Development of Species Identification in Ducklings: IV. Change in Species-Specific Perception Caused by Auditory Deprivation." *J. Comp. Physiol. Psychol.* **92** (1978): 375–387.

Gottlieb, G. "Comparative Psychology and Ethology." In *the First Century of Experimental Psychology,* edited by E. Hearst, 147–176. Hillsdale, NJ: Erlbaum, 1979.

Gottlieb, G. "Development of Species Identification in Ducklings: VI. Specific Embryonic Experience Required to Maintain Species-Specific Perception in Peking Ducklings." *J. Comp. Physiol. Psychol.* **94** (1980): 579–587.

Gould, S. J., and R. C. Lewontin. "The Spandrels of San Marco and the Panglossian Paradigm: A Critique of the Adaptationist Program." *Proc. R. Soc. Lond. B* **205** (1979): 581–598.

Gray, L. "Feeding Diversity in Deermice." *J. Comp. Physiol. Psychol.* **93** (1979): 1118–1126.

Gray, L. "Genetic and Experiential Differences Affect Foraging Behavior." In *Foraging Behavior: Ecological, Ethological, and Psychological Approaches,* edited by A. Kamil and T. D. Sargent, 455–473. New York: Garland STPM, 1981.

Gray, L., and R. R. Tardif. "Development of Feeding Diversity in Deermice." *J. Comp. Physiol. Psychol.* **93** (1979): 1127–1135.

Grouse, L., M. D. Chilton, and D. J. McCarthy. "Hybridization of Ribonucleic Acid with Unique Sequences of Mouse Deoxyribonucleic Acid." *Biochemistry* **11** (1972): 798–805.

Grouse, L. D., B. K. Schrier, E. L. Bennett, M. R. Rosenzweig, and P. G. Nelson. "Sequence Diversity Studies of Rat Brain RNA: Effects of Environmental Complexity on Rat Brain RNA Diversity." *J. Neurochem.* **30** (1978): 191–203.

Groves, S. "Age-Related Differences in Ruddy Turnstone Foraging and Aggressive Behavior." *Auk.* **95** (1978): 95–103.

Hainsworth, F. R. "Foraging Efficiency and Parental Care in *Colibri coruscans.*" *Condor* **79** (1977): 69–75.

Hall, W. G., C. P. Cramer, and E. M. Blass. "The Ontogeny of Suckling in Rats: Transitions Toward Adult Ingestion." *J. Comp. Physiol. Psychol.* **91** (1977): 1141–1155.

Harlow, H. F. "The Evolution of Learning." In *Behavior and Evolution*, edited by A. Roe and G. G. Simpson, 269–290. New Haven, CT: Yale University Press, 1958.

Heath, J. E. "Temperature Regulation and Diurnal Activity in Horned Lizards." *Univ. Calif. Publ. Zool.* **64(3)** (1965): 97–136.

Hinde, R. A. "Constraints on Learning—an Introduction to the Problems." In *Constraints on Learning: Limitations and Predispositions*, edited by R. A. Hinde and J. Stevenson-Hinde, 1–19. New York: Academic Press, 1973.

Hinde, R. A. "The Concept of Function." In *Function and Evolution in Behaviour: Essays in Honour of Professor Niko Tinbergen, FRS*, edited by G. Baerends, C. G. Beer, and A. Manning, 3–15. Oxford, UK: Clarendon, 1975.

Hinde, R. A., and J. Stevenson, eds. *Constraints on Learning*. New York: Academic Press, 1973.

Hodos, W., and C. B. G. Campbell. *"Scala naturae:* Why There is No Theory in Comparative Psychology." *Psychol. Rev.* **76** (1969): 337–350.

Houston, D.C. "Breeding of the White-Backed and Rüppell's Griffon Vultures, *Gyps africanus* and *G. rueppellii.*" *Ibis* **118** (1976): 14–40.

Hrdy, S. B. "Care and Exploitation of Nonhuman Primate Infants by Conspecifics Other than the Mother." In *Advances in the Study of Behavior*, edited by J. S. Rosenblatt, R. A. Hinde, E. Shaw, and C. Beer, Vol. 6, 101–158. New York: Academic Press, 1976.

Hull, C. L. "A Functional Interpretation of the Conditioned Reflex." *Psychol. Rev.* **36** (1929): 498–511.

Hull, C. L. "Mind, Mechanism, and Adaptive Behavior." *Psychol. Rev.* **44** (1937): 1–32.

Hull, C. L. *Principles of Behavior.* New York: Appleton, 1943.

Humphrey, G. *The Nature of Learning in Its Relation to the Living System.* London: Kegan Paul, 1933.

Immelmann, K. "Song Development in the Zebra Finch and Other Estrildid Finches." In *Bird Vocalizations*, edited by R.A. Hinde, 61–74. London and New York: Cambridge University Press, 1969.

James, W. *Principles of Psychology.* New York: Holt, 1890.

Jensen, C. "Generality of Learning Differences in Brain-Weight Selected Mice." *J. Comp. Physiol. Psychol.* **91** (1977): 629–641.

Jerison, H. J. "Brain Evolution: New Light on Old Principles." *Science* **170** (1970): 1224–1225.

Jerison, H. J. *Evolution of the Brain Intelligence.* New York: Academic Press, 1973.

Johnston, T. D. "Ecological Dimensions of Learning: A Study of Motor Skill Acquisition in Infant Primates." Unpublished doctoral dissertation, University of Connecticut, 1978.

Johnston, T. D. "Contrasting Approaches to a Theory of Learning." *Behav. Brain Sci.* **4** (1981): 125–173, (with commentary).

Johnston, T. D. "Learning and the Evolution of Developmental Systems." In *Readings and Essays in Evolutionary Epistemology*, edited by H. C. Plotkin. New York: Wiley, in press.

Johnston, T. D., and G. Gottlieb. "Development of Visual Species Identification in Ducklings: What Is the Role of Imprinting?" *Anim. Behav.* **29** (1981): 1082–1099.

Johnston, T. D., and M. T. Turvey. "A Sketch of an Ecological Metatheory for Theories of Learning." In *The Psychology of Learning and Memory*, edited by G. H. Bower, Vol. 14, 147–205. New York: Academic Press, 1980.

Kalat, J. W. "Biological Significance of Food Aversion Learning." In *Food Aversion Learning*, edited by N. W. Milgram, L. Krames, and T. M. Alloway, 73–103. New York: Plenum, 1977.

Kear, J. "Food Selection in Finches with Special Reference to Interspecific Differences." *Proc. Zool. Soc. London* **138** (1962): 163–204.

Kennedy, C., and L. Sokoloff. "An Adaptation of the Nitrous Oxide Method to the Study of the Cerebral Circulation in Children: Normal Values for Cerebral Blood Flow and Cerebral Metabolic Rate in Childhood." *Clin. Invest.* **36** (1959): 1130–1137.

Kimble, J. *Foundations of Conditioning and Learning*. New York: Appleton, 1967.

Krebs, J. "Optimal Foraging: Decision Rules for Predators." In *Behavioural Ecology* edited by J. R. Krebs and N. B. Davies, 23–63. Oxford: Blackwell's, 1978.

Krech, D., M. R. Rosenzweig, and E. L. Bennett. "Relations Between Brain Chemistry and Problem-Solving Among Rats Raised in Enriched and Impoverished Environments." *J. Comp. Physiol. Psychol.* **55** (1962): 801–807.

Kroodsma, D. E. "Reproductive Development in a Female Songbird: Differential Stimulation by Quality of Male Song." *Science* **192** (1976): 574–575.

Kroodsma, D. E. "A Re-Evaluation of Song Development in the Song Sparrow." *Anim. Behav.* **25** (1977): 390–399.

Kuo, Z. Y. "How Are Our Instincts Acquired?" *Psychol. Rev.* **29** (1922): 344-365.

Lack, D. *The Natural Regulation of Animal Numbers*. London and New York: Oxford University Press, 1954.

Lack, D. *Population Studies of Birds*. London and New York: Oxford University Press, 1966.

Lassen, N. A., D. H. Ingvar, and E. Skinhøj. "Brain Function and Blood Flow." *Sci. Am.* **239**(4) (1978): 62–71.

Leatherwood, P. "Influence of Early Undernutrition on Behavioral Development and Learning in Rodents." In *Studies on the Development of Behavior and the Nervous System. Vol. 4. Early Influences*, edited by G. Gottlieb, 187–209. New York: Academic Press, 1978.

Lewontin, R. C. *The Genetic Basis of Evolutionary Change*. New York: Columbia University Press, 1974.

Lewontin, R. C. "Adaptation." *Sci. Am.* **239**(3) (1978): 212–230.

Lewontin, R. C. "Sociobiology as an Adaptationist Program." *Behav. Sci.* **24** (1979): 5–14.

Lloyd, D. G. "Genetic and Phenotypic Models of Natural Selection. *J. Theor. Biol.* **69** (1977): 543–560.

Logue, A. W. "Taste Aversion Learning and the Generality of the Laws of Learning." *Psychol. Bull.* **86** (1979): 276–296.

LoLordo, V. M. "Constraints on Learning." In *Animal Learning: Survey and Analysis*, edited by M. E. Bitterman, V. M. LoLordo, J. B. Overmier, and M. E. Rashotte, 473–504. New York: Plenum, 1979.

Lorenz, K. Z. *Evolution and Modification of Behavior.* Chicago: University of Chicago Press, 1965.

Lorenz, K. Z. "Innate Bases of Learning." In *On the Biology of Learning*, edited by K. H. Pribram, 13–93. New York: Harcourt, 1969.

MacArthur, R. H., and E. O. Wilson. *The Theory of Island Biogeography.* Princeton, NJ: Princeton University Press, 1967.

McGraw, M. "Maturation of Behavior." In *Manual of Child Psychology*, edited by L. Carmichael, 332–369. New York: Wiley, 1946.

McKenna, J. J. "The Evolution of Allomothering Behavior Among Colobine Monkeys: Function and Opportunism in Evolution." *Am. Anthropol.* **81** (1979): 818–840.

Marler, P. "Comparative Study of Song Development in Sparrows." *Proc. XIVth. Int. Ornithol. Congr. 1967*, 231–244.

Marler, P. "A Comparative Approach to Vocal Development: Song Learning in the White-Crowned Sparrow." *J. Comp. Physiol. Psychol. Monogr.* **71** (No. 2, Pt. 2) (1970): 1–25.

Marler, P. "On Strategies of Behavioral Development." In *Function and Evolution in Behaviour: Essays in Honour of Professor Niko Tinbergen, FRS*, edited by G. Baerends, C. Beer, and A. Manning, 254–275. Oxford, UK: Clarendon, 1975.

Marler, P. "Sensory Templates in Species-Specific Behavior." In *Simpler Networks: An Approach to Patterned Behavior*, edited by J. Fentress, 314–329. New York: Sinauer, 1976.

Marler, P. "Perception and Innate Knowledge." In *The Nature of Life* edited by W. H. Heidcamp, 111–139. Baltimore, MD: University Park Press, 1977.

Marler, P., and P. Mundinger. "Vocal Learning in Birds." In *The Ontogeny of Vertebrate Behavior*, edited by H. Moltz, 389–450. New York: Academic Press, 1971.

Marler, P., and S. Peters. "Selective Vocal Learning in a Sparrow." *Science* **198** (1977): 519–521.

Marquis, D. G. "The Criterion of Innate Behavior." *Psychol. Rev.* **37** (1930): 334–349.

Mason, H. L., and J. H. Langenheim. "Language Analysis and the Concept *Environment.*" *Ecology* **38** (1957): 325–340.

Mayr, E. "Behavior Programs and Evolutionary Strategies." *Am. Scient.* **62** (1974): 650–659.

Miller, D. B., and R. Tallarico. "On the Correlation of Brain Size and Problem-solving Behavior of Ring Doves and Pigeons." *Brain Behav. Evol.* **10** (1974): 265–273.

Morgan, C. L. *Habit and Instinct*. London: Arnold, 1896.

Mueller, H. C., and D. D. Berger. "Prey Preferences in the Sharp-Shinned Hawk: The Roles of Sex, Age, Experience, and Motivation." *Auk.* **87** (1970): 452–457.

Mulligan, J. A. "Singing Behavior and Its Development in the Song Sparrow, *Melospiza melodia*." *Univ. Calif. Publ. Zool.* **81** (1966): 1–76.

Nottebohm, F. "Ontogeny of Bird Song." *Science* **167** (1970): 950–956.

Nottebohm, F. "The Origins of Vocal Learning." *Am. Nat.* **106** (1972): 116–140.

Oleson, J. "Contralateral Focal Increase of Cerebral Blood Flow in Man During Arm Work." *Brain* **94** (1971): 635–646.

Oppenheim, R. W. "Metamorphosis and Adaptation in the Behavior of Developing Organisms." *Dev. Psychobiol.* **13** (1980): 353–356.

Oppenheim, R. W. "Ontogenetic Adaptations and Retrogressive Processes in the Development of the Nervous System and Behaviour: A Neuroembryological Perspective." In *Maturation and Development: Biological and Psychological Perspectives*, edited by K. J. Connolly and H. F. R. Prechtl, 73–109. Philadelphia, PA: Lippincott, 1981.

Orians, G. H. "Age and Hunting Success in the Brown Pelican (*Pelecanus occidentalis*)." *Anim. Behav.* **17** (1969): 316–319.

Partridge, L. "Field and Laboratory Observations on the Foraging and Feeding Techniques of Blue Tits (*Parus caeruleus*) and Coal Tits (*Parus ater*) in Relation to Their Habitats." *Anim. Behav.* **24** (1976): 534–544.

Perrins, C. M. "Population Fluctuations and Clutch Size in the Great Tit, *Parus major*)." *J. Anim. Ecol.* **34** (1965): 601–647.

Perrins, C. *British Tits*. London: Collins, 1979.

Peterson, R. H., and C. L. Prosser. "The Effects of Cooling on Electrical Responses of Goldfish (*Carassinus auratus*) Central Nervous System." *J. Comp. Biochem. Physiol.* **42A** (1972): 1019–1037.

Petrinovich, L. "Probabilistic Functionalism: A Conception of Research Method." *Am. Psychol.* **34** (1979): 373–390.

Plotkin, H. C., and F. J. Odling-Smee. "Learning, Change, and Evolution: An Enquiry into the Teleonomy of Learning." In *Advances in the Study of Behavior* edited by J. S. Rosenblatt, R. A. Hinde, C. G. Beer, and M. C. Busnel, Vol. 4, 1–41. New York: Academic Press, 1979.

Pyke, G. H. "Are Animals Efficient Harvesters?" *Anim. Behav.* **26** (1978): 241–250.

Pyke, G. H., H. R. Pulliam, and E. L. Charnov. "Optimal Foraging: A Selective Review of Theories and Facts." *Q. Rev. Biol.* **52** (1977): 137–154.

Pyle, D. W. "Oviposition Site Differences in Strains of *Drosophila* Selected for Divergent Geotactic Maze Behavior." *Am. Nat.* **110** (1976): 181–184.

Pyle, D. W. "Correlated Responses to Selection for a Behavioral Trait in *Drosophila melanogaster*." *Behav. Genet.* **8** (1978): 333–340.

Randinsky, L. "Evolution of Brain Size in Carnivores and Ungulates." *Am. Nat.* **112** (1978): 815–831.

Raichle, M. "Sensori-Motor Area Increase of Oxygen Uptake and Blood Flow in the Human Brain During Contralateral Hand Exercise: Preliminary Observations by the O-15 Method." In *Brain Work. the Coupling of Function, Metabolism and Blood Flow in the Brain*, edited by D. H. Ingvar and N. A. Lassen, 372–376. Copenhagen: Munksgaard, 1975.

Razran, G. *Mind in Evolution.* Boston: Houghton Mifflin, 1971.

Recher, H. F., and J. A. Recher. "Comparative Foraging Efficiency of Adult and Immature Little Blue Herons (*Florida caerula*)." *Anim. Behav.* **17** (1969): 320–322.

Rensch, B. "Increase in Learning Capacity with Increase in Brain Size." *Am. Nat.* **90** (1956): 81–95.

Rensch, B. "Trends Towards Progress of Brains and Sense Organs." *Cold Spring Harbor Symp. Quant. Biol.* **24** (1959): 291–303.

Rescorla, R. A., and P. C. Holland. "Some Behavioral Approaches to the Study of Learning." In *Neural Mechanisms of Learning and Memory*, edited by M. R. Rosenzweig and M. L. Bennett, 165–192. Cambridge, MA: MIT Press, 1976.

Revusky, S. "The Role of Interference in Association Over a Delay." In *Animal Memory*, edited by W. K. Honig and P. H. R. James, 155–213. New York: Academic Press, 1971.

Revusky, S. "Learning as a General Process with an Emphasis on Data from Feeding Experiments. In *Food Aversion Learning*, edited by N. W. Milgram, L. Krames, and T. M. Alloway, 1–51. New York: Plenum, 1977.

Richardson, C. C., R. B. Inman, and A. Kornberg. "Enzymatic Synthesis of Deoxyribonucleic Acid. XVIII. The Repair of Partially Single-Stranded DNA Templates by DNA Polymerase." *J. Mol. Biol.* **9** (1964): 46–69.

Richdale, L. F. *A Population Study of Penguins.* Oxford, UK: Oxford University Press, 1957.

Ricklefs, R. E. "Energetics of Reproduction in Birds." In *Avian Energetics* edited by R. A. Paynter, 152–292. Cambridge, MA: Nuttall Ornithological Club, 1974.

Riddell, W. I., and K. G. Corl. "Comparative Investigation of the Relationship Between Cerebral Indices and Learning Abilities." *Brain Behav. Evol.* **14** (1977): 385–398.

Romanes, G. J. *Mental Evolution in Animals.* New York: Appleton, 1884.

Roots, B. I., and C. L. Prosser. "Temperature Acclimation and the Nervous System in Fish." *J. Exp. Biol.* **39** (1962): 617–629.

Rosenzweig, M. R. "Environmental Complexity, Cerebral Change, and Behavior." *Am. Psychol.* **21** (1966): 321–332.

Rozin, P. "The Evolution of Intelligence and Access to the Cognitive Unconscious." *Progr. Psychobiol. Physiol. Psychol.* **6** (1976): 245–280.

Rozin, P., and J. W. Kalat. "Specific Hungers and Poison Avoidance as Adaptive Specializations of Learning." *Psychol. Rev.* **78** (1971): 459–486.

Sasvári, L. "Observational Learning in Great, Blue and Marsh Tits." *Anim. Behav.* **27** (1979): 767–771.

Seligman, M. E. P. "On the Generality of the Laws of Learning." *Psychol. Rev.* **77** (1970): 406–418.

Seligman, M. E. P., and J. L. Hager, eds. *Biological Boundaries of Learning.* Englewood Cliffs, NJ: Prentice-Hall, 1972.

Shettleworth, S. J. "Constraints on Learning." In *Advances in the Study of Behavior* edited by D. S. Lehrman, R. A. Hinde, and E. Shaw, Vol. 4, 1–68. New York: Academic Press, 1972.

Shettleworth, S. J. "Function and Mechanism in Learning." In *Advances in Analysis of Behavior*, edited by M. Zeiler and P. Harzem, Vol. 3. New York: Academic Press, in press.

Sidman, R. L., M. C. Green, and S. H. Appel. *Catalog of Neurological Mutants of the Mouse.* Cambridge, MA: Harvard University Press, 1965.

Simmons, K. E. L. "Ecological Determinants of Breeding Adaptations and Social Behaviour in Two Fish-Eating Birds." In *Social Behavior in Birds and Mammals*, edited by J. H. Crook, 37–77. London: Academic Press, 1970.

Skinner, B. F. *About Behaviorism.* New York: Knopf, 1974.

Slobodkin, L. B. "Towards a Predictive Theory of Evolution." In *Population Biology and Evolution* edited by R. C. Lewontin, 187–205. Syracuse, NY: Syracuse University Press, 1968.

Slobodkin, L. B., and A. Rapoport. "An Optimum Strategy of Evolution." *Q. Rev. Biol.* **49** (1974): 181–200.

Sluckin, W. *Imprinting and Early Learning.* Chicago: Aldine, 1973.

Snow, D. W. "The Breeding of the Blackbird, *Turdus merula* at Oxford." *Ibis* **100** (1958): 1–30.

Sokoloff, L. "Metabolism of the Central Nervous System in vivo." In *Handbook of Physiology, Section 1: Neurophysiology*, edited by J. Field, H. W. Magoun, and V. E. Hall, Vol. III, 1843–1864. Washington, DC: American Physiological Society, 1960.

Sommerhoff, G. *Analytical Biology.* London and New York: Oxford University Press, 1950.

Staddon, J. E. R., ed. *Limits to Action: The Allocation of Individual Behavior.* New York: Academic Press, 1980.

Stearns, S. C. "The Evolution of Life History Traits: A Critique of the Theory and a Review of the Data." *Annu. Rev. Ecol. Syst.* **8** (1977): 145–171.

Stirling, I., and P. B. Latour. "Comparative Hunting Abilities of Polar Bear Cubs of Different Ages." *Can. J. Zool.* **56** (1978): 1768–1772.

Tebēcis, A. K., and J. W. Phillips. "Reflex Response Changes of the Toad Spinal Cord to Variations in Temperature and pH." *Comp. Biochem. Physiol.* **25** (1968): 1035–1047.

Thorndike, E. L. *Animal Intelligence.* New York: Hafner, 1911.

Thorpe, W. H. *Learning and Instinct in Animals.* London: Methuen, 1963.

Tinbergen, L. "The Natural Control of Insects in Pinewoods: I. Factors Influencing the Intensity of Predation by Songbirds." *Arch. Neerl. Zool.* **13** (1960): 265–343.

Tinbergen, N. *The Study of Instinct.* London and New York: Oxford University Press, 1951.

Tinbergen, N. "On Aims and Methods of Ethology." *Z. Tierpsychol.* **20** (1963): 410–429.

Tinbergen, N., F. Broekhuysen, F. Feekes, C. W. Houghton, H. Kruuk, and E. Szule. "Egg Shell Removal by the Blackheaded Gull, *Larus ridibundus* I.: A Behaviour Component of Camouflage." *Behaviour* **19** (1963): 74–117.

Tolman, E. C. "There is More Than One Kind of Learning." *Psychol. Rev.* **56** (1949): 144–155.

Tryon, R. C. "Genetic Differences in Maze-Learning Ability in Rats." In *Thirty-ninth Yearbook of the National Society for the Study of Education*, 111–119. Bloomington, IL: Public School Publishing, 1940.

Tschirgi, R. D. "Chemical Environment of the Central Nervous System." In *Handbook of Physiology, Section I: Neurophysiology*, edited by J. Field, H. W. Magoun, and V. E. Hall, Vol. III, 1865–1890. Washington, DC: American Physiological Society, 1960.

Verbeck, N. A. M. "Comparative Feeding Behavior of Immature and Adult Herring Gulls." *Wilson Bull.* **89** (1977): 415–421.

Wahlsten, D. "Genetic Experiments with Animal Learning: A Critical Review." *Behav. Biol.* **7** (1972): 143–182.

Warren, J. M. "Primate Learning in Comparative Perspective." In *Behavior of Non-Human Primates*, edited by A. M. Schrier, H. F. Harlow, and F. Stollnitz, Vol. 1, 249–281. New York: Academic Press, 1965.

Watson, J. B. *Behaviorism.* New York: Norton, 1924.

Watson, J. D. *The Molecular Biology of the Gene*, 2nd ed. New York: Benjamin, 1970.

Welker, W. I. "Brain Evolution in Mammals: A Review of Concepts, Problems, and Methods." In *Evolution of Brain and Behavior in Vertebrates*, edited by R. B. Masterson, M. E. Bitterman, C. B. G. Campbell, and N. Hotton, 251–344. Hillsdale, NJ: Erlbaum, 1976.

Wheeler, P. H. "Elaborate CNS Cooling Structures in Large Dinosaurs." *Nature (London)* **275** (1978): 441–443.

Williams, G. C. *Adaptation and Natural Selection.* Princeton, NJ: Princeton University Press, 1966.

Wooller, R. D., and J. C. Coulson. "Factors Affecting the Age of First Breeding of the Kittiwake *Rissa tridactyla.*" *Ibis* **119** (1977): 339–349.

Wright, S. "The Role of Mutation, Inbreeding, Crossbreeding, and Selection in Evolution." *Proc. XI Int. Congr. Genet.* **1** (1932): 356–366.

Yarczower, M., and L. Hazlett. "Evolutionary Scales and Anagenesis." *Psychol. Bull.* **84** (1977): 1088–1097.

New Work

Sharoni Shafir

Preface to Chapter 21

The relationship among computational tools, cognitive mechanisms, and evolution is a unifying theme of this book. Todd works within this framework, expanding the learning–evolution relationship into the realm of sexual selection. He employs simulations of a genetics-based model to investigate the evolution of mate-preference learning, and the role of such learning in the macroevolutionary process of speciation.

EVOLUTION OF LEARNING VIA SEXUAL SELECTION

A question that has attracted much attention in evolutionary biology is what conditions favor the evolution of learned, rather than inherited, behavior. Todd points out that this question has mostly been asked with regard to traits that are subject to viability selection. These are traits that allow an animal to exploit resources in its physical environment, such as finding shelter or food. Todd extends this question to a trait that allows an animal to successfully find a mate. Thus the trait is under selection pressure through sexual selection rather than through natural selection acting on viability or fertility. He asks whether a gene for learning to choose a mate, by imprinting on the phenotype of one's parents, can invade and fix in a population in which mate preferences are inherited.

Todd's model is an "idea model," not constructed around the biology of any specific organism and not intended to precisely mirror a particular genetic system. It borrows concepts from genetics, including mutation and recombination, and it shares properties of two main classes of population genetics models. Simple models assume discrete traits (e.g., to help or not to help) that are typically determined by two alleles, at each of up to three loci. Quantitative models treat traits that are expressed along a continuum (e.g., height), and assume them to be determined by many loci and alleles. Simple models track allelic frequencies and quantitative models track phenotypic trends.

Both classes of models can often be analyzed analytically. However, when a genetic mechanism is explicitly specified (as in simple models), but involves multiple alleles and loci, the model may soon become analytically intractable. Todd's model is of this kind. Whether mate preference is inherited or learned is determined by two alleles at one locus. Two other genes determine what an individual looks like, two genes determine the individual's mate preference, and another gene determines the degree of specificity (choosiness) of mate preference. The latter five genes are represented by 15 to 30 bits each. The complexity of such a model makes it well suited for exploration by simulation.

Simulations reveal that the gene for parental imprinting of mate preference does invade and fix, and that the rate at which it invades depends on how quickly individuals' phenotypes in the population change. These results offer two take-home messages. The first bears directly on the evolution of parental imprinting of mate preferences—the simulations provide a biologically testable hypothesis for the conditions under which such preferences should evolve. The degree of parental imprinting of mate preference in different species within a taxon could be correlated to the rate at which the species has evolved. As Todd points out, such a comparison has already been performed with some birds (Immelmann, 1972), and the findings agree with the model's predictions.

The second take-home message is that simulation models can complement verbal and analytical models and provide insight into understanding the evolution of sexually selected traits. Such an approach may prove useful, for example, in resolving the dispute surrounding Zahavi's handicap principle (Zahavi, 1975, 1977), which may well be one of the most controversial hypotheses for the evolution of sexually selected traits. The idea is that, through female choice, males evolve traits that handicap them. Numerous verbal and theoretical models reached contradictory conclusions about whether the principle can work or not. Grafen's (Grafen, 1990) recent genetic model that supports the handicap principle has contributed to the principle's greater acceptance. The model's complicated mathematics, however, limit its intelligibility to a very small group of people. Verbal arguments remain the most powerful tool of persuasion. It could prove rewarding to construct a mechanistic simulation model that "shows" how the handicap principle works.

THE EFFECT OF LEARNING ON EVOLUTION

Almost a century ago Baldwin introduced a "new factor in evolution." The Baldwin effect, by which learning can guide and accelerate the rate of evolution, has not been embraced by biologists as an important force in evolution, to say the least. Todd discusses how the perception that the Baldwin effect violates the most fundamental tenets of biology, Weismann's doctrine (somatic changes do not affect germ-line cells) and the Central Dogma (information flows from DNA to RNA to proteins), may contribute to the reluctance to accept it. This perception is, of course, unmerited. We should be seeing more models in biology that incorporate learning as a potential evolutionary force. It would be interesting, for example, to consider the effect of learned mate preferences on the evolution of signals by the handicap principle. Todd argues that due to a tight connection between mate-preference learning and reproductive fitness, the Baldwin effect may be particularly relevant in the realm of sexual selection. In the simulations presented, populations in which mate preferences are learned, in fact, evolve faster than those in which mate preferences are inherited. In the former, distinct, yet sympatric, subpopulations are formed; the isolating mechanism is mate preference. Reproductively virtually separating the subpopulations may culminate in a speciation event.

Sympatric speciation is considered extremely rare in animals, if it happens at all. The finding that learned mate preferences could lead to sympatric speciation is very exciting. It suggests a dominant role for sexual, rather than natural, selection. And even more revealing, it suggests that learning to choose mates, rather than inheriting mate preferences, is a key component in sympatric speciation. The potential for sympatric speciation may have been greatly underestimated by traditional population genetics models that failed to incorporate learning.

Todd's model prompts us to ask two main questions: Have we neglected to consider the potential role of learning as an important factor in evolution, particularly regarding sexually selected traits? Has that bias caused us to underestimate the importance of sympatric speciation, particularly where mate preference is the isolating mechanism? An idea model is not intended to supply answers. Rather, if successful, it contributes to guiding research, both theoretical and empirical, along new paths. Todd's chapter on sexual selection and the evolution of learning is a refreshing contribution toward that goal.

REFERENCES

Grafen, A. "Biological Signals as Handicaps." *J. Theor. Biol.* **144** (1990): 517–546.

Immelmann, K. "Sexual and Other Long-Term Aspects of Imprinting in Birds and Other Species." In *Advances in the Study of Behavior*, edited by D. S. Lehrman, R. A. Hinde, and E. Shaw, Vol. 4. New York: Academic Press, 1972.

Zahavi, A. "Mate Selection—A Selection for a Handicap." *J. Theor. Biol.* **53** (1975): 205–214.

Zahavi, A. "The Cost of Honesty (Further Remarks on the Handicap Principle)." *J. Theor. Biol.* **67** (1977): 603–605.

Peter M. Todd

Chapter 21:
Sexual Selection and the Evolution of Learning

1. INTRODUCTION

There are two realms in which every living organism must operate. The first is an economic realm of matter and energy, requiring the accrual of resources necessary for an organism's ongoing survival. The second is a genealogical realm of genetic information, requiring appropriate action to ensure an organism's ongoing genetic perpetuation. As Eldredge (1986, p. 351) puts it, "organisms seem to be both energy conversion machines and reproducing 'packages' of genetic information." Evolution, as "the primary channel of communication between living systems and their environments" (Plotkin & Odling-Smee, 1979, p. 13), clearly shapes organisms in response to both of these environmental realms. Less well appreciated is the fact that learning, as a secondary means of adapting an organism's internal organization to the external order in the environment, can similarly operate in both domains (see Table 1).

At first blush, the two realms seem to be manifested in distinctly separate dimensions of an organism's existence. Genetic information has primacy within an organism, internally controlling the construction and maintenance of the organism's phenotype in a one-way manner. Externally, matter/energy transfer becomes

TABLE 1 Economic versus genealogical realms in evolution, learning, and behavior.

	Economic	Genealogical
Main function:	metabolism, growth, survival	replication, reproduction
Evolution via:	natural selection	sexual selection, kin selection
	speciation through ecological specialization	speciation through mate-preference divergence
Learning of:	habitat/food/resource preferences and identities	mate preferences and identities, kin/offspring identities
Behaviors:	habitat/food/resource choice and use	mate choice, kin/offspring care

the major concern, with differential economic success between organisms being used by natural selection to control the reproductive success of the organism's genotype. Thus, according to Eldredge (1986, pp. 359–360): "Within organisms, information takes priority in causation over economics (even though most of that information is concerned with the construction and functioning of economic aspects of organisms, i.e., the soma). Among organisms, economics holds the upper hand: How well an organism does economically will materially affect its chances for reproductive success."

Since learning is also primarily concerned with the world external to the organism (while making changes internally in response), it is natural that the emphasis in studies of learning should be on its uses in the economic realm. Thus, texts on learning typically focus on such functions of learning as finding and obtaining food, selecting among types of food or other resources, navigating courses and remembering locations, choosing habitats, avoiding pain or aversive stimuli, predicting outcomes, adjusting to temporal patterns, learning language, etc. Seldom (e.g., Gallistel, 1990; Thorpe, 1956, 1963; Davey, 1989), if ever (e.g., Walker, 1987; Mackintosh, 1974) do these standard texts mention any functions relevant to the genetic information realm including mating, sex, and reproduction. Much of the work on the evolution of learning, particularly simulation studies, also tends to discuss only the role of natural selection in shaping learning mechanisms, again emphasizing how learning can pick up the economic slack that evolution can't manage to track (see, for example, Ackley & Littman, 1992; Belew, 1990; Todd & Miller, 1991a; Plotkin

& Odling-Smee, 1979). But learning has wider applications than these works would suggest, as we will see.

The two realms of energy economics and genetic information are not each limited to just the internal or the external worlds of organisms—both also cross the boundary of an organism's skin. Eldredge (1986, p. 360) mentions one direction of this interpenetration: economic concerns come into play internally in organisms as well, since the energy efficiency of an organism's physiology partly determines the size, strength, etc. of the phenotype that takes part in external naturally selected relations. But missing from Eldredge's description is the influence in the other direction: genealogical information is often of central concern in the *external* life of an organism. Deciding whom to mate with, protect, feed, exchange resources with, and otherwise interact with socially can crucially depend on the genetic relationship between the organisms involved. These are areas where inclusive (kin-based) fitness (Hamilton, 1964) rather than individual fitness, and sexual selection (Darwin, 1871; Fisher, 1930) rather than natural selection, predominate. Inclusive fitness concerns the propagation of one's own genetic information in the bodies of others; for instance, in sexual species, each offspring typically shares half its genes with each parent, so a parent can increase its inclusive fitness by promoting the (survival and) reproduction of its offspring and thereby spreading at least some of its own genes. Sexual selection via mate choice is also driven by the propagation of genes, but more circuitously. By choosing to mate with others whose genes are likely to benefit its offspring, an individual can further help to spread the 50% of its genetic makeup that the offspring also carry. For example, females (such as peahens) may select males (such as peacocks) with longer, showier tail-feathers that other females also find attractive and, as a result, have sons who themselves have long tails and attract more mates and further spread their mother's genes (the "sexy son" hypothesis—see Fisher, 1930; Lande, 1981; see Zahavi, 1975, for an alternative based on better genes for survival). Thus, both sexual selection and selection based on inclusive fitness join natural selection in operating in the external world between organisms, and both cause the genealogical realm also to have an impact in that external world.

Since genetic information can be so important not only within but also between organisms, learning can apply in the genealogical realm as well as the economic. In this case, rather than helping an individual to transfer matter and energy more adaptively, learning can help an individual transfer its genetic information more adaptively. By learning who its closest kin are, for instance, an organism can better decide whom to feed, whom to warn of approaching predators, whom to expect fair exchanges of resources with, and probably whom to *avoid* mating with (because of the dangers of inbreeding), so as to best promote its own genetic interests. This is the broad class of functions of learning with which we shall be concerned in this paper. In particular, we focus here on sexually selected learning mechanisms that aid an organism in choosing mates. These forms of learning are interesting because they can themselves influence sexual selection, and thereby have a strong impact on the course of evolution as a whole. The other forms of genealogical learning, used for kin selection and parental investment in offspring, are also interesting and deserve further study; however, these functions usually involve economic transfers of resources as the means of promoting related genes. Mate choice is situated more purely in the genealogical realm—there is often less (if any) transfer of resources involved—and so we will consider here only learning mechanisms in the service of this task.

In the rest of this paper, we will first consider the ways in which learning can evolve through sexual selection, including the sexually selected functions learning may have, the time scales at which these kinds of learning operate, and the differences from naturally selected evolution of learning. We will present a simulation model of the evolution of one particular form of sexually selected learning—parental imprinting—to illustrate these points. We then investigate how learning can affect sexual selection and evolution in turn, leading to such macroevolutionary effects as speciation and runaway selection. This is again illustrated through the simulation study of parental imprinting. Finally, we draw conclusions from this line of inquiry, and describe further directions for the study of sexually selected learning.

2. THE EVOLUTION OF SEXUALLY SELECTED FORMS OF LEARNING

Naturally selected forms of learning evolve because they help organisms to eke out an existence in the economic realm. They aid in acquiring the matter and energy necessary for survival by adjusting an organism's behavior to the changing patterns of material and energetic resources in the environment. Sexually selected forms of learning, in contrast, evolve because they help organisms in the genealogical realm, aiding in the transfer of genetic information to further generations by (among other functions) guiding the selection of appropriate mates with whom to have successful offspring. These types of learning adjust an organism's behavior to the changing patterns of genetic resources in the environment.

Mechanisms for learning mate preferences can evolve in the same way that other mate-choice mechanisms often do: they can genetically hitch-hike along with the good genes they help select when choosing mates, so that offspring inherit both the traits from the genes and the means of choosing them. Thus, the preference learning mechanisms are not sexually selected in quite the same way that, say, a peacock's extravagant tail is: there, more and more elaborate plumage has been chosen by generations of females for its attractive properties (and probably in direct opposition to its energetic costs). Mate preference learning mechanisms, on the other hand, are not typically selected by members of the opposite sex *per se*[1] (in part because they do not have readily perceivable external physical manifestations, such as tail-feathers); rather, they can *effect* the selection of potential mates themselves. Thus, they are not chosen, but choosing; and the extent to which they guide mate choice in an adaptive manner will determine their success in being passed on to successive generations. (In this sense the learning mechanisms are akin to the perceptual or decision-making mechanisms a female peahen might use to judge a peacock's tail, which can also evolve through sexual selection.) Since these learning mechanisms affect the reproduction of the organisms rather than their survival, it is sexual selection rather than natural selection which is responsible for their evolution.

2.1 ADAPTIVE FUNCTIONS OF MATE PREFERENCE LEARNING

The function most commonly ascribed to learning of any kind is the tracking of environmental change that occurs too quickly to be responded to by the primary adaptive process of evolution or the secondary adaptive process of individual development (Plotkin & Odling-Smee, 1979). An organism's behavioral system maps from momentary environmental conditions to appropriate actions; the various adaptive processes adjust that behavioral map at different time scales and using different information-storage mediums. Evolution can track changes in the environment requiring changes in the behavioral map that occur at time scales rather longer than the generation time of a given species. The evolved adaptations to environmental change are stored in the genes of an interbreeding population. Developmental processes can adjust behavioral maps to environmental changes that occur no faster than once per lifetime (which might be different from generation time). Multiple *different* kinds of change occurring within an individual's lifetime may be tracked by development (e.g., the environment getting warmer, and food plants changing color). But owing to development's one-way temporal nature, once a *particular* environmental state has been responded to, any further alteration in that state will be beyond development's adaptive ability (e.g., the environment getting colder again—see Plotkin & Odling-Smee, 1979, p. 25). The developed responses to environmental conditions are stored in the phenotype of an individual organism. Finally, learning can track environmental changes that happen *faster* than the lifetime (or generation time) of an individual, including multiple changes or reversals of the same environmental state. Learned adjustments in an individual's behavioral map are stored in the pattern of neural connectivity of its nervous system.

The types of environmental states that are typically tracked by naturally selected learning include things that change fairly frequently, and are frequently

---

[1] It is possible that individuals *could* prefer mates who display various learning abilities, and thus directly sexually select certain learning mechanisms; for instance, females of some bird species may prefer males who can learn and sing the most novel or widest variety of songs, and thereby directly select for song-learning (Aoki, 1989). We will not consider this possibility here.

TABLE 2 Rate of change versus frequency of use of information in sexually selected learning.

		Rate of Information Change	
		Slow	**Rapid**
	Once	learn once any time before mating	learn once just before mating
Frequency of Information Use			
	Repeatedly	learn once any time before first mating	learn multiple times, before each mating (most common?)

important to know about. For instance, the location of the best food patch might change from day to day, and since an organism typically needs to eat on a regular basis, a learning mechanism that can track changes in patch location rapidly (in time for dinner!) will prove useful. The aspects of the social environment that sexually selected learning typically tracks, on the other hand, might change repeatedly and rapidly within an individual's lifetime, or might not change at all, and might be useful repeatedly during a lifetime, or only once. We thus get a two-by-two table (see Table 2) that can tell us something about the time scales at which sexually selected learning might operate and be adaptive for a given species. The dimensions can be described as follows.

The mate choice mechanisms of different species may rely on sources of information with various rates of change relative to the lifetime of individuals of that species. On one end of the spectrum, the information that preferences are based on may be fixed throughout the individual's life. This could be the case, for instance, if mate choice were based on the genetic similarity between two individuals, to help avoid inbreeding or achieve "optimal outbreeding" (a balance between the costs and benefits of inbreeding and outbreeding—see Bateson, 1978, 1983). Since the genetic relationship between two specific individuals doesn't change over time, the "relatedness" or similarity between them has to be learned only once. On the other end of the rate-of-change spectrum, mate choice information might alter frequently within a lifetime. For example, choices based on the brightest plumage around, the individual guarding the best fruit-producing tree, or other superlative features, can quickly change as new individuals appear in the mating population or underlying resource distributions shift in the environment.

Just as the information that mate choice is based on may change a variable number of times, it can be *used* a variable number of times as well. In some species, mate choices are only made once, either because the species is semelparous (dying after its only breeding opportunity, as in many insects and other invertebrates, and some vertebrates such as the Pacific salmon—see Daly and Wilson, 1983), or because individuals form a monogamous pair-bond for life (as in some bird species in particular). In others, mate choices are made repeatedly, because the species is iteroparous (able to reproduce multiple times) and not strictly pair-bonding (as in most mammal species). The number of times preferences are used can also influence how they are learned.

We will now consider how these two dimensions interact to affect the nature of the sexually selected form of learning. If mate choice information does not change during an individual's lifetime and is only used once, then mate preferences need only be learned once, any time before the single mating event occurs. If the information does not change but is used *multiple* times, then again preferences need to be learned only once, any time before the first mating opportunity arises. Both of these cases can include preference learning early in life, which might be the optimal time for exposure to the relevant preference information. For example, if it is adaptive to choose mates similar to one's parents, then imprinting on the relevant parental characteristics should be done while the parents are still around to learn from, i.e., soon after birth. Gradual learning over an extended period also fits into both cases. For example, if it is adaptive to choose a mate who is representative of a stable population of potential mates, then learning an "average" set of characteristics through encounters with multiple individuals over some length of time will be a good strategy.

The situation is different, however, when the information upon which mate choices are based *does* change during an individual's lifetime. If the information is only used once, the preferences should be learned only once, but in this case they should be learned just before the mating event. Any earlier than this, and the learned preferences may no longer be adaptive once the single mating opportunity comes up. Thus, early parental imprinting will probably not be a viable learning strategy in this case. If multiple mate choices are made, then individuals should learn their preferences anew shortly before each mating opportunity. The majority of sexually selected learning mechanisms probably fall into this category. A particularly clear example of this sort of learning is provided by the Coolidge effect (Dewsbury, 1981), in which individuals (both males and females—see Lester & Gorzalka, 1988) who become sexually satiated with one partner will have rejuvenated interest upon encountering a new partner. Perhaps through a process of habituation and dishabituation (Havens & Rose, 1992), the individuals learn characteristics (or the identity) of their current partner and use these to influence their preference for their next partner (i.e., anyone else). Other forms of sexually selected learning for novelty seeking in mates similarly fall into this changing information/multiple mating category. Culturally influenced preferences (e.g., in humans—Fisher, 1992—and in birds—Aoki, 1989) and the mate-preference copying found in some species (in particular when there is a cost to making an independent choice—see Pruett-Jones, 1992; Dugatkin & Godin, 1992) can also involve this kind of learning.

Sexually selected learning mechanisms may thus be adapted to genealogical information sources in the environment that vary across a wide range of time scales. By looking at when these learning mechanisms are employed during an individual's life to help in the mate choice process, we can get a handle on the type and rate of change of information involved, using the categorization scheme just developed. Similarly, we can map from these characteristics of the mate choice information to the rough sort of learning that will be most adaptive. One example of this kind of time scale/learning mechanism relationship is the one-time parental imprinting learning used to track population changes that occur at about lifetime scales. We will now explore this particular case in more detail.

2.2 THE EVOLUTION OF SEXUAL (PARENTAL) IMPRINTING

To get a better understanding of how sexual selection alone can result in the evolution of a particular learning mechanism, we have developed a simple simulation of the evolution of one relevant form of sexually selected learning, sexual or parental imprinting. We present here a brief account of the simulation and results from this study; a much more detailed description can be found in an earlier paper (Todd & Miller, 1993). This section focuses on the evolution of the imprinting ability, while the reverse effects of imprinting on the course of evolution itself are presented in section 3.1.

In sexual (or parental, as opposed to filial) imprinting, the mate preferences exhibited by individuals later in life are learned through exposure to other individuals, usually a parent, at a very young age. Sexual imprinting is seen most frequently in certain birds, and has been studied extensively among those species (Immelmann, 1972). A vivid example is that of the lesser snow goose, which occurs in two color phases, blue and white. These geese seem to prefer mates that match the coloring of their parents, rather than their own (Cooch, 1961; Cooke & McNally, 1975). If an individual of one color, say blue, is raised by foster parents of the other color, i.e., white, its preference will usually follow that experience, and it will later seek white mates.

Sexually imprinting on one's parents could serve two different functions in tuning up one's mate preferences: learning about one's parents in particular, or learning about the species more generally. In the former case, the adaptive function of sexual imprinting could be to achieve greater fitness by getting the right balance between inbreeding and outbreeding, as Bateson (1978, 1983) suggests. However, optimal outbreeding theory relies on detailed assumptions about the naturally selected fitness effects of various genetic combinations in the economic realm (e.g., lethal homozygotes caused by inbreeding). Since we are interested here just in the sexually selected effects in the genealogical realm, we chose to investigate an alternate possible function of sexual imprinting: that it allows individual behavior to somehow track important environmental changes faster than is possible through evolution of the genotype. As mentioned earlier, this is the most commonly postulated adaptive function of learning in general. In the case of sexual imprinting, the environmental variable tracked by learning would be *the population itself*, specifically, the current frequency distribution of viable phenotypes. That is, sexually imprinting on the phenotypes of one's parents allows mate preferences to directly follow ongoing population dynamics with only one generation of lag time.

If one phenotype proves very successful in a particular generation, for example the blue form of the snow goose, sexual imprinting will allow all of the offspring of that phenotype to learn preferences for it. This allows them to avoid mating with lower-fitness individuals of different phenotypes, to prefer mating with higher-fitness individuals resembling their manifestly successful parents, and thereby to have offspring with higher-fitness genes. Genetically fixed mate preferences would evolve much more slowly in response to shifts in the population, and individuals with such fixed preferences would be less well calibrated in their mate choice and so would tend to have fewer offspring, and ones of lower viability or attractiveness. Immelmann (1975) highlights these advantages of imprinting: "Any changes in the environment of a species, including the appearance of the species itself, are followed automatically by a corresponding change in the relevant object preferences, if the latter are acquired through personal experience.... This is of special importance in rapidly evolving groups of animals as well as in unstable environments where a fair degree of opportunism will be favored as compared with absolutely rigid preferences for particular conditions" (p. 245).

THE SIMULATION MODEL. Simulating the evolution of sexual imprinting by sexual selection is somewhat simpler than simulating the evolution of other forms of learning by natural selection. In our previous studies of the evolution of learning in the economic realm (Miller & Todd, 1990; Todd & Miller, 1991a, 1991b), we used a genetic algorithm to evolve simple neural networks that controlled the feeding behavior of simulated organisms in simulated environments. These networks could learn food preferences over time through experience in the simulated environments, and thus guide the creatures to behave more and more adaptively. The model we develop here diverges from the previous studies in two major ways. First, to emphasize the role of sexual selection in the evolution of the populations we use in this study, we have completely eliminated nonsexual forms of selection—that is, we have a completely flat fitness landscape, with each individual equally viable. The forces that drive evolution will come through the number of mating opportunities that each individual is able to obtain and exploit. This means the standard genetic algorithm has been modified, both eliminating differential viability between individuals, and changing the way in which individuals are chosen for crossover and representation in the next generation. Second, we do not use neural networks to simulate the learning mechanisms of creatures in this scenario. Rather, sexual imprinting is modeled by simply making the child's ideal mate preference equivalent to the phenotype of one of its parents, as described below. Because this type of imprinting it so straightforward, the use of neural networks would unnecessarily complicate the simulation and cloud the theoretical issues.

Each individual in our simulations has a genotype made up of just six genes: two "trait" genes that code for phenotypic attributes, two "preference" genes that code for the preferred phenotypic attributes of potential mates, one "choosiness" gene that codes for the amount of deviation tolerated from the specified phenotypic mate preferences (i.e., how far an individual will generalize its sexual receptivity from its ideal during its mate search), and one "learning" gene that indicates whether the actual mate preference values are genetically determined or parentally imprinted. Using two phenotypic dimensions for the traits and preferences makes visualization easy: the phenotype of each individual can be represented as a single (x,y) point on a two-dimensional plot, the individual's mate preferences can be represented as a circular region (whose radius is specified by the choosiness gene) around another (x,y) point, and the phenotype frequency distribution of an entire population can be represented as a set of points in the same space. We interpret the two-dimensional phenotype space as a toroid of 1000 × 1000 arbitrary units (where the top edge connects to the bottom, and left edge to right). It is imperative to remember that positions in this abstract two-dimensional phenotype space are not spatial locations in physical space, and that separation in this space cannot be interpreted geographically. Mate preferences permitting, it is as easy for two individuals far apart in phenotype space to mate as for two individuals close in this space.

After the first two genes specify the individual's location in phenotype space, the rest of the genotype determines the individual's *mate preference function*. First, the location of the individual's ideal mate must be assigned. If this mate preference is inherited (i.e., the learning gene is "off"), then the ideal mate location is simply that coded in the individual's preference genes—it uses whatever preferences it inherited. But if this mate preference is to be learned (the learning gene is "on"), then parental imprinting occurs: the ideal mate location for this individual is assigned to be the actual phenotypic location of one of its parents. The two offspring created for every pair of parents each imprint on a different parent. Once this ideal mate location in phenotype space is decided, either by inheritance or learning, the choosiness gene tells just how far away from this ideal the individual would be willing to settle in choosing a mate.

The mate preference function maps the phenotypic distance between the individual's ideal mate location and some potential mate onto a probability that the individual would actually choose to indulge in this mating opportunity. The closer a potential mate is to the individual's ideal mate location, the better the chance that the individual will choose to mate. Thus we have implemented sexual imprinting so that offspring prefer an exact match to their parents, but can also allow a significant amount of deviation from that ideal, via the inherited choosiness parameter. We use a mate preference function in which the probability of mating falls off linearly with increasing distance between (1) the phenotypic location of the

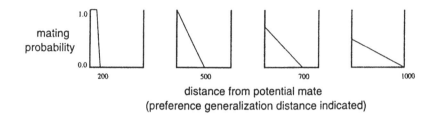

distance from potential mate
(preference generalization distance indicated)

FIGURE 1   The area-normalized mating probability functions corresponding to four different inherited preference generalization distances, showing how the maximum mating probability drops with larger preference generalization distances to maintain a roughly equal total mating probability under each of the four curves. From Todd and Miller, 1993 © MIT Press, 1993.

ideal preferred mate (as genetically specified or imprinted) and (2) the phenotypic location of the actual potential mate under consideration. Beyond the "critical generalization distance" determined by the choosiness gene, this probability of mating goes to zero (see Figure 1).

Implementing sexual selection with mating preferences is fairly straightforward. To create the next generation, a "mom" individual is first picked at random from the current population. (If natural selection were operating in this scenario, differential fitnesses would be imposed here to alter the probability of selecting different moms; but since our fitness landscape is flat, all potential moms are picked with equal probability.) Since it takes two to tango, a "dad" is next selected from the population in the same way. The phenotypic distance between each parent's ideal mate location and the other parent's actual location is computed (using the Euclidean metric in the two-dimensional toroidal phenotype space), and each individual figures its probability of mating with the other based on this distance and its individual mate preference function. The two probabilities are likely to be different: for example, the mom might like the dad more than the reverse, either because he is closer to her ideal than vice versa, or because she is less choosy than he, or both. This situation is shown in Figure 2, where the mom's preferences include the location of the potential mate dad, but the mom herself falls outside the dad's generalization region around his own ideal mate preference.

To accommodate this possible difference of opinion between the two potential parents, we multiply their probabilities together to yield a single probability that this pair will mate. Multiplying the two probabilities gives equal weight to each individual's choice in the matter—it is equivalent to each deciding independently (given their own probability) whether or not they will accept the other as a mate, and then mating only if they both want to (i.e., both accept each other). Mated pairs of individuals are crossed over in the usual genetic algorithm manner (using two-point crossover), and the resulting offspring are put into the next generation.

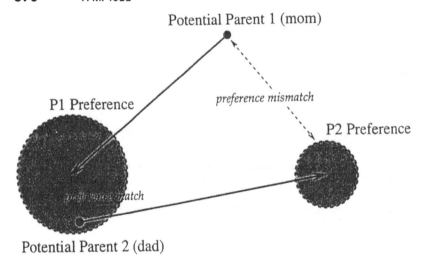

FIGURE 2   Two individuals and their mate preferences. Here potential parent 1 ("mom")
would like to mate with potential parent 2 ("dad"), because he lies within her mate
preference generalization region, but potential parent 2 will reject 1 because she is
outside his generalization region. From Todd and Miller, 1993 © MIT Press, 1993.

If the selected pair fails to mate, then a new dad is again chosen at random (with
replacement), and the process is repeated, until the initially selected mom finds
a mate. Once the mom mates, a new mom is picked (with replacement), and the
search for a suitable dad begins again. This process, with a mom-selection loop
wrapped around an inner dad-selection loop, continues until the next generation
is filled, that is, until 50 pairs have mated and produced 100 new offspring. If a
mom cannot find a suitable mate after five times through the population—i.e.,
after 500 mating attempts for our population size of 100—she is deemed hopelessly
picky, and a new mom is chosen. (Remember that there are no assigned sexes in
the population, so any individual can be a selected "mom," and given a mom, any
of the 100 individuals in the population—including another copy of herself!—can
then be chosen as a potential "dad" to see if this pair will mate.) Thus only those
individuals who are sexually selected will be genetically represented in the next
generation.

    We use traditional bit-wise mutation (with a mutation rate of 0.01 unless oth-
erwise stated) and two-point crossover. The crossover rate is 1.0—the mom and
dad always cross over—because obligate sexual recombination is called for with our
sexual selection scheme. To allow fairly fine-grained structure to emerge in the phenotype space, we use 15 bits to encode each trait and preference gene, 30 bits for
the choosiness gene, and a single bit for the learning gene, yielding a total genotype

length of 91 bits. (Because mutation and genetic drift can obviously have a substantial effect on the one-bit learning ability gene, we were careful to run controlled experiments with and without sexual selection, to disentangle the effects of drift from the effects of selection, as described below.) The population size for all runs in this paper is 100 individuals.

Since natural populations usually form a cluster somewhere in phenotype space, restricted to a small range of the total possible phenotypic variation, it would be inappropriate here to start out with a random initial population uniformly distributed throughout the entire phenotype space (the default for most GA applications). Instead, we give the initial population some phenotypic elbow room to spread into by starting it as a small random cluster in the middle of a much larger space of potential phenotypes. The entire space, as we mentioned earlier, is a 1000-by-1000 grid; initial trait genes are constrained to code for x and y phenotype traits in a range from 450 to 550, and the initial choosiness genes were constrained to code for generalization distances in a range from 100 to 200, to closely cover this phenotype range. This initial population (and all later generations) can be graphically displayed in a square grid with a dot at each individual's (x,y) phenotype position, an arrow leading from that location to the individual's ideal mate location, and a circle centered at the ideal with a radius equal to the individual's generalization distance, as shown in Figure 2 for just two individuals. In Figure 3 we show such a representation for every individual in Generation 0, with the centered square of initial phenotypic positions overlaid by the arrows and overlapping halos of the population's generalization regions (the arrows are not visible in this figure, but can be seen clearly in later figures in this paper).

SIMULATION RESULTS. Given this simulation setting, will sexual imprinting actually evolve in a population that begins without imprinters? To test this, we initialize a population of nonlearners (all having the learning gene set to "off") and start it cycling through generations. The learning gene evolves in the standard way through mutation and sexual selection, and determines whether the mate preferences are inherited or imprinted, as described earlier. Now if we monitor the frequency of the imprinting allele (i.e., individuals whose learning gene is "on") as the population evolves, we can see whether sexual imprinting will indeed prove adaptive and spread through the population. As illustrated by the four sample runs shown in Figure 4, the answer is affirmative.[2]

But what makes this imprinting adaptive, so that it evolves so steadily in the population? Surprisingly, part of the action is in the individuals who *never* get to mate—i.e., those individuals who in the "mom" role proved hopelessly picky or unattractive, and failed to reproduce even after 500 mating attempts.

---

[2]Because our learning gene is just a single bit, we had to be sure that the rise in the number of learners in the population was not simply due to genetic drift affecting that one-bit gene. When we tested this hypothesis, we found that imprinting was, in fact, evolving in a rapid and directed manner—see Todd and Miller (1993).

Since the moms are selected at random from the parent population, we would expect the proportion of imprinting moms who reproduce to match the proportion of imprinters in the parent population as a whole—*if* none of the moms failed in their attempts to find a mate. That is, we would expect the *number* of

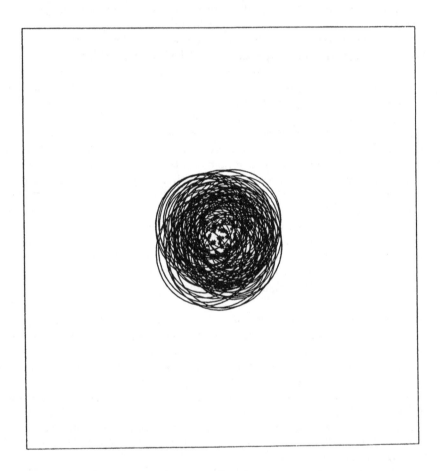

FIGURE 3 Graphic representation of the initial population at generation 0, showing the phenotypic location of each individual (marked with a "∗") and the circular outline of each individual's preference generalization region (i.e., a circle with radius equal to the individual's maximum preference generalization distance). From Todd and Miller, 1993 © MIT Press, 1993.

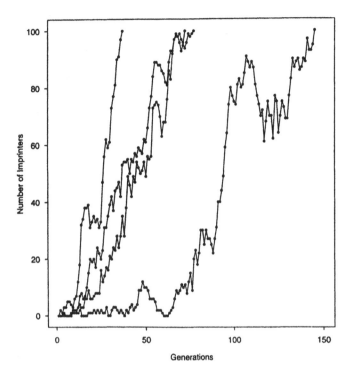

FIGURE 4 Number of sexual imprinters versus generations, showing steady upward evolution of imprinting, for four runs beginning with no imprinters. From Todd and Miller, 1993 © MIT Press, 1993.

imprinting moms in one generation to match *half* the number of imprinters in the previous generation (since moms in mated pairs make up only half the population). But when we compare these two numbers, we find that imprinters are slightly but significantly more common in the set of moms who successfully reproduce than random sampling from the previous generation would have produced, by about 0.3 individuals per generation.

However, the genes from the moms account for only *half* of what gets into the next generation; the selected dad individuals fill out the other half. And the dads, too, add a disproportionate number of imprinters, compared to the previous generation (about 0.6 new imprinters per generation). The genes of these dads get into the next generation based just on how well those dads' own preferences match the phenotypic locations of the moms and how well they match the preferences of the moms, but their imprinting only affects the former. This indicates that imprinters do, in fact, have mate preferences that are better adapted to the evolving population structure, accounting for their ability to take over the population as evolution progresses. Since the sexual selection on individuals in the "mom" role accounts for 0.3 new imprinters per generation, and selection on individuals in the "dad" role accounts for 0.6 new imprinters per generation, we can conclude that imprinting is favored through both channels.

The adaptive function of sexual imprinting may be described as tracking phenotypic variations in a population over time. These variations are caused by the forces of mutation and crossover, creating new distributions of phenotypes from one generation to the next. But if sexual imprinting evolves because it tracks phenotypic change, what happens when the rate of that change varies? We hypothesized that very slow change across successive generations would give imprinting no advantage over inherited preferences, while very rapid change would overwhelm the ability of imprinting to keep up. We realized that here was another case of a U-shaped relationship between the time to evolve a form of learning and the noisiness of the environmental signal that learning relies on—the same sort of relationship we have found earlier in our studies of both the evolution of associative learning (Todd & Miller, 1991a) and sensitization and habituation learning (Todd & Miller, 1991b; see also Littman and Ackley, 1991, for a similar result). In this case, the environmental condition that individuals must track is the current location of the population in phenotype space. The cue that an imprinting individual uses as an indication of that information is the phenotypic location of its own parent(s). And the noisiness of that cue corresponds to the mutation rate of the evolving population. High mutation rates tend to disturb the correlation between cue (parental phenotype) and environmental condition (population average phenotype), because each parent is more likely to have been mutated away from the population average. Thus, by manipulating the mutation rate, we manipulate the correlation between the perceptual cue and the environmental condition on which learning depends. Based on the same logic used in our previous studies of the evolution of learning (Todd & Miller, 1991a, 1991b), we hypothesized that imprinting would prove most useful, and evolve most quickly, at intermediate mutation rates.

The plot in Figure 5 shows that our suspicions were correct. Here we have controlled the effective rate of phenotypic change in the population by varying the mutation rate, from 0.0 to 0.1. The values plotted for comparison at the far right actually come from the random-walk case where the learning gene is not used and so is subject only to drift. We also seed the initial population differently for all of the runs included in this figure; rather than beginning with a population containing no imprinters, we start with a population containing half imprinters (50) and half nonimprinters (50). If we started with no imprinting individuals, the evolution of imprinters would be artificially impeded by the slow introduction of imprinting alleles at the very low mutation rates. We graph the number of generations taken to reach 100% imprinters in the population, or 999 generations if the population has not reached 100% by then. As expected, the average number of generations to evolve imprinting (indicated by the solid line) follows a U-shaped curve, growing large toward the extremes of high and low variation (mutation rate), and small for midaling values. (Note that a longer time to evolve imprinting signifies a lower adaptive value for learning; conversely, where the time to evolve imprinting is lowest, at the intermediate mutation rates, its adaptive value is apparently highest.) Thus, we have found once again that the rate at which a learning mechanism evolves can be affected by the "noisiness" of the environmental cue (or cues) that it employs, even in the case of a sexually selected form of learning.

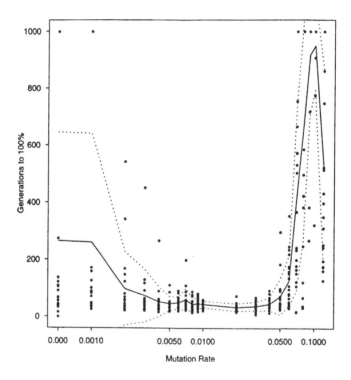

FIGURE 5 Number of generations to fixation of sexual imprinting, versus mutation rate. Each dot represents a single run out of 20 total runs; mean values are indicated by the solid line; and standard deviations around the mean are indicated by the dashed lines. The mutation rate varies from 0.0 to 0.1, in steps of .001 from 0.0 to 0.01, and in steps of .01 from 0.01 to 0.1. The x-axis is transformed logarithmically (but with mutation rate 0.0 added at the far left) to allow the small mutation rates to be seen more distinctly. The values for the random-walk case (no selection on the learning gene) are shown at the far right. From Todd and Miller, 1993 © MIT Press, 1993.

3. EFFECTS OF SEXUALLY SELECTED LEARNING ON EVOLUTION

The hierarchy of adaptive processes described in section 2.1 is most often talked about in terms of a one-way upward flow of control, from the most basic underlying process of evolution, to the next higher level of development, on upward to learning, and finally to behavior at the top end of the chain of command. This direction of flow is summarized in, and reinforced by, Weismann's doctrine and its cousin, the Central Dogma of molecular biology (Eldredge, 1986, p. 359). The former idea states that cells in the germ line are unaffected by what happens to cells in the soma during an organism's lifetime: changes in the body cannot cause (directed) changes in the gametes. The latter dogma is that information flows from DNA to RNA to proteins, and never the reverse: again, the genetic blueprint cannot be changed (in a directed manner). Together, both statements refute the inheritance of acquired characteristics; the evolved (inherited) genetic information store controls what develops and what is learned, but the reverse path is blocked.

At least directly. Indirectly, though, there are a variety of mechanisms of "downward causation" (Plotkin, 1988; Campbell, 1974), through which the higher-up adaptive processes can affect the courses of those lower down the chain of command. The clearest of these reverse-control effects is the "can't get there from here" constraint: the phenotype-building (developmental) machinery already in place in a genetic line essentially prevents evolution from suddenly exploring wildly different design options (e.g., wheels). This is because of the low likelihood of making successful sweeping alterations, all in one go, in a process aimed down a particular ontogenetic path. Thus, development exerts downward causation on evolution and affects its course, channeling its tinkering along lines similar to those it has explored before.

Of particular interest to us here is the realization that learning can also exert downward causation on the course of evolution. This is most widely described with regard to naturally selected forms of learning in terms of the *Baldwin effect* (Baldwin, 1896; Hinton & Nowlan, 1987; Belew, 1990; Schull, 1990). Here, the presence of learning ability in a population helps it to *genetically* climb up peaks in the adaptive landscape (Wright, 1932) more rapidly than evolution alone could achieve. Thus learning "guides" evolution to faster adaptation, by allowing a population to do some of the exploration of the adaptive landscape within the lifetime of the organisms. The Weismann-Central Dogma notions are not violated—the information acquired by learning during an individual's lifetime does not change its genetic information, but rather just makes it more likely that those with a genetic makeup that already puts them nearer the adaptive peak will be fitter and pass on more copies of those genes. But this downward causation does allow a kind of end-run around the blocking of the Weismann doctrine, letting acquired characteristics more strongly affect what is inherited.

With sexually selected forms of learning, the Weismann doctrine is faced with an even more direct challenge. Mate preference learning in particular is not only sexually selected, but also itself helps *produce* sexual selection through the choices it promotes. Mate preference learning in some sense allows an individual to say, "That's the kind of characteristics I want in *my* kids," and to choose a mate who'll provide genes more likely to yield those characteristics. So characteristics to be passed on genetically *are* in this sense acquired, not through personal soma change affecting one's own genes, but through recruitment of the genes of others. Mate preference learning thus comes face to face with Weismann's concerns; but, as Plotkin (1988, p. 153) assures us, "The Weismann maxim, of course, is never violated by any form of learning. What learning may do is change the frequency of genes in a gene pool. It never alters the genes qualitatively."

But sexually selected learning can still have a pronounced effect on the course of evolution. Here downward causation occurs even more strongly than in the Baldwin effect; rather than indirectly affecting the naturally selecting culling process of who's where on the adaptive peak, mate preference learning *directly* affects the sexually selecting choice process of who gets to mate with whom. The choosing of mates by learning individuals can provide a more direct influence on evolution than the competition for survival between learning individuals. With mate preference learning "the learning mechanism 'sees more clearly and directly' to the genetic level; in other forms of learning the 'vision' is obscured by a multitude of other phenotypic factors that are also contributing to individual fitness" (Plotkin, 1988, p. 153). Thus sexually selected forms of learning allow a much tighter loop to be formed with evolution in causing genetic change. As a result, major macroevolutionary effects are much more likely to emerge, as we will see next, returning to our simulation model of the evolution of sexual imprinting.

3.1 SEXUAL IMPRINTING AND SPECIATION

We saw in section 2.2 that sexual imprinting learning can evolve in a population to facilitate individual mate selection. But sexual imprinting can have more than just adaptive effects at the individual level—it can also have important effects on the structure and evolution of the population as a whole. In particular, sexual imprinting of mate preferences can often lead to speciation, the splitting of the population into phenotypically isolated subpopulations between which mating rarely occurs. Speciation is one of the most important of macroevolutionary phenomena, leading to the variety of life we find on our planet today. That sexually selected individual learning can have an effect on such a major population-level process clearly shows the importance of this path of downward causation between learning and evolution in the genetic information realm.

The sexual selection model presented here demonstrates the viability of *sympatric* speciation, the splitting apart of a geographically coherent lineage into reproductively isolated populations without the intervention of spatial barriers to mating (sympatric means "same fatherland"). The most widely accepted theory of speciation, Mayr's (1942, 1947) *allopatric* ("other fatherland") model, denies the likelihood of such sympatric speciation: geographic features such as mountains and rivers (or just being on different geographic margins of a parent population) are seen as the primary forces that split populations apart. Dobzhansky (1937) advocated the possibility of sympatric speciation, but proposed that it would rely on an analogous splitting force: a low-fitness valley separating populations on neighboring high-fitness peaks in an adaptive landscape. In contrast to both of these speciation mechanisms, our model of sympatric speciation (originally described in Todd and Miller, 1991c), requires neither spatial restrictions on mating, nor any natural selection at all: speciation happens on a geographically undifferentiated flat fitness landscape.

In our model, a species appears as a cluster of individuals in phenotype space, all having mate preferences centered on or near that cluster. We begin our simulations with a single initial species, as shown in Figure 3. From there, different things can happen to the population, depending on the types of mate preferences implemented. First, if we only allow inherited mate preferences, then the population tends to be fairly stable. It may drift slightly from its initial central position, and may spit out mutant individuals from time to time, but in general it will remain a single species, sometimes getting even more tightly clustered in phenotype space (provided the mutation rate is not too high—here and in the following run we use a mutation rate of 0.005). This phenomenon can be seen in the view of a typical population over several generations in Figure 6, which shows the locations of the individuals in the population along with their preferences and generalization regions. In generation 10 we see the initial cluster, with one individual's choosiness gene mutated to create a very large generalization distance (the large circle extending off the edges of the plot). This central cluster remains intact for the 700 generations we have shown here, growing a little tighter over time as evolution discovers that choosier individuals can do better at finding mates. At generation 50, we see a couple of individuals spat out from the central cluster, but with preferences pointing back into that cluster, and we also see one individual inside the cluster whose preference has mutated to lie outside. All three of these individuals will be unable to find acceptable or willing mates, and so will not have offspring in the next generation. At generation 200, a few more individuals and preferences have mutated away from the majority of the population. There is even the beginning of a small potential species at the far right edge, but this is short-lived (though another attempted species reappears at this location in generation 500). Overall, though, the original central species holds sway, shifting slightly "eastward" by generation 700, but remaining largely stationary and coherent. Evolved mate preferences can thus be interpreted as a somewhat conservative force, tending to keep the population together phenotypically by dampening population movement and splitting.

Learned mate preferences, sexually imprinted from the parents, lead to quite different results. If we look at a population with all imprinted preferences, we see relatively frequent speciation events, as shown in Figure 7.

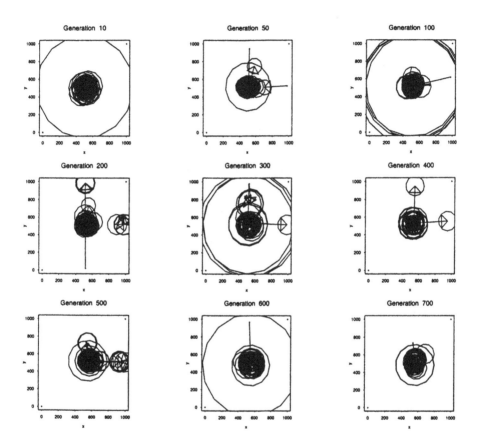

FIGURE 6   Population dynamics with evolved (inherited) mate preferences over 700 generations, showing in each instance the phenotypic location of each individual, an arrow extending from the individual's phenotypic location to its phenotypic preference (most arrows are not visible due to crowding, however), and a circular preference region around each phenotypic preference, with size (radius) corresponding to the individual's preference generalization distance. From Todd and Miller, 1993 © MIT Press, 1993.

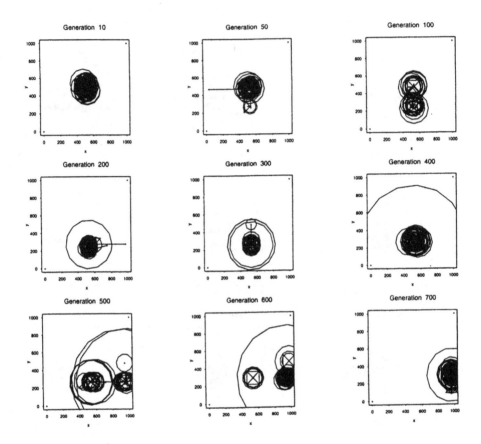

FIGURE 7 Population dynamics with learned (imprinted) mate preferences over 700 generations, using the same display conventions as Figure 6. From Todd and Miller, 1993 © MIT Press, 1993.

Here we can see species splitting off at generations 100 and 500, with only one species winning out and surviving in between, probably because of our fixed population size. In addition, the population as a whole moves around in phenotype space considerably more than in the inherited-preference run in Figure 6. Finally, the species clusters are for the most part more tightly clumped than with inherited preferences, because of greater evolved individual choosiness. This probably allows smaller changes in the distribution of population phenotypes and preferences to lead to the isolation and splitting off of new species, compared to the changes necessary to overcome the cohesive force of inherited preferences and cause speciation in that case.

Preliminary analysis of many simulation runs indicates that speciation is approximately three times more common in imprinting populations than in nonimprinting ones. This suggests that sexual imprinting functions at the population level as a more fluid, less conservative adaptive force, keeping single populations in cohesive clusters that are more susceptible to being split by the random jiggling forces of mutation and crossover. Once the newly split species begin to diverge in phenotype space, sexual imprinting will again tend to consolidate each cluster, keeping individual mate preferences local so that little interbreeding will occur between the subpopulations. The clusters will then continue to drift further apart until they are reproductively well isolated, effectively constituting new species.

Based mainly on empirical findings rather than simulation results, Immelmann (1972) has proposed that the main function of sexual imprinting may be, in fact, to facilitate speciation. Immelmann has stated that "imprinting may be of special advantage in any rapidly evolving group, as well as wherever several closely related and similar species occur in the same region [i.e., sympatric situations]. Interestingly enough, both statements really do seem to apply to all groups of birds in which imprinting has been found to be a widespread phenomenon..." (Immelmann 1972, p. 167). Indeed, if we combine the finding that populations composed entirely of imprinters tend to speciate relatively frequently, with the result from section 2.2 that imprinting evolves and spreads through our populations, we can conclude that the tendency to speciate itself evolves in our model. In this case, speciation emerges as a side effect following the evolution of individually adaptive, parent-imprinted mate preferences. But in nature, if Immelmann is correct, the ability to speciate could sometimes prove so adaptive that it imposes higher-level (e.g., species, genus...) selection for imprinting. Certainly the individual-level advantage of imprinting for tuning up mate preferences will generally outweigh higher-level pressures, but selection among more frequently speciating lineages is an intriguing possibility.

3.2 OTHER FORMS OF SEXUALLY SELECTED LEARNING AND THEIR EVOLUTIONARY EFFECTS

To consider what other types of sexually selected learning might exist in the service of mate choice, we should first elaborate the functions of mate choice in general. The initial cut that mate choice mechanisms should make in dividing the reproductively suitable from the unsuitable is selection of the proper species, and of the proper sex. Selection of mates based on genetic viability relative to oneself is important, allowing individuals to avoid deleterious inbreeding and the danger of recessive lethals, and deleterious outbreeding and the breakup of co-adapted gene complexes. Absolute genetic quality is also valuable, as assessed by direct viability indicators or indirectly through the preferences of others. Economic quality can also be a basis of choice, as individuals select among prospective mates with different short-term nuptial provisioning behaviors and long-term parental investment abilities. The social relationship to potential mates can matter as well: neophobia can lead to pair-bonding and long-term monogamous relationships, while neophilia can result in promiscuous behavior as exemplified by the Coolidge effect. These forms of mate choice are listed roughly by increasing rate of change of the underlying information on which preferences will be based, from the stable, long-term indicators of a particular species, to the highly variable identity of a promiscuous individual's latest mate. The kinds of learning appropriate to each type of mate choice function will vary as a result of this rate of information change, as we discussed in section 2.1.

Long-term imprinting learning seems the most natural mechanism to use for selection of mates of the appropriate species and sex (if indeed any form of learning is used for this aspect of mate choice in a particular species). For inbreeding and outbreeding avoidance, imprinting on parents and siblings is also reasonable. Learning the appearance and identity of kin that are born later in one's life could also be useful to maintain the proper balance of inbreeding and outbreeding, and learning one's *own* appearance could be adaptive for choosing others who share the same sorts of traits, to avoid breaking up one's own co-adapted sets of genes. To help select mates with good absolute genetic quality, an individual could learn the correlations between different traits and the viabilities of those who bear them. Alternatively, individuals can learn the preferences of others and copy these in choosing a mate; imprinting on one's parents may actually serve this purpose, since each parent reflects (at least to some extent) the choice criteria of the other, and thus can be used as a preference template. Assessing the economic quality of individuals via their nuptial gifts can involve learning the current distribution of gifts from all potential suitors, so that those bearing the best gifts can be given top billing. Judging long-term parental investment ability may involve learning about the types of territories and resources various potential mates control, and again comparing this to the (learned) range of possibilities. Finally, pair-bonding will likely involve learning the appearance, call, scent, etc. of one's long-term mate for ongoing identification purposes. Promiscuity, on the other hand, will involve at least short-term memory for the characteristics of one's most recent mate(s), and probably habituation to old mates and sensitization or dishabituation to new ones (Havens & Rose, 1992).

The possible evolutionary effects of these different kinds of mate preference learning are varied as well. Imprinting on stable, species-level characteristics of appropriate mates is likely to lead to stabilizing selection and relatively little evolutionary movement of the population. Imprinting on more variable traits, such as those peculiar to one's parents, can lead to population movement in phenotype space and the generation of new species, as we saw in section 3.1. Other trait preferences, as opposed to individual preferences, could similarly lead to phenotypic change and speciation. Learning through repeated environmental experience, as opposed to one-shot imprinting learning, can generally increase the accuracy of the mate preferences it affects, since information can be averaged and combined over multiple trials. This can result in more directed, less noisy (or stochastic) evolution of the traits involved, in contrast to the often highly perturbed path of lone natural selection (see Miller and Todd, in press, for further discussion of this point). Certain general features of the learning processes involved in mate preference learning could also have effects on trait evolution. For instance, the peak-shift phenomenon often observed in associative learning, in which individuals exhibit the greatest response to stimuli that are *more* positive than the positive training examples they saw, could help lead to moderate sexual dimorphism via peak-shifted mate preferences (Weary, Guilford, & Weisman, 1993). Finally, if the kinds of preferences that are learned are directional rather than absolute—that is, if they are for differences in traits relative to the individual, rather than for specific values of the traits themselves—then runaway selection processes can be started, resulting in evolving populations shooting off in phenotype space (ten Cate & Bateson, 1988; Miller & Todd, 1993; Kirkpatrick, 1987).

This brief list of some of the functions of mate choice mechanisms, some of the types of sexually selected learning that can bear on those choice mechanisms, and some of the evolutionary effects that can result from the learned preferences, is not meant to be exhaustive. It is merely illustrative of the further range of possibilities involved in sexually selected forms of learning, beyond the single case of sexual imprinting we have explored here in detail. In future work, we intend to consider others of these mate preference learning mechanisms in similar detail, to explore further the conditions under which they can evolve, and the effects they can have in turn on the course of evolution.

4. CONCLUSIONS

Learning can be adaptive not only for energy and resource capitalization in the economic realm of survival, but also for exploiting genetic resources in the genealogical realm of reproduction. Sexually selected forms of learning for use in mate choice are a particularly clear example of adaptive processes that gather and exploit this genetic information. Mate preference learning evolves when learned preferences give individuals an advantage in finding appropriate mates, as we saw in our simulation of the evolution of parental imprinting learning. Furthermore, the presence of such preference learning can affect the course of evolution itself in a direct way: "mate-choice learning presents us with the most direct example of learning's being potentially able to alter the composition of the gene pool" (Plotkin, 1988, p. 153). The resulting macroevolutionary effects include population-wide phenotypic change and the eventual splitting of the population into new species. Sexually selected learning can thus exhibit a more powerful form of the Baldwin effect, previously described only in terms of naturally selected forms of learning.

There are a variety of reasons that more attention should be focused on the phenomena of sexually selected learning. The fact that this kind of learning can have a marked impact on the evolutionary process itself makes it significant to those interested in macroevolutionary phenomena and the interactions of learning and evolution. Sexually selected forms of learning are likely to be quite widespread in nature, because the sorts of traits that they track—sexually selected characteristics, the identities and relationships of particular individuals, the average characteristics of populations and subpopulations—are generally highly variable and rapidly changing. Thus evolution alone will be unlikely to keep up with the rate of alteration of these traits, and learning will prove an adaptive addition. Sexual selection in general is currently a hot topic in biology, after decades of neglect, and the opportunities to add to our understanding in this area, not least in the area of learning, are great. And because of the complicated dynamic nature of the interactions between evolution and learning, in the sexual selection realm in particular, computer simulation models may be one of the best ways to push forward theoretically.

Finally, after starting by considering learning in the economic realm, and then exploring the sexually selected functions of learning in the genealogical realm, we may come full circle: some forms of learning that originally evolve in the service of mate choice may be co-opted by natural selection for functions in the economic realm again. Sexual selection might be a source of innovation not only morphologically and behaviorally (Miller & Todd, in press), but in the sphere of learning mechanisms as well. For instance, mechanisms that originally evolved for use in memorizing the location and distribution of potential mates (Gaulin & Fitzgerald, 1989) could be further adapted for use in finding food in patchy environments. If such an evolutionary relationship is possible, then the importance of studying sexually selected forms of learning may be further increased, as the source of some forms of naturally selected learning as well. Of course, the reverse may also hold; and so, in the end, we must balance our attention between both domains, and explore the evolution of plastic individuals and their learning mechanisms in the economic *and* the genealogical realms, via natural *and* sexual selection.

ACKNOWLEDGMENTS

The ideas in this paper were developed in collusion with Geoffrey Miller. The author takes responsibility for any errors.

REFERENCES

Ackley, D., and M. Littman. "Interactions Between Evolution and Learning." In *Artificial Life II*, edited by C. G. Langton, C. Taylor, J. D. Farmer, and S. Rasmussen, 487–509. Redwood City, CA: Addison-Wesley, 1992.

Aoki, K. "A Sexual-Selection Model for the Evolution of Imitative Learning of Song in Polygynous Birds." *Am. Naturalist* **134** (1989): 599–612.

Baldwin, J. M. "A New Factor in Evolution." *Am. Naturalist* **30** (1896): 441–451.

Bateson, P. "Sexual Imprinting and Optimal Outbreeding." *Nature* **273** (1978): 659–660.

Bateson, P. "Optimal Outbreeding." In *Mate Choice*, edited by P. Bateson. Cambridge: Cambridge University Press, 1983.

Belew, R. K. "Evolution, Learning, and Culture: Computational Metaphors for Adaptive Algorithms." *Complex Systems* **4** (1990): 11–49. Also Technical Report CS89-156, Computer Science and Engineering Department, University of California at San Diego, La Jolla, CA.

Campbell, D. T. "'Downward Causation' in Hierarchically Organized Biological Systems." In *Studies in the Philosophy of Biology*, edited by F. Ayala and T. Dobzhansky. London: Macmillan, 1974.

Cooch, G. "Ecological Aspects of the Blue Snow Goose Complex." *Auk* **78** (1961): 72–89.

Cooke, F., and C. M. McNally. "Mate Selection and Colour Preferences in Lesser Snow Geese." *Behaviour* **53** (1975): 151–170.

Daly, M., and M. Wilson. *Sex, Evolution, and Behavior*, 2nd ed. Belmont, CA: Wadsworth, 1983.

Darwin, C. R. *The Descent of Man and Selection in Relation to Sex*. London: Murray, 1871.

Davey, G. *Ecological Learning Theory*. London: Routledge, 1989.

Dewsbury, D. A. "Effects of Novelty on Copulatory Behavior: The Coolidge Effect and Related Phenomena." *Psychol. Bull.* **89** (1981): 464–482.

Dobzhansky, T. *Genetics and the Origin of Species*. New York: Columbia University Press, 1937.

Dugatkin, L. A., and J.-G. J. Godin. "Reversal of Female Mate Choice by Copying in the Guppy (*Poecilia reticulata*)." *Proc. Roy. Soc. London* Series B (Biological Sciences) **249(1325)** (1992): 179–184.

Eldredge, N. "Information, Economics, and Evolution." *Ann. Rev. Ecology & Systematics* **17** (1986): 351–369.

Fisher, H. *Anatomy of Love: The Natural History of Monogamy, Adultery, and Divorce*. New York: Simon & Schuster, 1992.

Fisher, R. A. *The Genetical Theory of Natural Selection*. Oxford: Clarendon Press, 1930. (Reprinted 1958; New York: Dover.)

Gallistel, C. R. *The Organization of Learning*. Cambridge, MA: MIT Press/Bradford Books, 1990.

Gaulin, S. J. C., and R. W. Fitzgerald. "Sexual Selection for Spatial-Learning Ability." *Animal Behav.* **37(2)** (1989): 322–331.

Hamilton, W. D. "The Genetical Evolution of Social Behavior, I and II." *J. Theor. Biol.* **7** (1964): 1–32.

Havens, M. D., and J. D. Rose. "Investigation of Familiar and Novel Chemosensory Stimuli by Golden Hamsters: Effects of Castration and Testosterone Replacement." *Hormones & Behav.* **26(4)** (1992): 505–511.

Hinton, G. E., and S. J. Nowlan. "How Learning Guides Evolution." *Complex Systems* **1** (1987): 495–502.

Immelmann, K. "Sexual and Other Long-Term Aspects of Imprinting in Birds and Other Species." In *Advances in the Study of Behavior*, Vol. 4, edited by D. S. Lehrman, R. A. Hinde, and E. Shaw. New York: Academic Press, 1972.

Immelmann, K. "The Evolutionary Significance of Early Experience." In *Function and Evolution in Behavior: Essays in Honour of Professor Niko Tinbergen*, edited by G. Baerends, C. Beer, and A. Manning. Oxford, UK: Clarendon Press, 1975.

Kirkpatrick, M. "The Evolutionary Forces Acting on Female Preferences in Polygynous Animals." In *Sexual Selection: Testing the Alternatives*, edited by J. W. Bradbury and M. B. Andersson, 67–82. New York: Wiley, 1987.

Lande, R. "Models of Speciation by Sexual Selection on Polygenic Traits." *Proc. Natl. Acad. Sci. USA* **78** (1981): 3721–3725.

Lester, G. L. L., and B. B. Gorzalka. "Effect of Novel and Familiar Mating Partners on the Duration of Sexual Receptivity in the Female Hamster." *Behav. & Neural Biol.* **49(3)** (1988): 398–405.

Littman, M. L., and D. H. Ackley. "Adaptation in Constant Utility Non-stationary Environments." In *Proceedings of the Fourth International Conference on Genetic Algorithms*, edited by R. K. Belew and L. B. Booker, 136–142. San Mateo, CA: Morgan Kaufmann, 1991.

Mackintosh, N. J. *The Psychology of Animal Learning*. London: Academic Press, 1974.

Mayr, E. *Systematics and the Origin of Species*. New York: Columbia University Press, 1942.

Mayr, E. "Ecological Factors in Speciation." *Evolution* **1** (1947): 263–288.

Miller, G. F., and P. M. Todd. "Exploring Adaptive Agency I: Theory and Methods for Simulating the Evolution of Learning." In *Proceedings of the 1990 Connectionist Models Summer School*, edited by D. S. Touretzky, J. L. Elman, T. J. Sejnowski, and G. E. Hinton, 65–80. San Mateo, CA: Morgan Kaufmann.

Miller, G. F., and P. M. Todd. "Evolutionary Wanderlust: Sexual Selection With Directional Mate Preferences." In *From Animals to Animats 2: Proceedings of the Second International Conference on Simulation of Adaptive Behavior*, edited by J.-A. Meyer, H. L. Roitblat, and S. W. Wilson, 21–30. Cambridge, MA: MIT Press/Bradford Books, 1993.

Miller, G. F., and P. M. Todd. "The Role of Mate Choice in Biocomputation: Sexual Selection as a Process of Search, Optimization, and Diversification."

In *Evolution as a Computational Process*, edited by W. Banzhaf and F. H. Eeckman. Berlin: Springer-Verlag, in press.

Plotkin, H. C. "Learning and Evolution." In *The Role of Behavior in Evolution*, edited by H. C. Plotkin, 133–164. Cambridge, MA: MIT Press/Bradford Books, 1988.

Plotkin, H. C., and F. J. Odling-Smee. "Learning, Change, and Evolution: An Enquiry Into the Teleonomy of Learning." *Adv. Stud. Behav.* **10** (1979): 1–41.

Pruett-Jones, S. "Independent Versus Non-independent Mate Choice: Do Females Copy Each Other?" *Am. Naturalist* **140** (1992): 1000–1009.

Schull, J. "Are Species Intelligent?" *Behav. & Brain Sci.* **13** (1990): 63–108.

ten Cate, C., and P. Bateson. "Sexual Selection: The Evolution of Conspicuous Characteristics in Birds by Means of Imprinting." *Evolution* **42(6)** (1988): 1355–1358.

Thorpe, W. H. *Learning and Instinct in Animals.* Cambridge, MA: Harvard University Press, 1956.

Thorpe, W. H. *Learning and Instinct in Animals*, 2nd ed. Cambridge, MA: Harvard University Press, 1963.

Todd, P. M., and G. F. Miller. "Exploring Adaptive Agency II: Simulating the Evolution of Associative Learning." In *From Animals to Animats: Proceedings of the First International Conference on Simulation of Adaptive Behavior* edited by J.-A. Meyer and S. W. Wilson, 306–315. Cambridge, MA: MIT Press/Bradford Books, 1991a.

Todd, P. M., and G. F. Miller. "Exploring Adaptive Agency III: Simulating the Evolution of Habituation and Sensitization." In *Proceedings of the First International Conference on Parallel Problem Solving from Nature*, edited by H.-P. Schwefel and R. Maenner, 307–313. Berlin: Springer-Verlag, 1991b.

Todd, P. M., and G. F. Miller. "On the Sympatric Origin of Species: Mercurial Mating in the Quicksilver Model." In *Proceedings of the Fourth International Conference on Genetic Algorithms*, edited by R. K. Belew and L. B. Booker, 547–554. San Mateo, CA: Morgan Kaufmann, 1991c.

Todd, P. M., and G. F. Miller. "Parental Guidance Suggested: How Parental Imprinting Evolves Through Sexual Selection as an Adaptive Learning Mechanism." *Adap. Behav.* **2(1)** (1993): 5–47.

Walker, S. *Animal Learning, an Introduction.* London: Routledge and Kegan Paul, 1987.

Weary, D. M., T. C. Guilford, and R. G. Weisman. "A Product of Discriminative Learning May Lead to Female Preferences for Elaborate Males." *Evolution* **47(1)** (1993): 333–336.

Wright, S. "The Roles of Mutation, Inbreeding, Crossbreeding, and Selection in Evolution." *Proc. Sixth Int. Cong. Genetics* **1** (1932): 356–366.

Zahavi, A. "Mate Selection: A Selection for a Handicap." *J. Theor. Biol.* **53** (1975): 205–214.

Melanie Mitchell

Preface to Chapter 22

Like Hinton and Nowlan's classic paper "How Learning Guides Evolution," the following chapter by Miglino, Nolfi, and Parisi describes a simple "idea model" that nonetheless makes an important point about evolution. The point is that, while genetic material is the substrate of evolutionary changes, organisms possess many levels of organization, and the repercussions of evolutionary change are felt in different ways by different levels.

Miglino, Nolfi, and Parisi study these differences via a computer model in which organisms can be described at four different levels: genetic, neural, behavioral, and fitness, with each giving rise to the next higher level. Adaptive change is that which leads to higher fitness; ultimately fitness is the only criterion for selection. Changes at other levels sometime give rise to higher fitness, sometimes give rise to lower fitness, and sometimes have no effect on fitness at all (i.e., are "neutral"). There are a number of ways in which neutral changes can occur. For example, a mutation in the "nonfunctional" part of the genetic level (e.g., in a gene that is never expressed in creating the neural level) will have no effect on the eventual fitness of the mutated organism; likewise a mutation that changes the neural level in a way that is not related to its function will have no effect on fitness.

The purpose of Miglino, Nolfi, and Parisi's model is to study, in a simple system, how evolution proceeds at the different levels. In this way, their model is more realistic than the "flat" (single-level) model of Hinton and Nowlan. The reason for using a computer model is, of course, that such an undertaking would be very difficult with real organisms. For example, the mappings from genes to nervous-system structures to even the simplest behaviors is only beginning to be understood. Also, the vastly different time scales of real-world evolution at the different levels makes it nearly impossible to perform satisfactory controlled experiments. In a computer model, the mappings can be laid out by the programmer, and many simulations can be run, allowing a study of all the levels simultaneously. Of course, this and all the models described in this book face the charge of being too simplistic and not capturing enough of the real world to be meaningful. Answering this charge is the ever-present task faced by would-be modelers.

In Miglino, Nolfi, and Parisi's model, simple simulated organisms controlled by neural networks perform a food-searching task in a simple environment. An organism's genetic level consists of a bit string that encodes developmental instructions for a neural network. The neural level consists of the particular neural network created by the instructions in the genome—it maps sensory inputs to motor responses. The behavioral level is the organism's movement in the environment in response to stimuli. The fitness level is the amount of food eaten over the organism's lifetime. Each level rests upon the immediately lower level, and each lower level has "functional" and "nonfunctional" components. Only the functional components give rise to the next level.

The authors let such systems evolve via mutation at the genetic level and selection on the fitness level from a population of organisms that were allowed to navigate individually for a number of time steps. This process produced highly fit organisms within one thousand or so generations. The authors then traced the "lineages" of the most highly fit organisms over many runs, looking at differences between parents and offspring at each level over evolutionary time. They found that evolution across these lineages consists of many nonadaptive yet retained changes at each level; the lower the level, the higher the percentage of retained changes that were nonadaptive. In fact, at the lower levels the vast majority of retained changes were nonadaptive. However, some of these nonadaptive changes were used later on (via additional mutations) to produce new structures that were adaptive. The authors interpret this as a demonstration of the argument made by Stephen Jay Gould and other evolutionists against "ultra-Darwinists" who believe that change is gradual and continuous, and that, under natural selection, all retained changes are adaptive in some way. Gould and others have argued that many, if not most retained changes are, in fact, nonadaptive, at least at the time at which they first occur.

To many readers of this book, Miglino, Nolfi, and Parisi's demonstration may seem to be merely preaching to the choir. They are making a point that may seem obvious in retrospect: the mappings from lower levels to higher levels are typically many-to-many, so there is a lot of room in the space of, say, genes in which evolution can "tweak things at random" with impunity. Eventually, such tweaking is bound to happen on a state from which actual adaptive mutations can be had. Most retained changes are a result of such neutral tweaking. But making this (possibly) obvious point in a computer model opens the possibility of understanding these various adaptive and nonadaptive processes in more detail. If we are to understand evolution in nature (or in our computer simulations) and to have any means of predicting its course, we must understand the effects it has on various levels of organization in organisms. Some questions that could potentially be addressed by computer models are: What is the relative importance of adaptive versus nonadaptive changes at different levels? What are the mechanisms at different levels underlying both types of change? What are the mechanisms for retaining nonadaptive changes? What exactly is the role of hierarchy in evolution? What are the relative rates of changes, the relative rates at which different levels evolve? Does evolution naturally lead to states of higher "evolvability" (i.e., potential for adaptive change)? Answering such questions will, I think, also be essential for making sense of the vast quantities of data coming from genetic sequencing efforts such as the Human Genome project. Miglino, Nolfi, and Parisi's model is a first stab at addressing such questions.

The model as it stands involves no learning or plasticity during the lifetimes of organisms. However, questions about the effect of evolution on different levels of organization are very relevant to questions about the interaction of individual plasticity and evolution. The major theme of this book is that, while learning and other forms of phenotypic plasticity occur on different organizational levels and at different rates than genetic evolutionary change, the two processes nonetheless can have (and have had) profound effects on each another. One possible symbiotic effect is suggested by Miglino, Nolfi, and Parisi's results. The fact that higher levels of organization are often immune to genetic mutations is often a good thing for a species, since it confers a degree of stability to higher levels. However, in rapidly changing or not-easily-predictable environments, this stability can be a drawback, since it limits the rate at which changes can be made at the higher, behaviorally significant levels. Learning (or, more generally, phenotypic plasticity) is a mechanism that acts directly on a higher level of organization, and thus allows an organism to change quickly at these higher levels in response to changes in the environment for which genetic evolution alone would be too slow to track.

Orazio Miglino, Stefano Nolfi, and Domenico Parisi

Chapter 22:
Discontinuity in Evolution: How Different Levels of Organization Imply Preadaptation

1. INTRODUCTION

Although sequences of nucleotide bases in DNA and amino acids in proteins appear to mutate at approximately constant rates, evolution seems to have been anything but steady. In fact, most dramatic changes in evolution appear to have occurred rather abruptly (Eldredge and Gould; 1972; Gould and Eldredge, 1993).

Gould (1991; see also Gould and Vrba, 1982) claims that, in order to explain such radical changes, one should abandon the "adaptationist" or "ultra-Darwinist" perspective according to which all phenotypic traits are selected because they have a well-defined function and they increase the fitness of the individual exhibiting them. The idea that all change is gradual and continuous is a consequence of interpreting phenotypic changes as primarily adaptive. Gould proposed the classification of phenotypic traits in three different categories: *adaptations* (traits that have a well-defined function), *preadaptations* (traits built for one function and then adapted to another), and *exaptations* (traits not built as adaptations at all but later adapted for some function), and he claims that the last two categories must greatly exceed adaptations in number and importance.

Adaptive Individuals in Evolving Populations, Ed. R. K. Belew & M. Mitchell,
SFI Studies in the Sciences of Complexity, Vol. XXVI, Addison-Wesley, 1996   **399**

Of course, no biologist has ever advocated a complete equivalence between retained changes and adaptations. Adaptive changes in some trait may entail correlated but not necessarily functional alterations of other traits (e.g., the human chin appears to be a by-product of functional changes in other parts of the human face during hominization). The architecture of genetic and embryological systems defines "channels" of possible change. Selection may be required to push an organism down a channel but the channel itself, though not necessarily an adaptation, acts as a major determinant of the direction taken by evolution. The real questions then are the relative importance of adaptive versus adaptively neutral changes and a better understanding of the mechanisms underlying both types of change.

Using a computational model we will try to show that adaptively neutral evolutionary changes, i.e., changes that are not correlated with a change in fitness, may largely outnumber adaptive ones. This fact appears mainly due to the hierarchical organization of organisms, to the internal complexity of each level of organization, and to the complex nonlinear mapping between levels. We will also show how some of these adaptively neutral changes may subsequently become the basis for further changes that do prove adaptive and, therefore, how exaptive phenomena may arise producing sudden evolutionary changes in the behavior of organisms.

2. DESCRIBING ORGANISMS AT MULTIPLE LEVELS

It is obvious that biological systems can and should be described at various hierarchical levels. For example, an organism can be described at the genetic level, at the neural level, at the behavioral level, and at the level of fitness. A description of the organism at the genetic level is a description of the genotype of the organism, that is, of the genetic material that the organism has inherited from its parent(s) and that directs the construction of the phenotypic organism. A description at the neural level is a description of the nervous system of the organism, which is just one aspect of the phenotypic organism. A description at the behavioral level is a description of the behavior that is exhibited by the organism, given its nervous system, in a particular environment. Finally, a description at the level of fitness is a measurement of the fitness obtained by the organism given its behavior in that environment.

A critical role in our understanding of organisms is played by how these different levels interact and how changes at one level are related to changes at other levels. However, a multilevel approach to the study of real organisms is very difficult to realize given the different conceptual and methodological tools traditionally used by geneticists, neuroscientists, behavioral scientists, and evolutionary biologists. As a consequence, organisms tend to be studied separately at these different levels and by scientists who belong to different disciplines.

Computational models of evolution (Holland, 1975; Langton, 1992) allow us to analyze the same system, for example a simulated organism, at various levels simultaneously and to investigate how changes at one level are related to changes at other levels. As we will show in the next section one can simulate within the same experiment the genotype, the nervous system, the behavior, and the environment of a particular artificial organism and one can examine what evolutionary changes at one of these different levels accompany changes at other levels.

3. THE MODEL

We developed artificial organisms (Orgs) that perform a simple navigation task in a simulated environment (Treves, Miglino, and Parisi, 1992). To each Org corresponds a string of genetic material or genotype which specifies a set of developmental instructions. These instructions generate a certain number of neurons and control the growth and branching process of the axons of the neurons (Nolfi and Parisi, in press). The result of this growing process is a neural network that represents the nervous system of the Org. The architecture and connection weights of such a network determine the way in which the Org responds to environmental stimuli, i.e., the Org's behavior. Such behavior, through interaction with the environment to which the Org is exposed, determines its fitness, i.e., the Org's reproductive chances.

3.1 GENETIC MODEL

An Org's genotype is represented as a string of 0s and 1s (see Figure 1).

The string has a fixed length and is divided up into blocks (see Figure 2), each block corresponding to a single neuron that may or may not get expressed during the development process. Each block contains instructions that determine the developmental properties of the corresponding neuron (see Figure 3). There are

```
0110010100110011001101010100111
1101010100010100011100100101101
1100010000111000110010101000110
1001100100001101010111010100110
1001101110101011101011110001101
1100100110010100011100........11
```

FIGURE 1 An Org's genotype.

genotype   genes

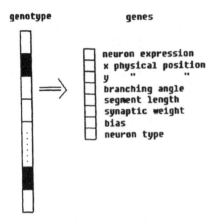

neuron expression
x physical position
y   "   "
branching angle
segment length
synaptic weight
bias
neuron type

FIGURE 2 Developmental instructions specified in an Org's genotype. Inactive blocks which correspond to unexpressed neurons are represented as black cells while active blocks are represented as empty cells.

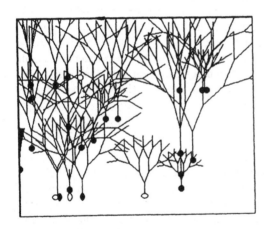

FIGURE 3 Developmental growth of neural axons resulting from an evolved genetic string.

three different types of neurons: sensory neurons, internal neurons, and motor neurons. Genotypes are 40 blocks in length; i.e., the nervous system of each Org can include a maximum number of 40 neurons. The first 8 blocks correspond to sensory neurons, the last 5 blocks to motor neurons, and the 27 intermediate blocks to internal neurons. The total genome length is 1600 bits.

3.2 DEVELOPMENTAL MODEL

Each block specifies the following instructions ("genes") (see Figure 2):

a. The "neuron expression gene" determines if the other instructions contained in the block will be executed or not, i.e., if the corresponding neuron will be present or not in the Org's nervous system.

b. The two "physical position genes" specify the Cartesian coordinates of the neuron in the bidimensional nervous system of the corresponding Org.

c. The "branching angle gene" and the "segment length gene" determine the angle of branching of the neuron's axon and the length of the branching segments.

d. The "synaptic weight gene" determines the synaptic weights of all the connections that will be established by the neuron. (All connections departing from the same neuron have the same weight.)

e. The "bias gene" determines the activation bias of the neuron.

f. The "neuron type gene" determines if a sensory neuron codifies the angle (relative to the Org's direction) or the distance of one of the two fixed landmarks present in the environment and if a motor neuron codifies the first or the second motor neuron (i.e., the first or the second bit of the binary representation of the four possible motor actions that can be performed by the Org). This implies that each Org can have four different types of sensory neurons and two different types of motor neurons. More neurons of the same type may be present. When output neurons of the same type are present, the actual motor response is computed by averaging the activation levels of the corresponding neurons.

The result of the execution of the genotypic instructions of an evolved Org is shown in Figure 3.

Neurons of different types and in different parts of the Org's nervous system are created and connections between neurons are established through the growth process of neurons' axons. When the growing axonal branch of a particular neuron reaches another neuron, a connection between the two neurons is established. The resulting neural network is shown in Figure 4. In Figure 5 is shown the functional part of the same network, i.e., the same network after isolated (nonfunctional) neurons and groups of interconnected neurons have been removed. Notice that only the functional neural network determines how the corresponding Org responds to environmental input.

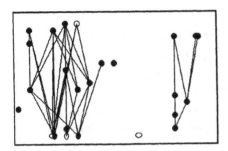

FIGURE 4 Connections established during the growth process shown in Figure 3. The bottom layer contains sensory neurons, the upper layer motor neurons, and the remaining layers internal neurons.

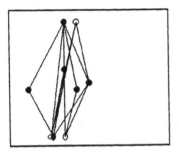

FIGURE 5 Resulting functional network after elimination of isolated (nonfunctional) neurons and groups of interconnected neurons.

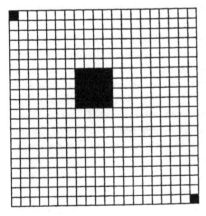

FIGURE 6 Environment. The two black cells represent the two landmarks and the shaded square in the center of the environment represents the target area of 4×4 cells that must be reached by an Org in order to increase its fitness. The environment is actually a bidimensional space of infinite size; i.e., Orgs can go outside the 20×20 environment displayed in the figure. Yet, the starting position of Orgs is fixed at the edge of the 20×20 environment.

3.3 NEURAL MODEL

Through interaction with the environment, the functional network determines an Org's behavior. At each time step, the Org receives some activation value in the sensory neurons, depending on its position in the environment and its direction. Such an input determines in turn, through a spreading activation process, the activation value of the internal and output neurons. These last neurons determine the Org's motor reaction to the current input stimulus, i.e., the Org's behavior. Internal and output neurons have a Heavyside activation function.

3.4 BEHAVIORAL MODEL

Each Org lives in a simulated environment which is a two-dimensional square divided up into 20 × 20 cells (see Figure 6). The environment contains a small central target area of 4 × 4 cells. Each Org has a facing direction and a rudimentary sensory system that allows it to receive as input the angle (relative to where the Org is currently facing) and the distance of two fixed landmarks situated in the environment. Each Org is also equipped with a simple motor system that provides it with the possibility, at each time step, to move a cell forward, to turn left or right, or to remain still. We decided that in order to increase its fitness an Org should use the sensory information from the two landmarks to reach the target area and remain there.

It is useful to distinguish between the potential behavior and the actual behavior of an Org. The way in which an Org reacts to all possible input stimuli is the Org's potential behavior while the way in which the Org reacts to the stimuli it actually experiences during its life is the Org's actual behavior. In ecological networks (cf. Parisi, Cecconi, and Nolfi, 1990) the stimuli an Org experiences during its life are partially determined by the Org itself through its motor behavior. (In Figure 7 we show that a typical Org actually visits most of the world's cells in some of the four possible orientations only). As a consequence, only the way in which an Org reacts to a given (self-selected) subset of all potential input stimuli has a role in determining that Org's fitness (Nolfi and Parisi, 1993).

While the actual behavior of an Org is determined by observing how the Org spontaneously behaves in its environment, its potential behavior can be determined by testing it in artificial conditions, i.e., by exposing it to all possible stimuli that can be experienced in the 20 × 20 environment (see Figure 8).

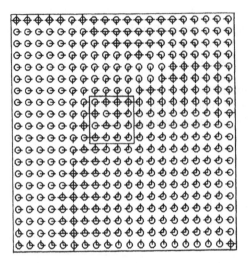

FIGURE 7 Actual behavior of an evolved Org. We show with which orientations an evolved Org visits each cell of the environment. The trace is the result of the Org's natural movements after being placed in all the peripheral cells of the world. As can be seen, very few cells are visited in all four possible orientations.

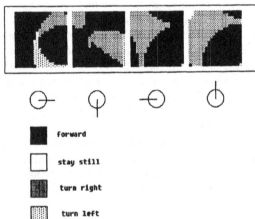

- forward
- stay still
- turn right
- turn left

FIGURE 8 Potential behavior of the same Org of Figure 7. Each of the four pictures shows the reactions of this Org in all the world cells for each of four possible orientations.

3.5 HIERARCHICAL RELATION AMONG LEVELS

The overall picture (see Figure 9) is a system organized in four hierarchical levels: genotype, nervous system, behavior, and fitness. The genotype determines the nervous system that in turn determines the behavior that in turn determines the fitness. The mapping from one level to the next higher level is complex and nonlinear. It is not the case that each feature of a lower level corresponds to a single feature of the next higher level and vice versa. The mapping is not one-to-one but many-to-many. For example, many "genes" (see Figure 2) determine the existence of a single neural connection and, vice versa, a single "gene" can have a role in determining the existence of many connections. Or, many internal neurons may contribute to determining the activation level of a single output neuron and, vice versa, a single internal neuron can determine the activation level of more than one output neuron.

Another important feature of our model is that at each of the first three levels (genotype, nervous system, behavior) a functional component can be distinguished from a nonfunctional component. At the genetic level, only the blocks that are actually expressed, i.e., that have their expression gene set to 1, are functional in determining the nervous system (see Figure 2). Similarly, at the nervous system level, only the neurons that are interconnected and contribute to determining the motor responses to environmental stimuli are functional in determining the Org's behavior (see Figure 4 and 5). And finally, only the motor responses that an Org produces in response to stimuli it actually experiences during its lifetime are functional in

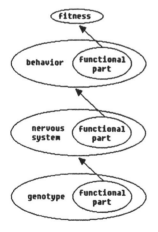

FIGURE 9 Hierarchical organization of an Org. The functional component of the genotype is the subpart of the genotype that is expressed and that determines the nervous system. The functional component of the nervous system is the subpart of the developed neural structure that contributes to determining the Org's motor reactions. The functional component of behavior represents the motor responses to the stimuli that the Org actually experiences during its lifetime.

determining the Org's fitness (see Figure 7 and 8). In other words, it is only the functional component of each level that determines both the functional and the nonfunctional component of the next higher level.

The nonlinear mapping between levels and the existence of nonfunctional components at all levels imply that individuals that have the same structure at some level can differ at some lower level of their hierarchical organization for two reasons: (a) they can have two different functional components at one level that result in the same structure at the next higher level; (b) they can have different nonfunctional components at one level because, by definition, these differences do not affect the higher levels.

4. RESULTS OF SIMULATIONS

We have run a set of ten simulations using the model described above. Each simulation begins in generation 0 (G0) with 100 randomly generated genotypes resulting in 100 Orgs with different neural and behavioral characteristics. An Org's nervous system does not develop during its "life" in the environment; in other words, the development of the nervous system is supposed to be instantaneous. (For a variant of the present model in which genotypes map into phenotypes in time, see Nolfi and Parisi, 1995). G0 networks are allowed to "live" for a total of 2400 actions divided into 80 epochs. Each epoch consists of 30 actions starting from a randomly chosen cell located at the edge of the 20 × 20 environment. Orgs are placed in individual copies in the environment; i.e., they live in isolation. The fitness criterion is chosen to favor the selection of individuals that reach the central 4 × 4 target area as quickly as possible and then remain inside the target area. Fitness is calculated by counting +10 points for each cycle spent in the 4 × 4 target area, +1 point for each movement forward, and −1 point for each cycle spent outside the 20 × 20 cell environment.

At the end of their lives (2400 actions) the 20 Orgs that have accumulated the most fitness are allowed to reproduce by generating 5 copies of their genotypes. The new set of 20 × 5 Orgs is the next generation (G1). Twenty mutations are applied during the copying process (crossover is not used). For each mutation a randomly chosen bit of the genetic string is replaced by a new bit value. Since there is a 50% chance that the mutation event will not cause a change, the expected number of actual changes is 10.

After 1400 generations, Orgs are able to reach the target area efficiently starting from almost any peripheral cell and to remain inside the target area after having reached it.

In order to analyze in detail evolutionary change at various levels of organization we have concentrated on a subset of all the individuals constituting a population. If we take the best individual of the last generation we can reconstruct the entire lineage of this individual until the single originator of the lineage in the first generation is reached. The lineage is constituted by a total of 1400 individuals, one for each generation, i.e., by the best individual of the last generation, by its (single) parent in the preceding generation, by its grandparent, its grand grandparent, etc. We call this lineage the "winning lineage." Since the individual representing the winning lineage in one generation is the offspring of the individual representing the lineage in the preceding generation, the winning lineage is composed of a total of 1399 parent/offspring pairs.

If we look at how fitness changes from parent to offspring in the winning lineage across generations by testing all the individuals of the lineage in the same environmental conditions (we place all the individuals in all the peripheral cells of the environment once), we obtain an interesting pattern: periods of stasis, which can last hundreds of generations, are followed by rapid increases in fitness (see Figure 10).

We then compared the two members of each of the 1399 parent/offspring pairs of the winning lineage at the different levels of organization of these organisms. More specifically, we measured the percentage of parent/offspring pairs in which the offspring differs from its parent in each of the descriptive dimensions we have identified. What we found is that the percentage of changes decreases when one goes up from lower to higher levels of organization. In Figure 11 is shown the percentage of cases in which an offspring differs from its parent in (G) its genotype, (FG) its functional genotype, (N) its overall neural network, (FN) its functional neural network, (B) its potential behavior, (AB) its actual behavior, and (F) its fitness. More than 99% of the individuals that make up the winning lineage differ from their parents at the level of the functional part of their genotype, i.e., the portion of the genotype that is expressed in the phenotype. On the other hand, only 50% of these same individuals differ from their parents at the functional nervous system level, 20% at the potential behavior level, and only 10% at the fitness level. This implies that most of the genetic mutations do not affect the fitness level and, as a consequence, are adaptively neutral.

Mutations that do not affect the fitness level do not have any role from the perspective of the individual's reproductive chances. However, these adaptively neutral mutations produce a great number of changes at the lower levels and this can have long-term consequences for the evolutionary process. In order to understand the role of mutations that do not affect the fitness level, we can examine the changes which occur in Orgs at different levels of their organization during an evolutionary period in which the fitness level is stable. If fitness is not observed to increase generation after generation, by definition no adaptive mutations arise, and then changes in Orgs reflect only the effect of neutrally adaptive mutations.

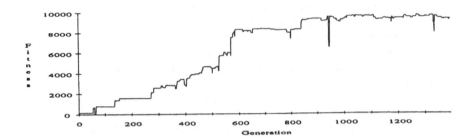

FIGURE 10 Fitness of the 1400 individuals constituting the winning lineage in a typical run tested in the same environmental conditions. Orgs are placed once in each of the 76 peripheral cells of the 20 × 20 cell environment facing the target area. The total number of startings is 80 because Orgs are placed in each of the four corner cells with two different orientations.

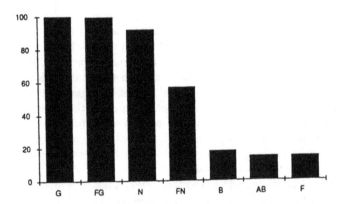

FIGURE 11 Percentage of parent/offspring pairs in the winning lineage in which the offspring differs from its parent at various levels of organization: (G) genotype, (FG) functional genotype, (N) overall nervous system, (FN) functional nervous system, (B) potential behavior, (AB) actual behavior and (F) fitness. Each black bar represents the average result of ten different simulations.

In the simulation of Figure 10 we have a quite long stretch of evolutionary time (from generation 154 to generation 272) in which the fitness level of the winning lineage does not change at all (cf. Figure 10). During this period of 118 generations, which is stable from the point of view of fitness, 54 of the 118 Orgs (remember that a single individual represents the winning lineage in each generation) differ at the functional neural level and 17 Orgs differ at the level of their potential behavior (as measured in artificial conditions; cf. above) with respect to their parents. In Figure 12 is shown the functional neural network and the potential behavior of eight of these Orgs that most significantly differ from their parents. (The Orgs represent Generations 154, 168, 188, 190, 225, 234, 246, and 257, respectively.) The first Org (representing Generation 273) in which an adaptive mutation occurred and that, as a consequence, obtained a higher fitness value than its 118 predecessors, is also shown.

If we examine the behavior of these eight Orgs (leaving aside for the moment the Org of Generation 273), we observe some changes in how they respond to the input in some areas of the environment (more precisely, in the NW area) but these changes in behavior do not lead to any change in fitness. These Orgs oscillate between two alternative solutions (e.g., the behavior of the Org representing Generation 154 is different from the behavior of the Org representing Generation 168 but it is almost identical to the behavior of the Org representing Generation 257) but these solutions are equivalent from the point of view of fitness.

At the level of the functional nervous system, on the other hand, we see an interesting process going on. If we look at the neural architectures of the two Orgs representing Generations 154 and 257, respectively, we see that these architectures are very different despite the fact that they generate the same behavior. However, if we look at the architecture of the Org of Generation 273, i.e., the first Org that received an adaptive mutation, we observe that its architecture is almost identical (except for a single connection) to an architecture already present in the population since Generation 246. This means that the architecture of the Org of Generation 273 was preselected or preadapted at least 27 generations before. (In Gould's terminology [Gould, 1991] we should use the term "exaptation" instead of "preadaptation" because this is an example of a trait not built as an adaptation but later adapted to a function. However, we prefer to use the term "preadaptation" in its more general sense which includes both preadaptation and exaptation phenomena.).

This preadapted architecture was not obtained in any purposeful way. It was not selected against other architectures because it allowed some further advantage in adaptive terms. It was selected and retained by chance until it was able to generate a more fit Org (in generation 273) through a single or a few mutations. Furthermore, the fact that more fit Orgs did not arise for a long period of more than 100 generations implies that adaptive mutations in these Orgs were very improbable. Therefore, the adaptively neutral changes that affected the nervous system level were crucial in determining the successive adaptive ones. Without such neutral changes it might have been impossible for the evolutionary process to generate more adaptive Orgs.

5. DISCUSSION

Biological organisms can be described at various hierarchical levels, for example, at the genetic level, at the neural level, at the behavioral level, and at the fitness level. Each level determines the structure of the organism at the successive level but not all changes that occur at a lower level cause corresponding changes at higher levels. Each level maps into the next higher level in complex ways. Therefore, a change at a lower level does not necessarily translate into some change at the next higher level. Furthermore, since only part of the structure at each level (the functional part) is responsible for determining the structure at successive levels, only changes that affect these functional parts can cause changes at successive levels.

If we examine the lineage of the best individual of the last generation in an evolutionary simulation with growing neural networks and we compare the two Orgs in each parent/offspring pair, we see that different levels change at different rates. While almost all offspring differ from their parents at the genetic level (because of mutations), the probability that the two members of a pair differ at higher levels gradually decreases as one ascends levels. This implies that most of the mutations do not affect the fitness level and, as a consequence, should be considered nonadaptive or neutral. However, these mutations may produce changes at other levels and this can have long-term consequences for the evolutionary process.

Since most mutations do not affect fitness and may be maintained in the population for purely chance reasons with no adaptive advantage, evolution is, to a significant extent, neutral. These mutations and the changes they determine at levels lower than the fitness level determine the future course of evolution, but are initially selectively neutral.

Some neutral changes, as we have shown in our simulations, may subsequently become the basis for further changes that do prove adaptive. Hence, preadaptation phenomena may arise. The population may turn out to be preadapted to these further adaptive changes. These preadaptation phenomena may explain the discontinuous and abrupt changes observed in our simulations. Neutral mutations can accumulate in the nonfunctional components of a particular level without affecting in any way the next higher level. However, due to some new mutation a portion of the changed nonfunctional component of the lower level can become suddenly functional and, as a consequence, visible at the next higher level.

We have shown that higher levels of organization have a high degree of stability in that they are mostly immune to genetic mutations. This fact can become a drawback in rapidly changing environments. Having the chance to directly affect higher levels of organization (for example, the neural and behavioral level through learning [Nolfi, Elman, and Parisi, 1994]), phenotypic plasticity can be seen as a way to overcome this problem. In order to obtain phenotypic plasticity, we modified

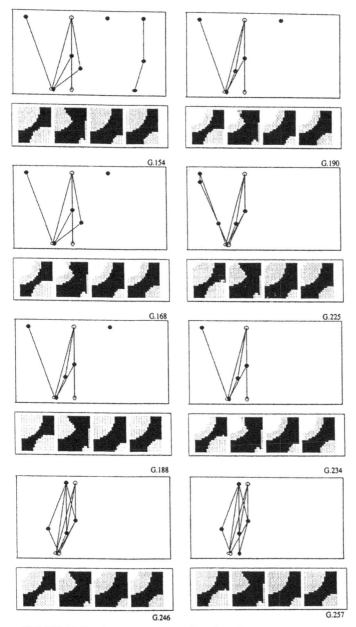

FIGURE 12 Caption on next page. (continued)

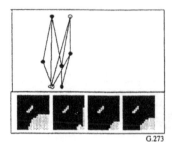

FIGURE 12 (continued) Functional neural architecture and potential behavior (see Figure 7) of 8 Orgs of the successful lineage in a phase of the evolutionary process in which fitness is stable. The ninth Org is the first Org in which an adaptive mutation caused a sudden increase in fitness after this stable period. The number at the bottom of each picture represents the generation of the corresponding Org.

the model described in this paper by allowing the genotype to determine the development of the nervous system during an Org's lifetime (Nolfi and Parisi, 1995) and by making the developmental process sensitive to the external environment (Nolfi, Miglino, and Parisi, 1994). We believe that this is a very promising direction of research for simulations with evolutionary neural networks. Our hope is that such extensions will help us to shed light on a fundamental question: how and in which conditions has phenotypic plasticity evolved?

ACKNOWLEDGMENTS

We would like to thank Rik Belew, Melanie Mitchell, and Charles Taylor for their useful comments.

REFERENCES

Eldredge, N., and S. J. Gould. "Punctuated Equilibria: An Alternative to Phyletic Gradualism." In *Models in Paleobiology*, edited by T. J. Schopf. San Francisco: Freeman, 1972.

Gould, S. J., and E. S. Vrba. "Exaptation—a Missing Term in the Science of Form." *Paleobiology* **8** (1982): 4–15.

Gould, S. J. "Exaptation: A Crucial Tool for an Evolutionary Psychology." *J. Soc. Issues* **3** (1991): 43–65.

Gould, S. J., and N. Eldredge. "Punctuate Equilibrium Comes of Age." *Nature* **366** (1993): 227–233.

Holland, J. J. *Adaptation in Natural and Artificial Systems.* Ann Arbor, MI: University of Michigan Press, 1975.

Langton, C. G. "Artificial Life." In *1991 Lectures in Complex Systems*, edited by L. Nadel and D. L. Stein, 189–242. Reading, MA: Addison-Wesley, 1992.

Nolfi, S., J. Elman, and D. Parisi. "Learning and Evolution in Neural Networks." *Adaptive Behavior* **2** (1994): 5–28.

Nolfi, S., O. Miglino, and D. Parisi. "Phenotypic Plasticity in Evolving Neural Networks." In *Proceedings of the International Conference From Perception to Action*, edited by D. P. Gaussier, and J-D. Nicoud, 146–157. Los Alamitos, CA: IEEE Press, 1994.

Nolfi, S., and D. Parisi. "Genotypes for Neural Networks." In *The Handbook of Brain Theory and Neural Networks*, edited by M. A. Arbib. Cambridge, MA: MIT Press, in press.

Nolfi, S. and D. Parisi. "Self-Selection of Input Stimuli for Improving Performance." In *Neural Networks in Robotics*, edited by G. A. Bekey. Boston, MA: Kluwer Academic, 1993.

Nolfi, S., and D. Parisi. "Evolving Neural Networks that Develop in Time." Technical Report, Institute of Psychology C.N.R., Rome, 1995.

Parisi, D., F. Cecconi, and S. Nolfi. "Econets: Neural Networks that Learn in an Environment." *Network* **1** (1990): 149–168.

Treves, A., O. Miglino, and D. Parisi. "Rats, Nets, Maps, and the Emergence of Place Cells." *Psychobiology* **1** (1992): 1–8.

William E. Hart

Preface to Chapter 23

The following chapter provides an overview of some recent experiments that examine the relationship between evolution and learning. These experiments were motivated, in part, by the work of Hinton and Nowlan (Chapter 25) which lends plausibility to the notion that there may be indirect causal influences of learning on evolution. The authors examine the evolutionary performance of a population of organisms that gather food. Each organism collects food from a different environment, and organisms reproduce according to the amount of food that they gather. In addition, the organisms may perform a learning task that involves predicting how the organism's motor actions change the position of food in the environment. To evaluate the effect of the learning task, the authors compare the evolutionary performance of populations that learn to that of populations that do not learn. They observe that the average number of food elements gathered by the organisms is greater for populations of organisms that learn.

The authors emphasize three differences between their experiments and those of Hinton and Nowlan. The authors' experiments use a learning method that makes deterministic modifications to the organism, while the learning method used by Hinton and Nowlan is stochastic (i.e., each individual learns by making random guesses). Second, the fitness of an organism is not solely a function of its genotype, since the environment in which an organism gathers food can affect the fitness of

Adaptive Individuals in Evolving Populations, Ed. R. K. Belew & M. Mitchell,
SFI Studies in the Sciences of Complexity, Vol. XXVI, Addison-Wesley, 1996   **417**

the organism. Since organisms collect food from different environments, the environment is not a constant factor in the fitness function. Third, the learning task that they consider is not directly related to the evolutionary task. The evolutionary performance of organisms depends on their ability to gather food, which is not necessarily improved by their performance in the prediction task. Hinton and Nowlan's experiments use a learning task that improves the fitness of organisms directly.

These differences are important because they lend biological plausibility to the results reported by the authors. The learning task considered by Hinton and Nowlan is directly related to the fitness of an organism. Organisms in real biological systems are constrained to perform certain tasks like food collection that are directly related to their fitness. However, Parisi and Nolfi's experiments illustrate how a learning task that is *not* directly related to an organism's fitness can be modeled. Further, their experiments indicate that performing a learning task that is indirectly related to an organisms fitness can still be used to guide evolution.

This conclusion also has implications for evolutionary algorithms used to solve engineering problems. Consider the process of an organism's food-gathering as a function of the initial conditions specified by the organism's genotype and the initial environment. From this perspective, indirect learning can be equated with learning algorithms that use a crude estimate of the direction of the local optimum. For example, suppose learning is performed using a gradient-based local search method. The authors' results suggest that coarse approximations to the gradient may provide enough directionality to allow learning to successfully guide evolution. This conclusion is similar to the results obtained by Hart and Belew (27), which show that stochastic, nongradient local search algorithms can be used to guide evolution.

Domenico Parisi and Stefano Nolfi

Chapter 23:
The Influence of Learning on Evolution

1. EVOLUTION AND LEARNING

Evolution and learning are two different ways in which the behavior, and other traits, of organisms can change. Evolution is change at the population level. Organisms reproduce selectively and subject to mechanisms (mutation, sexual recombination) which maintain inter-individual variability. This causes changes in the population from one generation to the next. Learning, on the other hand, is change at the individual level. By interacting during its life with a specific environment, an organism can change its behavior by incorporating, through its experience, aspects of the environment in its internal structure.

Evolutionary change is generationally cumulative. The changes which occur in a particular generation are superimposed upon changes that have occurred in previous generations. Learning is individually (but not generationally) cumulative. The changes that occur in an individual at a particular time of the individual's life are influenced by changes at preceding times but no changes due to learning are inherited by the individual's offspring.

Adaptive Individuals in Evolving Populations, Ed. R. K. Belew & M. Mitchell,
SFI Studies in the Sciences of Complexity, Vol. XXVI, Addison-Wesley, 1996

2. EVOLUTION'S INFLUENCE ON LEARNING

Although evolution and learning are two distinct kinds of change which occur in two distinct types of entities (populations and individual organisms), they may influence each other. The influence of evolution on learning is not surprising. Evolutionary change leaves its trace in the genotype. Hence, each individual inherits a genome which is the cumulative result at the level of the individual of the past evolutionary changes which have occurred at the level of the population. Since an individual's genome partially specifies the resulting phenotypic individual and it constrains how the individual will behave and what it will learn, the way is open for an influence of evolution on learning.

Simulations applying genetic algorithms to populations of neural networks have shown this to occur. In these simulations a strictly Darwinian framework is adopted: the changes that occur in an individual network as a consequence of learning are not inherited by the network's offspring. (For an example of a Lamarckian inheritance of learned changes, cf. Ackley and Littman, 1994). However, evolution progressively selects for networks that incorporate a predisposition to learn some specific task. Although the networks of later generations do not perform better in the task at birth, i.e., prior to learning, in comparison to networks of earlier generations they do learn more quickly. Therefore, what is genetically inherited is not a congenital ability to perform the task but only a predisposition to learn the task if given the appropriate experience.

A predisposition to learn some particular task can be incorporated in a neural network in a variety of different ways, for example as the network's initial matrix of connection weights or the network's architecture. Evolution may select initial weight matrices or network architectures that cause better learning. This has been shown to happen both in the case where the learning task and the evolutionary task are the same (for weight matrices, cf. Belew, McInerney, and Schraudolph, 1991; for network architectures, cf. Miller, Todd, and Hedge, 1989) and in the case where they are different (for weight matrices, cf. Parisi, Nolfi, and Cecconi, 1991; Nolfi, Elman, and Parisi, in press). The learning task and the evolutionary task are the same if the terminal performance in the learning task is identical with the fitness in terms of which individuals are selected for reproduction. The two tasks are different if the individual learns task A but its fitness is independently evaluated based on its performance on task B.

A predisposition to learn created by evolution can also be expressed in other ways, for example as the inheritance of an appropriate learning rate or momentum for learning some particular task (cf. Belew, McInerney, and Schraudolph, 1991) or, in more ecological conditions, as an inherited tendency to behave in such a way that the individual is exposed to the appropriate learning experiences. For example, assuming that the teaching input for some supervised learning task is only physically available in some specific locations in the environment, networks can be shown to evolve and, therefore, to inherit genetically a tendency to move in the environment in such a way that they tend to approach and stay close to those locations (Denaro and Parisi, 1994).

---

3. LEARNING'S INFLUENCE ON EVOLUTION

The influence of evolution on learning can be easily understood within a Darwinian framework because this framework provides a mechanism through which the influence can be realized: the inherited genotype is simultaneously the result of evolution and a partial cause of the phenotypic individual and its learning tendencies. A Darwinian framework makes the opposite influence, of learning on evolution, much more difficult to explain. Learning is realized as changes in the phenotypic individual but these changes are not inherited. Therefore, learning should have no direct effect on the course of evolution.

This orthodoxy has been challenged by some simple simulations done by Hinton and Nowlan (1987; reprinted in Chapter 27; herafter H&N). They have shown that if evolution must search for a very specific result (genome) among many useless alternative results (genomes), learning can aid this search even if it takes the simple form of random changes during an individual's life. Genomes that are close to the desired genome but which evolution alone would have a hard time finding can be discovered if inherited genomes are subject to random changes during life and if they reproduce as a function of how quickly the random changes transform them into the desired genome. In this way, evolution can converge on the desired genome more quickly than if learning is absent, although it remains true that learned changes are not inherited. If evolution is unaided by learning, its chances of success are restricted to the case that the single desired genome suddenly emerges because of the chance factors operating at reproduction (random mutations and sexual recombination). Learning solves half of the problem and therefore it makes the task of the chance factors much easier.

The H&N simulations have been influential in giving a new plausibility to the idea that, in addition to the more biologically plausible causal influence of evolution on learning, there may be a more indirect causal influence of learning on evolution. However, the H&N simulations have various limitations and it remains to be seen if learning continues to influence evolution when these limitations are removed.

A first limitation of the H&N simulations is that learning is equated with random changes. Changes due to learning appear to be nonrandom; they have a direction, a goal. There is a task to be learned. A second limitation is that in H&N's simulations there is no independent physical environment in which individual organisms live and, therefore, evolve and learn. Evolution and learning are changes which result from the interactions between organisms and environments. Evolution is based on fitness. But fitness is not a property of genotypes; it is a property of genotype/environment pairs (Odling-Smee, 1988).

The third limitation is perhaps the most serious one because it could be interpreted as restricting the significance of the H&N results. In their simulations the evolutionary task and the learning task coincide (see above). What determines the reproductive chances of a particular individual (its fitness) is its performance on the learning task. Individuals are more likely to reproduce if they happen to find a solution to their learning task early in their life. Therefore, one might object that it is this equating of fitness with learning performance that explains why learning can influence evolution in the H&N simulations.

In fact, the identity of evolutionary task and learning task in H&N and other simulations can represent a limitation of these simulations. In real organisms it is not clear that a good learning performance automatically translates into high fitness. Learning tasks, insofar as they are evolved, are likely to converge with fitness but, especially in more advanced organisms, the relationship between the two can be very complex and indirect.

Simulations in which the evolutionary task and the learning task are distinct tasks have also been done (e.g., Nolfi, Elman, and Parisi, in press). In these simulations an attempt is also made to overcome the other two limitations of the H&N simulations. Learning is not random changes but is directed learning (using backpropagation) and the organisms (networks) evolve and learn in a physical (simulated) environment. Hence, these simulations may represent a further and more robust test of an influence of learning on evolution.

4. HOW LEARNING TO PREDICT CAN HAVE AN EFFECT ON EVOLUTION

In these simulations, organisms reproduce as a function of their performance in the task of capturing food present in the environment. This is the evolutionary task. The organisms are modeled by neural networks living in an environment that contains randomly distributed pieces of food. Each network has input units encoding the position of the nearest food element relative to the organism. The network's output units encode various possible motor actions that allow the organism to move and turn. The network architecture is fixed. The problem for evolution is to find weight matrices for this fixed network architecture that cause an organism to respond to sensory information about food location with motor actions allowing the organism to capture food efficiently.

Furthermore, during its life each organism learns to predict how the position of food changes with its motor actions. This is the learning task. The learning task is a conceptually distinct task with respect to the evolutionary task although it may be causally related to fitness. The individual's reproductive chances are determined by its performance in the evolutionary task (capturing food), not in the learning

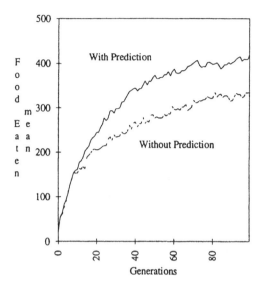

FIGURE 1 Average number of food elements eaten by successive generations of organisms evolved with and without learning during life. Each of the two curves represents the average performance of ten different simulations with different random assignment of weights.

task (predicting food position). In principle, an individual that learns quickly how to predict could have poor fitness and not reproduce, or an individual can be a slow learner but it can have high fitness and reproduce.

We have observed that in these conditions, learning has a positive influence on evolution (for a demonstration of the same effect on other tasks, cf. Cecconi and Parisi, 1991). Populations in which both evolution and learning occur exhibit a better growth of fitness across generations than populations that evolve but do not learn (cf. Figure 1).

In these simulations, reproduction is nonsexual (single parent) and the addition of random variability at reproduction is restricted to mutations. If we assume that the portion of weight space which is explored by learning overlaps significantly with the portion explored by mutations, we can advance the following explanation for the positive influence of learning on evolution. (The explanation is similar to the one given by H&N for their results but is extended to cover the more complex situations studied in our simulations).

In a population which only evolves, the decision regarding which individuals should reproduce can only be based on the fitnesses of genomes. If what is genetically inherited is a matrix of connection weights, each individual genome can be represented as a particular location in weight space which corresponds to a given height on the fitness surface. This height represents the fitness value of the particular genome. (Phenotypes, not genomes, have fitness. However, for simplicity we are identifying genomes and phenotypes.) To determine this value the individual is tested during its life.

The situation is different in a population that both evolves and learns. Learning causes changes in weights. This means that the individual moves in weight space and, therefore, its total fitness is a function of the various fitnesses (heights on the fitness surface) associated with the various locations traversed during learning. Since learning determines this path and the fitness of an individual is a function of the path traversed during learning, this appears to be the mechanism that allows learning to influence evolution.

But why does learning have a *beneficial* effect on evolution? Consider the decision whether a particular individual should reproduce or not. In a population without learning this decision can be based only on the fitness of a single location in weight space, i.e., the location occupied by the genome of the individual. By testing the individual, the fitness associated with this specific location can be known. However, the region surrounding the individual's location remains unknown. In other words, the selective mechanism ignores what fitnesses are associated with the locations that are near the given location.

Consider now two individuals, a and b, which have two rather different genomes and therefore are located in two distant locations in weight space (cf. Figure 2). Assume, however, that the genome of a and the genome of b have the same fitness (case 1); alternatively, assume that the genome of a has a higher fitness than the genome of b (case 2). In a population without learning a and b would have the same reproductive chances (case 1) or a would reproduce rather than b (case 2).

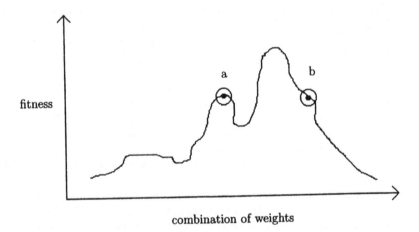

FIGURE 2   Fitnesses of all possible weights matrices. Point b has a better surrounding region than point a even though the fitnesses of the two points are identical. For practical reasons the N dimensions of the weight space are represented as a single dimension.

Because of mutations the offspring of a reproducing individual will not be located in precisely the same location as their parent, but rather somewhere in the region surrounding it. Imagine now that the region surrounding individual b is better than the region surrounding individual a; i.e., it contains more locations with a higher fitness than the region surrounding a. This implies that the offspring of b will have a higher fitness than the offspring of a. Therefore, instead of leaving the decision to chance (case 1) or favoring individual a (case 2), it would be more effective for evolution to select individual b for reproduction.

Learning is a mechanism which allows evolution to find out about the regions in weight space surrounding the location of candidates for reproduction. Learning involves movement in weight space. This implies that the total fitness of an individual which learns will be a function of both (i) the fitness of the starting point (inherited genome), and (ii) the fitnesses of the locations traversed during learning. If we assume that at least some of these locations are situated in the surrounding region explored by mutations, we can see that in populations that both evolve and learn, reproductive decisions can be based on knowledge of surrounding regions. In other words, evolution is based on the fitnesses of the currently living individuals but it would be more effective if it could be based also on the fitnesses of the offspring of these individuals. Learning allows evolution "to look into the future." Therefore, evolution with learning can be more effective than evolution alone.

The argument so far implies that even random movement in weight space during life can benefit evolution. And, in fact, this is what the H&N simulations demonstrate. As a matter of fact, a population of networks which are selected for the ability to capture food and at the same time "learn" during life on the basis of randomly generated teaching input, has a better evolutionary fitness curve than a population with no learning at all (Parisi, Nolfi, and Cecconi, 1991). (The H&N simulations are a pure example of learning as random movement in weight space. Backpropagation learning on the basis of a randomly generated teaching input, as described by Parisi, Nolfi, and Cecconi, 1991, may involve movement which has some directionality.)

However, the effect of learning on evolution is likely to vary as a function of the learning task and how the learning task is related to the evolutionary task. Learning has a tendency to push evolution toward particular kinds of solutions, and different types of learning (i.e., different learning tasks) can have different effects on the evolutionary process.

In order to understand how learning affects the course of evolution, we should consider two different surfaces, the fitness surface that represents the performance of each individual with respect to the evolutionary task and the learning surface that represents the performance of each individual with respect to the learning task. The fitness surface determines which individuals will reproduce while the learning surface determines how individuals move in weight space during their lifetime. But, as we have seen, the way in which individuals move in weight space during their

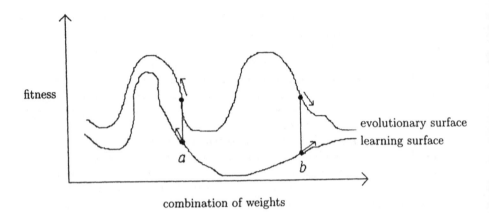

FIGURE 3 Fitness surface for the evolutionary task and performance surface for the learning task for all possible weight matrices. Movements due to learning are represented as arrows. Point a is in a region in which the two surfaces are dynamically correlated. Even if a and b have the same fitness on the evolutionary surface at birth, a has more probability to be selected than b since it is more likely to increase its fitness during life than b.

lifetime affects their fitness. The learning task and the corresponding learning surface determine the nature of the movement and, as a consequence, have an influence on which individuals are selected for reproduction.

Consider two individuals, a and b, which are located in two distant locations in weight space but have the same fitness; i.e., the two locations correspond to the same height on the fitness surface (cf. Figure 3). However, individual a is located in a region in which the fitness surface and the learning surface are dynamically correlated; i.e., a region in which movements that result in increases in height with respect to the learning surface tend to cause increases also with respect to the fitness surface. Individual b, on the other hand, is located in a region in which the two surfaces are not dynamically correlated. If individual b moves in weight space, it will go up in the learning surface but not necessarily in the fitness surface. Because of learning, the two individuals will move during their lifetime in a direction that improves their learning performance, i.e., in a direction in which their height on the learning surface tends to increase. This implies that individual a, which is located in a dynamically correlated region, will end up with a higher fitness than individual b and, therefore, will have more chances to be selected. The final result is that evolution will have a tendency to progressively select individuals which are located in dynamically correlated regions. Learning represents an evolutionary pressure to select individuals which improve their performance with respect to both the learning and the evolutionary task.

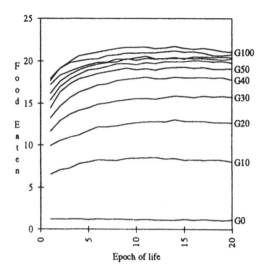

FIGURE 4 Average number of food elements eaten by organisms during their lifetime in successive generations. Each curve represents the average performance of ten different simulations with different random initial assignment of weights. For practical reasons, only generations 0, 10, 20, 30, 40, 50, and 100 are shown.

We have observed this influence of learning on evolution in our simulations with networks that are selected for the ability to collect food elements and at the same time learn during their life how to predict changes in sensory input which are caused by their movements. As we show in Figure 4, while the individuals of generation 0 do not increase their ability to capture food during life, the individuals of successive generations eat more food elements in successive epochs of their life. All individuals are subject to prediction learning during their life and all individuals, including those of generation 0, do learn how to predict. (The error on the prediction task has more or less the same decrease across epochs for all individuals.) However, learning has a beneficial effect on the fitness of an individual only in later generations. Thus, although learning to predict does not by itself increase the food collecting ability (generation 0), it is able to guide evolution so that it progressively selects individuals for which learning does have a beneficial effect on the evolutionary task, i.e., individuals located in regions of dynamic correlation between the evolutionary surface and the learning surface.

Although the direction in which learning guides evolution depends on the particular learning task, some effect may be expected whatever the learning task. For example, if one trains networks to do the XOR task and the networks belong to a population selected for the ability to capture food, one obtains the same result (cf. Parisi, Nolfi, and Cecconi, 1991). After a certain number of generations, individuals that are learning the XOR task show an improvement in their food collecting ability during life which parallels their improved performance on the XOR task; i.e., learning forces evolution to find individuals that are located in dynamically correlated regions. However, the strength of the effect is likely to be a function of the relation between the learning and evolutionary tasks. If the two tasks are correlated, i.e., a weight matrix which has a high (low) value on one surface tends to have a high value (low) on the other surface, the effects will be stronger. In the extreme case in which the learning task and the evolutionary task are the same, their performance surfaces will coincide and, therefore, by definition, an increase in performance with respect to one surface will correspond to an increase in performance with respect to the other surface.

REFERENCES

Ackley, D. E., and M. L. Littman. "A Case for Lamarkian Evolution." In *Artificial Life III*, edited by C. G. Langton, 3–10. Santa Fe Institute Studies in the Sciences of Complexity, Proc. XVII. Reading, MA: Addison-Wesley, 1994.

Belew, R. K., J. McInerney, and N. N. Schraudolph. "Evolving Networks: Using the Genetic Algorithm with Connectionist Learning." In *Artificial Life II*, edited by C. G. Langton, J. D. Farmer, S. Rasmussen, and C. E. Taylor, 511–548. Santa Fe Institute Studies in the Sciences of Complexity, Proc. X. Reading, MA: Addison-Wesley, 1991.

Cecconi, F., and D. Parisi. "Evolving Organisms that Can Reach for Objects." In *From Animals to Animats*, edited by J. A. Meyer, and S. W. Wilson. Cambridge, MA: MIT Press, 1994.

Denaro, D., and D. Parisi. "Cultural Transmission in a Population of Neural Networks." Technical Report, Institute of Psychology, National Research Council, Rome, 1994.

Hinton, G. E., and S. J. Nowlan. "How Learning Guides Evolution." *Complex Systems* 1 (1987): 495–502.

Miller, G. F., P. M. Todd, and S. U. Hedge. "Designing Neural Networks Using Genetic Algorithms." In *Proceedings of the Third International Conference on Genetic Algorithms*. San Mateo, CA: Morgan Kaufmann, 1989.

Nolfi, S., J. Elman, and D. Parisi. "Learning and Evolution in Neural Networks." *Adaptive Behavior* (in press).

Odling-Smee, R. "Niche-Constructing Phenotypes." In *The Role of Behavior in Evolution*, edited by H. C. Plotkin. Cambridge, MA: MIT Press, 1988.

Parisi, D., S. Nolfi, and F. Cecconi. "Learning, Behavior, and Evolution." In *Toward a Practice of Autonomous Systems*, edited by F. Varela and P. Bourgine. Cambridge, MA: MIT Press, 1991.

COMPUTER SCIENCE

Richard K. Belew, Melanie Mitchell, and David H. Ackley

Chapter 24:
Computation and the Natural Sciences

This book concerns issues usually discussed in the biology and psychology literature: the adaptation of individual organisms, the evolution of species, and interactions between these two processes. It is therefore clear why many of our workshop participants came from the biology and psychology communities. Where do the computer scientists fit into this interdisciplinary group's efforts? In this chapter, we discuss some of the contributions computer science can make to the natural sciences (especially biology and psychology) and, conversely, what contributions the natural sciences can make to computer science.

1. WHAT CAN COMPUTER SCIENCE OFFER THE NATURAL SCIENCES?

As a practical matter, computation is all but inescapable in the pursuit of natural science in this day and age. Computers are ubiquitous in nearly every aspect of science—data storage and retrieval, scientific calculations, solution of symbolic equations, statistical analysis of data, scientific visualization, and so on. Here we discuss two roles of computation that go beyond such essentially domain-neutral tasks: computer programs as models of natural systems, and computation itself as a framework for describing the mechanisms underlying natural systems.

COMPUTER PROGRAMS AS MODELS

A basic task of science is to build models—simplified and abstracted descriptions— of natural systems. Traditionally such models are embodied in mathematical equations—for example, a system of differential equations interpreted as modeling interactions between two competing populations in a particular ecology. Such models are used both to frame hypotheses about the mechanisms underlying a system's behavior and to make predictions about its future behavior. The success or failure of such predictions, along with less obvious aesthetic criteria (parsimony, elegance, etc.), are used to evaluate the quality of the model.

The development of high-speed computers has given scientists the alternative of modeling natural phenomena using executable computer programs, in addition to passive mathematical equations and natural language descriptions. A computer program is essentially a chunk of automated mathematics, and consequently many of the goals and methods of computational modeling are shared with traditional mathematical modeling. The fundamental goal of both is to elucidate mechanisms and to make predictions. The specification of algorithms requires the same precision as mathematical models, and can expose the same implicit, *cæteris paribus* (other things being equal), assumptions that often lurk inside natural language descriptions. Formal modeling, whether mathematical or computational, tends to force such assumptions into the light, allowing researchers to challenge their legitimacy, to explore their consequences, and to identify conflicts and compatibilites with other assumptions.

At the same time, computer models generally, and those of adaptive systems in particular, have a number of advantages over traditional mathematical approaches, and over purely empirical approaches. For example:

- Computer models can capture complications that are difficult to treat in a mathematical model. Most of the computer models described in this book explicitly simulate each individual organism in a population and all of the interactions between organisms. Formal mathematical models are typically limited to describing aggregate characteristics of a system (e.g., population size or average fitness of an evolving population) rather than individual interactions, even though the interesting dynamics of a system often depends as much on the specifics of individual interactions as on general trends. In that sense, computer models are a natural complement to formal mathematical modeling, providing as *observables* the very aggregate properties that traditional mathematical models take as givens.

- Mathematical models of population dynamics are often solvable only when infinite populations, time, or are assumed. This is clearly neither necessary nor possible with detailed computer simulations, and in this sense as well, computer modeling and mathematical modeling have complementary strengths.

- Computer models are ideally suited to the study of *transient* behavior—the changing behaviors of a system that has not yet reached an equilibrium. Capturing transient behavior using mathematical models is typically difficult if not impossible. Especially for evolving and adapting systems, however, the equilibrium states—death, extinction, no learning, no evolving—are the least interesting. John Holland has nicely captured this in the phrase "adaptive systems far from equilibrium" (Holland, 1994).
- Computer models support completely controlled experiments in which parameters, initial conditions, and boundary conditions are varied and the effects are studied. This level of control is essentially impossible to attain in natural world experiments.

Computer models have disadvantages as well. Highly realistic simulations of natural systems are often hard to program, slow to run, and produce results that are difficult to interpret. Overly simplified computer models may either produce meaningless results or (as one of our workshop participants put it) "say the obvious with fancy graphics." It is a challenge to find effective models that are as simple as possible while still being informative—a challenge taken up by the models described in this book.

The distinctions among computer models made by Roughgarden et al. (Chapter 2)—minimal models for ideas, minimal models for systems, and synthetic models of systems—are useful here. Most of the computer models in this book, like most in the artificial life literature, fall in the "minimal models for ideas" category. That is, they do not make reference to any particular species or place. Instead, they are simple, abstract models for exploring particular ideas such as sexual selection, the evolution of learning, or the Baldwin effect. These models—by themselves—do not make particular predictions that can be falsified with data; rather they explore the plausibility of a particular mechanism for producing a particular kind of behavior in a system.

By contrast, Shafir and Roughgarden's "The Effect of Memory Length on Individual Fitness in a Lizard" (Chapter 12) falls into the "minimal model for a system" category: It deals with a particular species and it attempts to predict actual data. A huge advantage enjoyed by this type of model is the ease of its evaluation: to what extent does it accurately and parsimoniously predict actual data?

Idea models are harder to evaluate. Rather than being judged on predictive success, a model must be judged on its ability to make a point, to demonstrate the plausibility of an idea, to communicate an idea precisely. Our discussions during the workshop reflected the debates going on at large in the artificial intelligence and artificial life communities, where such minimal idea models are common.

Many at our workshop felt that Hinton and Nowlan's model of the Baldwin effect (Chapter 25) is a successful example of an idea model. It was meant to demonstrate the plausibility of the Baldwin effect in the most simple scenario. Hinton and Nowlan's model makes Baldwin's somewhat vague argument algorithmically precise, and thus comprehensible to workers in a number of fields. Moreover, by virtue both of its strengths and its limitations, it has stimulated useful debate and illuminated important aspects of the Baldwin effect. Minimal idea models such as those developed by the workshop participants, and described in this book, may well make similar contributions.

MODELS AS COMMUNICATIVE DEVICES

Science is fundamentally a social process, and the evolution of scientific thinking depends critically on the effective dissemination of highly abstract concepts. Computer programs have a special significance in this context, providing, in effect, a new class of languages for communicating scientific ideas.

Getting one's theories understood and then accepted by one's colleagues has traditionally depended on *arguments*, captured in written documents. These key artifacts of scientific culture are expressed using a refined, highly stylized pidgin of natural language and formal notation. Depending on the specific field of study, scientific discourse also makes heavy use of logical-mathematical theorem proving, statistics, and multitudes of visual representations of data, such as figures, tables, and graphs. All are designed to propose a theory/model, demonstrate its behavior, buttress claims of relevance of the model to the phenomenon under study, defend it against alternatives, and so on.

Expressing a given scientific model in a *computational* language has several consequences. The fact that computational models share with traditional mathematics the ability to state scientific theories precisely was mentioned above. As programming languages improve, they are also beginning to offer scientists a more expressive medium than mathematics. The models can therefore be used to summarize large amounts of data. Given the widespread availability of (nearly!) standardized computers, researchers can readily provide everything necessary to observe the phenomena in question or perform an experiment. In short, computer programs stand to provide a new and qualitatively different artifact for scientific communication.

Since running a computer program requires relatively little domain-specific expertise on the part of the user (i.e., it requires little knowledge of population genetics to run a genetic algorithm program), the model and its behavior become more accessible to other scientists, particularly to those in disciplines other than that of the program's author. In some cases, the result may simply be the recapitulation of science known in one community by practitioners in another. Hinton and Nowlan's demonstration of the Baldwin effect (Chapter 25), for example, does not identify new phenomena; rather, it demonstrates a subtle relationship that was known only in evolutionary biology. Fundamental scientific advances frequently emerge from the discovery of unsuspected connections between disciplines. To the degree that computational models can "export" results from one discipline to another in a comprehensible fashion, their value is significant.

Computational models promise to do more than communicate existing knowledge, especially when applied to broad, difficult questions of interdisciplinary interests. They can also catalyze the integration of research from many different disciplines. A preeminent example is the set of "Parallel Distributed Processing" (PDP) volumes, which describe a variety of neural network models (Rumelhart & McClelland, 1986; McClelland & Rumelhart, 1986; McClelland & Rumelhart, 1988). The third volume in particular contains executable software for the models, and with those programs readers can replicate the basic results described in the first two volumes, as well as create variations relevant to their particular interests. Since such extensions share a common language—effectively the *language of PDP*—the results can readily be communicated across disciplines. The availability of this software has therefore been instrumental in facilitating communication among and integrating research efforts of neuroscientists, psychologists, linguists, anthropologists, computer scientists, and others studying facets of human thought and behavior.

In addition to providing a set of core concepts and language, the use of computational models allows even more direct commerce among interdisciplinary participants, at least in principle: The computational models themselves can be combined together—composed—as programs. For example, a neuroscientist's neural network (NNet) can be composed with an evolutionary biologist's genetic algorithm (GA) to model the behaviors of individuals in an evolving population. Such divisions of labor (e.g., neuroscientists validating the NNet and evolutionary biologists validating the GA) allow individual practitioners to develop models reflecting their own area of expertise. The success of this methodology, of course, depends on finding a conceptual correspondence bridging between models. That is, the evolutionary biologist and the neuroscientist must agree on critical shared features of their models (how the food is sensed by the NNet and related to selective fitness in the GA, for example). Technologies to support compositional modeling exist (cf. Zeigler, 1976), but this too requires a shared understanding of the relationship among models. The real problem, then, is "essentially the same as it always has been": adopting a consistent language that can be comprehended across disciplines. The glossary at the end of this book is offered as assistance in this regard.

COMPUTATION AS A FRAMEWORK FOR NATURAL SCIENCE

Although computer science is typically viewed as a subdiscipline of engineering, devoted to the practical tasks of building better hardware and software rather than understanding natural systems, this is only partially true. Turing, Ulam, von Neumann, Weiner, and other early pioneers of computer science were motivated as much by scientific as engineering goals in the development of their ideas about computation. They wanted not only to build computers but to understand the *computational basis* of natural systems—for example, looking at the brain as a computational device and trying to understand how it computes. In recent years this motivation has been recaptured by increasing numbers of computer scientists who see their mission as understanding computation in all its manifestations, natural and artificial. The computer scientists who attended our meeting (admittedly a skewed sample of the field!) tend to view themselves and their work in this way.

There are good reasons to believe that a computational framework will eventually be seen as essential for not only the cognitive sciences, but other natural sciences as well; we can only sketch these arguments here. First, a computational framework brings some powerful ideas that address the fundamental mechanisms, behaviors, and limits to behavior in natural systems, in a way that other descriptive frameworks do not. To the degree that adaptation in nature involves performing computations of some kind, then the notions of *computability* (what can and cannot be computed in principle) and *computational complexity* (how much time or memory it takes to perform particular computations) provide constraints on what kinds of mechanisms may be at work. In short, computer science offers not only its technological feats (in the form of powerful computers) to the natural sciences; it can also lend ideas from its theoretical side—what it means to perform a computation, ways in which systems can perform computations, what is computable in principle, and so on.

Broadly speaking, computer science can make two powerful types of statements about the difficulty of certain problems. The limits of computability, or *decidability*, demonstrate that there exist well-specified classes of questions for which it is, in general, impossible to compute the answer, even though the answer definitely exists and even if we have an infinite amount of time to compute. Turing's famous "Halting Problem" (deciding whether an arbitrary program on arbitrary input will halt or will continue indefinitely) is an example of a broad class of problems that are "undecidable" in just this sense.

One may wonder whether such unbounded-resource results can possibly have much relevance to the properties of natural systems. Stahl has observed that a number of "unsolvable problems...have interesting biological counterparts." He suggests, for example, that our:

> ...inability to predict symmetry properties of chains of symbols [shown by Post's combinatorial problem]...may be compared with...prediction of molecular chain symmetries in collagen or myosin, or study of RNA-DNA 'unzipping' during mitosis, etc. following action of new enzymes. (Stahl, 1965)

Whether or not the informal connections to which Stahl alludes can be substantiated, the simple fact that they can be plausibly made indicates the potential power of computational thinking applied to the natural world.

While computability concerns itself with what can and cannot be computed at all, computational complexity deals with what can and cannot be produced in a reasonable amount of computing time and space. An example often cited in this regard is the "Traveling Salesman" problem: Given a set of cities and distances between them, what "tour" minimizes the number of miles a traveling salesman would need to travel and yet visits all the cities? While very simple methods can solve this problem when the number of cities involved is small, all known algorithms increase *exponentially* in the amount of time and/or memory they require as the number of cities increases. Practically speaking, then, such problems are computationally "intractable." Further, there is now a large body of computer science theory strongly suggesting that such computational questions, though computable (in the formal sense just given above) are intractable regardless of how the computation is performed—simply building faster computers will not make this issue go away.

For present purposes, the key point is that to the degree that natural systems *are* computational systems, they are bound by the limits of computational complexity just as surely as any engineered algorithm. The problems of limited computational resources—especially time—have obvious relevance to natural evolutionary systems. "The quick and the dead" may well be Nature's way of distinguishing computationally efficient "programming" from looser code. Selection for space efficiency is less obvious, as it is unclear just how much more of a "load" an organism with more memory (genes, neurons, synapses, etc.) faces. In the context of models such as that of Zhivotovsky, Bergman, and Feldman (Chapter 10), where the effect of increased memory is directly addressed, these issues become of paramount importance. It is a major goal for many of us in computer science (and, increasingly, those in the natural sciences) to generalize the concepts and theories of computation to better understand how nature computes, and thus to better understand nature.

WHAT CAN THE NATURAL SCIENCES OFFER COMPUTER SCIENCE?

In the previous section we highlighted some emerging influences from the computational to the natural sciences. What about the reverse? How is computer science—especially its technological aspects—changed by virtue of its interactions with the biological and psychological themes considered here?

One connection is obvious: Good ideas can come from anywhere. Significant advances in computing have emerged from interpreting natural phenomena as computations. (Note that this source of inspiration remains available even if the "strong" computational stance toward natural systems just discussed proves fruitless.) Phenomena ranging from evolving populations to adapting individuals, from developmental processes to nervous systems to immune systems, all have sparked useful computer science research. This flow of ideas is similar to what has occurred in the field of artificial intelligence (AI). For several decades now, careful observers of cognitive processes have attempted to describe them in computational terms. In artificial life (Alife) the sources are typically more biological and less cognitive, but again novel computational approaches are the result.

A pragmatic difference between Alife and AI, at least to date, is that AI's deliverables have been *prima facia* more useful than those of Alife. For example, if a knowledge engineer converts the expertise of a clinical diagnostician (or portfolio manager, oil exploration "dip stick" analyst, etc.) into a competent expert computer program, we can reasonably believe the expert system has significant economic value. But what is the economic value of an Alife *model*, of an evolving species, or an adapting individual?

While not nearly as advanced as AI's expert systems, some aspects of ALife work are proving practically important. Neural networks and genetic algorithms—both originally conceived as models—are beginning to have important practical application. Recent results applying immunological research to the problem of computer viruses promises to provide another source of economically important technology arising from Alife research (Forrest et al., 1994; Kephart, 1994).

And so, like AI the interpretation of natural processes as computation and their realization as practical algorithm is part of what computer science can gain from natural science.

The natural sciences also offer *modes of analysis* to computer science that have proven useful in the understanding of complex natural phenomena. Computational systems generally, and those involved in Alife research in particular, are becoming extraordinarily complex and difficult to analyze. Biologists and psychologists have dealt with complicated natural phenomenon for a long time, and techniques that they have developed to understand their systems' behavior may be of use to computer scientists. A clear example of this transfer is provided in a paper by Ackley and Littman (1992). In their experiments they analyzed change in a gene they knew to be unfunctional as a calibration of evolutionary change occurring throughout their system. Further, they acknowledged explicitly the appropriation of this technique from population genetics.

Perhaps the most basic argument for the relevance of the natural sciences to computer sciences is simply that computer systems are getting more lifelike every day. The stereotypical properties of von Neumann machines are becoming less and less realistic as computer systems become larger, more complex, and more interconnected. Supercomputers involving many parallel processing elements are declining in cost. Distributed networks of computers, connected by local and wide area networks and running asynchronously, are everywhere. More and more resources are devoted to fault tolerance. Replicated processors and other redundant hardware help make systems more "fail soft" and robust. Databases are replicated across networks—"mirrored"—to increase availability and access speed. Individual hardware components are increasingly autonomous and self-reliant. In software, object-oriented programming is now a fundamental methodology for isolating behavior into largely autonomous components; client-server network models are another example of the same basic trick. All these properties—robustness, fault tolerance, redundancy, local autonomy, "fail soft" operation—are properties often associated with natural systems.

Furthermore, as declining costs and improving human-computer interfaces make computer systems accessible to more people, the once sparsely populated and elite world of networked computers is experiencing a remarkable population explosion. With the spread of networking, computers are no longer isolated and sterile "number crunchers"—more and more they are becoming *venues*, "cyberspace" oases supporting multifarious interactions between myriads of real people. This mass influx is creating phenomena with which that computer scientists are just beginning to grapple with—but which seem quite familiar to observers of natural systems: People flock to newly announced locations on the network, overgraze at sites containing popular information until that system is unable to function, horde data they might need later, squabble over resources like phone lines and modem pools, fight for dominance in discussion groups, and so on.

In short, more and more real biological and cultural processes are being recapitulated on the Net. As students of biological, cultural as well as computational phenomena, we are both witnesses and participants as this new growth medium begins to fill with life. For better and for worse, it is also the case that this electronic medium also makes extremely fine-grained data about users' behaviors available for scientific analysis.[1]

Most broadly, then, the infusion of so much life science into computer science dovetails with the use of computation as a framework for natural systems. It portends a "naturalization" of our fundamental notions of computation. Although traditional computer science has developed in fairly rigid correspondence to formal models like Turing's automata, and as typically implemented in von Neumann-style architectures, the range of other formal characterizations of computation—e.g., real-valued computation (Blum, Shub, & Smale, 1989; Moore, 1993), quantum computation (Feynman, 1986; Lloyd, 1993) and the "physics of information" (Zurek, 1990)—suggests that there is still a great deal more we can learn about the phenomenon of computation. The new demands created by attempts to incorporate biological phenomena into our computational systems, from the unpredictable and vexing behaviors of our all-too-natural users, to robustness in the face of very natural "acts of God," to the most basic problems of coordinating increasingly complex systems, means that our science of computation must be stretched a great deal further. It is no wonder, then, that seminal scientists such as Turing, Ulam, von Neumann, and Weiner were as concerned at the very beginning with biology's role in computation—and vice versa—as we are today.

---

[1]It is important to note, especially to technologists unfamiliar with the special rigors of collecting data from human subjects, that the current capabilities to eavesdrop in this manner do not automatically give us the right to do so.

REFERENCES

Ackley, D. H., and M. L. Littman. "Interactions Between Learning and Evolution." In *Artificial Life II*, edited by C. G. Langton, C. Taylor, J. D. Farmer, and S. Rasmussen, Santa Fe Institute Studies in the Sciences of Complexity, Proc. Vol. X. 487–507. Reading, MA: Addison-Wesley, 1992.

Blum, L., M. Shub, and S. Smale. "On a Theory of Computation and Complexity over the Real Numbers: NP-Completeness, Recursive Functions, and Universal Machines." *Bulletin of the American Mathematical Society* 21: (1989): 1–46.

Forrest, S., A. S. Perelson, L. Allen, and R. Cherukuri. "Self-Nonself Discrimination in a Computer." In *Proceedings of the 1994 IEEE Symposium on Research in Security and Privacy*. Los Alamitos, CA: IEEE Computer Society Press, 1994.

Feynman, R. "Quantum Mechanical Computers." *Foundations of Physics* 16(6) (1986): 507–531.

Holland, H. H. "Echoing Emergence: Objectives, Rough Definitions, and Speculations for ECHO-Class Models." In *Complexity: Metaphors, Models, and Reality*, edited by G. A. Cowan, D. Pines, and D. Meltzer, 309–342. Santa Fe Institute Studies in the Sciences of Complexity, Proc. Vol. XIX. Reading, MA: Addison-Wesley, 1994.

Kephart, J. O. "A Biologically Inspired Immune System for Computers." In *Artificial Life IV*, edited by R. A. Brooks and P. Maes, 130–139. Cambridge, MA: MIT Press, 1994.

Lloyd, S. "A Potentially Realizable Quantum Computer." *Science* 261 (1993): 1569–1571.

McClelland, J. L., D. E. Rumelhart, and the PDP Research Group, eds. *Parallel Distributed Processing*, Vol. 2. Cambridge, MA: MIT Press, 1986.

McClelland, J. L., and D. E. Rumelhart. *Explorations in PDP: A Handbook of Models, Programs and Exercises*. Cambridge, MA: MIT Press, 1988.

Moore, C. *Real-Valued, Continuous-Time Computers: A Model of Analog Computations, Part I*. Working Paper 93-04-018, Santa Fe Institute, Santa Fe, NM, 1993.

Rumelhart, D. E., J. L. McClelland, and the PDP Research Group. *Parallel Distributed Processing*, Volume 1: Foundations. Cambridge, MA: MIT Press, 1986.

Stahl, J. *Journal of Theoretical Biology* 8 (1965): 371–394.

Zeigler, B. P. *Theory of Modeling and Simulation*. New York: John Wiley, 1976.

Zurek, W. H., ed. *Complexity, Entropy, and the Physics of Information*. Santa Fe Institute Studies in the Sciences of Complexity, Proc. Vol. VIII. Reading, MA: Addison-Wesley, 1990.

Reprinted Classics

Melanie Mitchell and Richard K. Belew

Preface to Chapter 25

Despite its relatively recent publication (1987), we have included Hinton and Nowlan's "How Learning Can Guide Evolution" in our collection of "classic papers" (Chapter 25) because of its significant influence on the evolutionary computation and artificial life communities. For many in these communities, it was the first introduction to the Baldwin effect.

Hinton and Nowlan provide a simple demonstration of how the Baldwin effect might operate. Their demonstration is appealing because its simplicity and analytical tractability stands in sharp relief against the subtle relationship between learning and evolution that it elucidates. Ironically, despite the fact that both Hinton and Nowlan have been and continue to be seminal contributors to neural network research, this model includes a model of learning that is only remotely related to neural networks. Hinton and Nowlan present this problem in terms of neural networks so as to keep in mind the possibility of extending the example to more standard learning tasks and methods.

And they were successful: the work was among the first to connect the genetic algorithm (GA) and neural network (NNet) communities, a connection that is now flourishing (e.g., see Schaffer, Whitley, and Eshelman, 1992). More generally, it was among the earliest applications of computer models to the question of how learning and evolution can interact without Lamarckianism. Hinton and Nowlan's work has generated a good deal of interest in this area, resulting in a number of papers either extending their original model (e.g., Belew, 1990; Harvey, 1993; French and Messinger, 1994), or using other simplified models to demonstrate and measure the Baldwin effect (e.g., Ackley and Littman, 1992). Hinton and Nowlan's work therefore provides a paradigm case of using a simple "idea" model (cf. Chapter 2) to better understand a complex natural phenomenon.

Hinton and Nowlan claim that their results demonstrate that the Baldwin effect is, in principle, possible. When a fitness landscape is hard to search by evolution alone (for example, with the extreme "impulse" function considered by Hinton and Nowlan), learning can help organisms adapt to the environment by producing useful behaviors that have not yet been "hard-coded" by genetic evolution. Eventually differential survival allows these adaptations to become genetically encoded via random genetic variation. In essence, learning allows evolution to gain information about fitness in environments where evolution could not gain such information on its own, thereby allowing evolution to perform selection using that information.

An evolutionary hypothesis such as the Baldwin effect is very difficult to verify by direct experiment. Some have hypothesized that the Baldwin effect is at least in part responsible for such phenomena as bird songs (which start out being learned but later become innate) (Simpson, 1953). But direct evidence and controlled experiments are difficult, given (1) the evolutionary time scales required to see the effect; (2) the difficulty in neatly dividing learned and innate traits in animals; and (3) the complexity of the mapping from genes to phenotypic traits.

Given these experimental difficulties, Hinton and Nowlan's approach is to rely on simple computer simulations rather than direct observation. However, as was discussed in the Introduction, see Chapter 1, such "minimal idea" demonstrations are often controversial. For example, there are several ways in which Hinton and Nowlan's results might depend on unrealistic assumptions. First, the "learning" method used—random guessing—might be too simple to reflect anything about interactions between evolution and learning as found in nature. Hinton and Nowlan argue that "a more sophisticated learning procedure only strengthens the argument for the importance of the Baldwin effect." This is true insofar as a more sophisticated learning procedure would further broaden the "zone of increased fitness" shown in Figure 1 of their paper. However, if the learning procedure were *too* sophisticated—that is, capable of learning the necessary trait virtually independent of genetic predisposition—there would be little selection pressure for evolution to move from the ability to learn a trait to genetic hard-wiring of the same trait. Models similar to Hinton and Nowlan's but incorporating more sophisticated learning mechanisms may give insight into such tradeoffs.

There are other issues as well. The model's fitness function critically assumes that organisms are able to recognize just when they have guessed the right answer and can stop searching. In addition, many of the model's critical parameters (genome length, initial allele ratios, number of guesses during a lifetime) must be nearly perfectly balanced in order for the model to display Baldwin-like effects (Belew, 1990).

Hinton and Nowlan also play somewhat fast and loose with the metaphysics of individual and specic search. Consider: "Each learning trial can be almost as helpful to the learning search as the production and evaluation of a whole new organism." While it may well be true that guessing experiments by individuals are "cheap" versions of genetic experiments by species in their simulation, the biological reality of performing the two experiments, their relative rates of change, and "remembering" their outcomes are all quite different.

But the most fundamental simplification in Hinton and Nowlan's model is the direct, one-to-one correspondence it creates between genotypic alternatives and phenotypic ones. Lamarck's and Baldwin's stories differ only in the *directness* of the process by which phenotypic adaptations are converted into genotypic adaptations. Piaget's analytic device—the "phenocopy" (cf. Chapter 19)—was proposed to consider this exact question. In Hinton and Nowlan's model, the specification of a genetic trait and the guessing of it by a cognitive phenotype are directly interchangeable. This obviates the need for, indeed the possibility of, the complex biological processes of *development* by which genotype is transformed into phenotype. The entire model may well, therefore, turn out to be somewhat irrelevant to the real relation between learning at the individual level and evolution at the species level.

Even if this turns out to be the case, the paper has already played an important role. Hinton and Nowlan have boiled down a characterization of the Baldwin effect into perhaps its most austere computational model. They created an initial spark between two communities of NNet and GA modelers that continues to foment interesting work. And they stated their hypotheses with enough clarity that others can constructively disagree. That the paper accomplishes all of this in only seven pages is a lesson to all authors.

REFERENCES

Ackley, D., and M. Littman. "Interactions Between Learning and Evolution." In *Artificial Life II*, edited by C. G. Langton, C. Taylor, J. D. Farmer, and S. Rasmussen, 487–507. Santa Fe Institute Studies in the Sciences of Complexity, Proc. X. Reading, MA: Addison-Wesley, 1992.

Belew, R. K. "Evolution, Learning, and Culture: Computational Metaphors for Adaptive Algorithms." *Complex Systems* 4 (1990): 11–49.

French, R. M., and A. Messinger. "Genes, Phenes, and the Baldwin Effect: Learning and Evolution in a Simulated Population." In *Artificial Life IV*, edited by R. A. Brooks and P. Maes, 277–282. Cambridge, MA: MIT Press, 1994.

Harvey, I. "The Puzzle of the Persistent Question Marks: A Case Study of Genetic Drift." In *Proceedings of the Fifth International Conference on Genetic Algorithms*, edited by S. Forrest, 15–22. San Mateo, CA: Morgan Kaufmann, 1993.

Schaffer, J. D., D. Whitley, and L. J. Eshelman. "Combinations of Genetic Algorithms and Neural Networks: A Survey of the State of the Art." In *International Workshop on Combinations of Genetic Algorithms and Neural Networks*, 1–37. Los Alamitos, CA: IEEE Computer Society Press, 1992.

Simpson, G. G. "The Baldwin Effect." *Evolution* 7 (1953): 110–117.

Geoffrey E. Hinton and Steven J. Nowlan

Chapter 25:
How Learning Can Guide Evolution

This paper originally appeared in *Complex Systems* **1** (1987): 495–502. Reprinted by permission.

Abstract. The assumption that acquired characteristics are not inherited is often taken to imply that the adaptations that an organism learns during its lifetime cannot guide the course of evolution. This inference is incorrect (Baldwin, 1896). Learning alters the shape of the search space in which evolution operates and thereby provides good evolutionary paths towards sets of co-adapted alleles. We demonstrate that this effect allows learning organisms to evolve *much* faster than their nonlearning equivalents, even though the characteristics acquired by the phenotype are not communicated to the genotype.

1. INTRODUCTION

Many organisms learn useful adaptations during their lifetime. These adaptations are often the result of an exploratory search which tries out many possibilities in order to discover good solutions. It seems very wasteful not to make use of the exploration performed by the phenotype to facilitate the evolutionary search for good genotypes. The obvious way to achieve this is to transfer information about the acquired characteristics back to the genotype. Most biologists now accept that the Lamarckian hypothesis is not substantiated; some then infer that learning cannot guide the evolutionary search. We use a simple combinatorial argument to show that this inference is incorrect and that learning can be very effective in guiding the search, even when the specific adaptations that are learned are not communicated to the genotype. In difficult evolutionary searches which require many possibilities to be tested in order to discover a complex co-adaptation, we demonstrate that each learning trial can be almost as helpful to the evolutionary search as the production and evaluation of a whole new organism. This greatly increases the efficiency of evolution because a learning trial is much faster and requires much less expenditure of energy than the production of a whole organism.

Learning can provide an easy evolutionary path towards co-adapted alleles in environments that have no good evolutionary path for non-learning organisms. This type of interaction between learning and evolution was first proposed by Baldwin (1896) and Lloyd Morgan (1896) and is sometimes called the Baldwin effect. Waddington (1942) proposed a similar type of interaction between developmental processes and evolution and called it "canalization" or "genetic assimilation." So far as we can tell, there have been no computer simulations or analyses of the combinatorics that demonstrate the magnitude of the effect.

2. AN EXTREME AND SIMPLE EXAMPLE

Baldwinism is best understood by considering an extreme (and unrealistic) case in which the combinatorics are very clear. Imagine an organism that contains a neural net in which there are many potential connections. Suppose that the net only confers added reproductive fitness on the organism if it is connected in exactly the right way. In this worst case, there is no reasonable evolutionary path toward the good net and a pure evolutionary search can only discover which of the potential connections should be present by trying possibilities at random. The good net is like a needle in a haystack.

The evolutionary search space becomes much better if the genotype specifies some of the decisions about where to put connections, but leaves other decisions to learning. This has the effect of constructing a large zone of increased fitness around the good net. Whenever the genetically specified decisions are correct, the genotype falls within this zone and will have increased fitness because learning will stand a chance of discovering how to make the remaining decisions so as to produce the good net. This makes the evolutionary search much easier. It is like searching for a needle in a haystack when someone tells you when you are getting close. The central point of the argument is that the person who tells you that you are getting close does not need to tell you anything more.

3. A SIMULATION

We have simulated a simple example of this kind of interaction between learning and evolution. The neural net has 20 potential connections, and the genotype has 20 genes[1], each of which has three alternative forms (alleles) called 1, 0, and ?. The 1 allele specifies that a connection should be present, 0 specifies that it should be absent, and ? specifies a connection containing a switch which can be open or closed. It is left to learning to decide how the switches should be set. We assume, for simplicity, a learning mechanism that simply tries a random combination of switch settings on every trial. If the combination of the switch settings and the genetically specified decisions ever produce the one good net we assume that the switch settings are frozen. Otherwise they keep changing.[2]

The evolutionary search is modeled with a version of the genetic algorithm proposed by Holland (1975). Figure 1 shows how learning alters the shape of the search space in which evolution operates. Figure 2 shows what happens to the relative frequencies of the correct, incorrect, and ? alleles during a typical evolutionary search in which each organism runs many learning trials during its lifetime. Notice that the total number of organisms produced is far less than the 2^{20} that would be expected to find the good net by a pure evolutionary search. One interesting feature of Figure 2 is that there is very little selective pressure in favor of genetically specifying the last few potential connections, because a few learning trials is almost always sufficient to learn the correct settings of just a few switches.

The same problem was never solved by an evolutionary search without learning. This was not a surprising result; the problem was selected to be extremely difficult for an evolutionary search, which relies on the exploitation of small co-adapted sets of alleles to provide a better than random search of the space. The spike of fitness in our example (Figure 1) means that the only co-adaptation that confers

---

[1]We assume, for simplicity, that each potential connection is controlled by its own gene. Naturally, we do not believe that the relationship between genes and connections is so direct.

[2]This implicitly assumes that the organism can "recognize" when it has achieved the good net. This recognition ability (or an ability to tell when the switch settings have been improved) is required to make learning effective and so it must precede the Baldwin effect. Thus, it is possible that some properties of an organism which are currently genetically specified were once behavioral goals of the organism's ancestors.

improved fitness requires simultaneous co-adaptation of all 20 genes. Even if this co-adaptation is discovered, it is not easily passed to descendants. If an adapted individual mates with any individual other than one nearly identical to itself, the co-adaptation will probably be destroyed. The crux of the problem is that only the one good genotype is distinguished, and fitness is the only criterion for mate selection. To preserve the co-adaptation from generation to generation it is necessary for each good genotype, on average, to give rise to at least one good descendant in the next generation. If the dispersal of complex co-adaptations due to mating causes each good genotype to have less than one expected good descendant in the next generation, the co-adaptation will not spread, even if it is discovered many times. In our example, the expected number of good immediate descendants of a good genotype is below 1 without learning and above 1 with learning.

4. DISCUSSION

The most common argument in favor of learning is that some aspects of the environment are unpredictable, so it is positively advantageous to leave some decisions to learning rather than specifying them genetically (e.g. Harley, 1981). This argument is clearly correct and is *one* good reason for having a learning mechanism, but it is different from the Baldwin effect which applies to complex co-adaptations to *predictable* aspects of the environment.

To keep the argument simple, we started by assuming that learning was simply a random search through possible switch settings. When there is a single good combination and all other combinations are equally bad a random search is a reasonable strategy, but for most learning tasks there is more structure than this and the learning process should make use of the structure to home in on good switch configurations. More sophisticated learning procedures could be used in these cases (e.g. Rumelhart, Hinton, and Williams, 1986). Indeed, using a hillclimbing procedure as an inner loop to guide a genetic search can be very effective (Brady, 1985). As Holland (1975) has shown, genetic search is particularly good at obtaining evidence about what confers fitness from widely separated points in the search space. Hillclimbing, on the other hand, is good at local, myopic optimization. When the two techniques are combined, they often perform much better than either technique alone (Ackley, 1987). Thus, using a more sophisticated learning procedure only strengthens the argument for the importance of the Baldwin effect.

For simplicity, we assumed that the learning operates on exactly the same variables as the genetic search. This is not necessary for the argument. Each gene could influence the probabilities of large numbers of potential connections and the learning would still improve the evolutionary path for the genetic search. In this more general case, any Lamarckian attempt to inherit acquired characteristics would run into a severe computational difficulty: To know how to change the genotype in order to generate the acquired characteristics of the phenotype it is necessary to invert the forward function that maps from genotypes, via the processes of development and learning, to adapted phenotypes. This is generally a very complicated, non-linear, stochastic function and so it is very hard to *compute* how to change the genes to achieve desired changes in the phenotypes even when these desired changes are known.

We have focused on the interaction between evolution and learning, but the same combinatorial argument can be applied to the interaction between evolution and development. Instead of directly specifying the phenotype, the genes could specify the ingredients of an adaptive process and leave it to this process to achieve the required end result. An interesting model of this kind of adaptive process is described by Von der Malsburg and Willshaw (1977). Waddington (1942) suggested this type of mechanism to account for the inheritance of acquired characteristics within a Darwinian framework. There is selective pressure for genes which facilitate the development of certain useful characteristics in response to the environment. In the limit, the developmental process becomes *canalized*: The same characteristic will tend to develop regardless of the environmental factors that originally controlled it. Environmental control of the process is supplanted by internal genetic control. Thus, we have a mechanism which as evolution progresses allows some aspects of the phenotype that were initially specified indirectly via an adaptive process to become more directly specified.

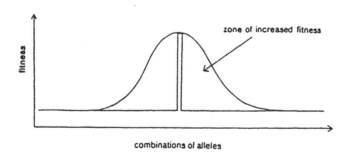

FIGURE 1   The shape of the search space in which evolution operates. The horizontal axis represents combinations of alleles and so it is not really one-dimensional. Without learning, the search space has a single spike of high fitness. One cannot do better than random search in such a space. With learning, there is a zone of increased fitness around the spike. This zone corresponds to genotypes which allow the correct combination of potential connections to be learned.

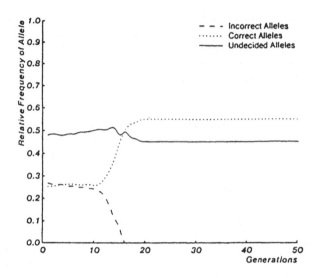

FIGURE 2   The evolution of the relative frequencies of the three possible types of allele. There are 1000 organisms in each generation, and each organism performs 1000 learning trials during its lifetime. The initial 1000 genotypes are generated by selecting each allele at random with a probability of 0.5 for the ? allele and 0.25 for each of the remaining two alleles. A typical genotype, therefore, has about ten decisions genetically specified and about ten left to learning. Since we run about 2^{10} learning trials for each organism, there is a reasonable chance that an organism which has the correct genetic specification of ten potential connections will learn the correct specification of the remaining ten. To generate the next generation from the current one, we perform 1000 matings. The two parents of a mating are different individuals which are chosen at random from the current generation. Any organism in the current generation that learned the good net has a much higher probability of being selected as a parent. The probability is *proportional* to $1 + 19n/1000$, where n is the number of learning trials that remain after the organism has learned the correct net. So organisms which learn immediately are 20 times as likely to be chosen as parents than organisms which never learn. The single offspring of each mating is generated by randomly choosing a cross-over point and copying all alleles from the first parent up to the cross-over point, and from the second parent beyond the cross-over point.

Our simulation supports the arguments of Baldwin and Waddington, and demonstrates that adaptive processes within the organism can be very effective in guiding evolution. The main limitation of the Baldwin effect is that it is only effective in spaces that would be hard to search without an adaptive process to restructure the space. The example we used in which there is a single spike of added fitness is clearly an extreme case, and it is difficult to assess the shape that real evolutionary search spaces would have if there were no adaptive processes to restructure them. It may be possible to throw some light on this issue by using computer simulations to explore the shape of the evolutionary search space for simple neural networks that do not learn, but such simulations always contain so many simplifying assumptions that it is hard to assess their biological relevance. We therefore conclude with a disjunction: For biologists who believe that evolutionary search spaces contain nice hills (even without the restructuring caused by adaptive processes) the Baldwin effect is of little interest,[3] but for biologists who are suspicious of the assertion that the natural search spaces are so nicely structured, the Baldwin effect is an important mechanism that allows adaptive processes within the organism to greatly improve the space in which it evolves.

ACKNOWLEDGMENTS

This research was supported by grant IST-8520359 from the National Science Foundation and by contract N00014-86-K-00167 from the Office of Naval Research. We thank David Ackley, Francis Crick, Graeme Mitchison, John Maynard-Smith, David Willshaw, and Rosalind Zalin for helpful discussions.

[3]One good reason for believing the search space must be nicely structured is that evolution works. But this does not show that the search space would be nicely structured in the absence of adaptive processes.

REFERENCES

Ackley, D. H. "Stochastic Iterated Genetic Hillclimbing." Doctoral dissertation, Carnegie-Mellon University, Pittsburgh, PA, 1987.

Baldwin, J. M. "A New Factor in Evolution." *American Naturalist* **30** (1896): 441–451.

Brady, R. M. "Optimization Strategies Gleaned From Biological Evolution." *Nature* **317** (1985): 804–806.

Harley, C. B. "Learning the Evolutionary Stable Strategy." *J. Theoretical Biology* **8** (1981): 611–633.

Holland, J. H. *Adaptation in Natural and Artificial Systems.* Ann Arbor, MI: University of Michigan Press, 1975.

Lloyd Morgan, C. "On Modification and Variation." *Science* **4** (1896): 733–740.

Rumelhart, D. E., G. E. Hinton, and R. J. Williams. "Learning Representations by Back-Propagating Errors." *Nature* **323** (1986): 533–536.

Von der Malsburg, C., and D. J. Willshaw. "How to Label Nerve Cells so that They Can Interconnect in an Ordered Fashion." *Proc. Natl. Acad. Sci. U.S.A.* **74** (1977): 5176–5178.

Waddington, C. H. "Canalization of Development and the Inheritance of Acquired Characters." *Nature* **150** (1942): 563–565.

John Maynard Smith

Natural Selection:
When Learning Guides Evolution

Reprinted with permission from *Nature* **329** (29 October 1987): 761–762.
Copyright © 1987 Macmillan Magazines Limited. Referencing style has
been altered to be consistent with the other chapters in this volume.

A perennial problem for evolutionary biologists is how natural selection can
give rise to a complex structure that is of value to the organism when fully formed,
but useless until it is complete. The orthodox answer, and I think often the correct
one, is that it doesn't, and that complex structures have passed through a number
of stages, each a little better than the last. For example, even random sequences
of amino acids can have some, rather nonspecific, catalytic activity. The vertebrate
eye could not arise in a single step, but a single light-sensitive cell is better than
nothing, a light-sensitive cell with a layer of pigment to one side is better still,
and so on. That is to say, evolution is a hill-climbing process. In a recent paper,
G. E. Hinton and S. J. Nowlan (1987) suggest how a combination of evolution and
learning may make possible the evolution of a structure that is of value only when
perfect.

An idea that has always had attractions is that adaptations acquired during an
individual's lifetime, by learning or in other ways, are passed on to its offspring. For

Lamarck, such inheritance of acquired characters was a main cause of evolution. For Darwin, natural selection was primary, but he also ascribed a role to "the effects of use and disuse." Weismann rejected the whole idea, essentially because he could not think of a mechanism whereby the structures acquired by an individual—for example, the blacksmith's muscles—could be translated into information in the gametes. With the development of modern genetics, and in particular of molecular biology, we have come to reject the idea that individual adaptation can alter the information in the gametes.

Even if we accept Weismann's view completely, however, it is still possible for individual learning to facilitate evolution. If individuals vary genetically in their capacity to learn, or to adapt developmentally, then those most able to adapt will leave most descendants, and the genes responsible will increase in frequency. In a fixed environment, when the best thing to learn remains constant, this can lead to the genetic determination of a character that, in earlier generations, had to be acquired afresh each generation. The idea goes back to J. M. Baldwin (1896), C. Lloyd Morgan (1896), and C. H. Waddington (1942). It has not always been well received by biologists, partly because they have suspected it of being Lamarckist (a suspicion that Waddington was curiously reluctant to allay), and partly because it was not obvious that it would work.

What Hinton and Nowlan have done is to answer these objections. They consider the following simple model, of whose biological unreality they are well aware. Imagine an organism with a neural net in which 20 connections must be correctly set ('on' and 'off') it its fitness is to be increased: if even one connection is wrong, it is no better than if all were wrong. This is the worst type of fitness surface, as there is no slope leading to the summit. There are 20 gene loci, each with three alleles, 0, 1 and ?. The first two specify on and off, and the third that the connection can be varied. During learning, an individual tries out a succession of settings for these variable connections. If, by chance, it tries out a correct set, and its genetically specified connections are also correct, it is rewarded, and does not alter its settings again. An individual makes 1,000 trials in its life. Its Darwinian fitness is increased by a factor of $1+19n/1,000$, where n is the number of trials after it has learnt the correct settings. Thus, an individual that is born with the correct settings genetically fixed has a fitness 20 times that of one that never learns.

Hinton and Nowlan simulated a population of 1,000, with initial gene frequencies of 0.5 of ? and 0.25 of 0 and 1 at each locus. Because on average 10 loci were fixed, about one individual in 1,000 would have all the fixed settings correct. Also, with 10 variable settings, it would have a reasonable chance of learning the correct settings in 1,000 trials. Individuals were chosen as parents with probabilities proportional to their fitnesses. Reproduction was sexual. The response to selection was rather slow for the first 10 generations, because most individuals never learnt, but the response then accelerated, and after 20 generations the frequency of correct alleles was high and learning was rapid.

To what extent has learning accelerated evolution in this model? The absence of learning would be represented by the case in which only 0 and 1 alleles were present.

There are then approximately 10^6 genotypes, of which one has a fitness of 20 and all the rest of 1. In a sexual population of 1,000, with initial allele frequencies of 0.5, a fit individual would arise about once in 1,000 generations (I ignore the fact that some wrong alleles would, in time, be fixed by genetic drift). Mating would disrupt the optimum genotype, however, and its offspring would have lost the adaptation. In effect, a sexual population would never evolve the correct settings.

What of an asexual population? An asexual population of 1,000 would be unlikely to include an optimal individual, and, unless one assumes an unreasonably high mutation rate, would never give rise to one. But a population of many millions would include optimal individuals that would breed true, and the correct settings would soon be established by selection.

Two comments are worth making. First, an asexual population, without learning, would try out more than 10^6 individuals before solving the problem, compared with 20,000 for the simulated sexual population with learning. Second, in the absence of learning, a large asexual population can evolve the adaptation, whereas a sexual one cannot (or does so excessively slowly). This illustrates the general fact that, when there are epistatic fitness interactions, sexual reproduction can actually slow down evolutionary progress, by breaking up co-adapted groups of genes as soon as they arise. (Epistasis arises in the present example because if, for example, A and B are correct alleles, and a and b incorrect ones, then A is fitter than a in the presence of B, but not in the presence of b.) Because sexual reproduction is almost universal, I wonder whether epistatic interactions can be as widespread as is sometimes thought.

Hinton and Nowlan show there are contexts in which learning (or developmental flexibility) speeds up evolution. It does so by altering the search space in which evolution operates, surrounding the optimum by a slope which natural selection can climb. To use their analogy, finding the optimal neural set in the absence of learning is like searching for a needle in a haystack. With learning, it is like searching for the needle when someone tells you when you are getting close.

REFERENCES

Baldwin, J. M. *Am. Nat.* **30** (1896): 441.
Hinton, G. E., and S. J. Nowlan. *Complex Systems* **1** (1987): 495.
Lloyd Morgan, C. *Science* **4** (1896): 733.
Waddington, C. H. *Nature* **150** (1942): 563.

New Work

Melanie Mitchell

Preface to Chapter 26

Michael Littman's chapter is a brief survey of ways in which evolution and learning can be combined in computer simulations. He asks an interesting question with relevance to both biology and computation: what kinds of problems can be solved by learning alone, what kinds by evolution alone, and what kinds require a combination of the two?

Many people have drawn analogies between learning and evolution, some going so far as to classify evolution as a type of learning. For example, "evolutionary algorithms"—computational search techniques inspired by evolution—are considered by many to be part of the field of machine learning. In this light, what does it mean to ask, "are there problems that evolution can solve that learning cannot, and vice versa?"

To make such questions meaningful, one has to define one's terms—"evolution," "learning," and "problem"—carefully. Littman defines evolution as "changes in the genome that alter inborn behavior" and learning as "changes in an individual's behavior as a result of interactions with its environment." The word "behavior" is presumably being construed rather broadly here to mean *any* changes in the individual's phenotype, but Littman uses this term because he is particularly interested in an individual's ability to act appropriately (i.e., in a way as to increase its probability of survival) in response to sensory input. Littman's notion of "problem" in this chapter is the task of optimizing one's fitness over time in an *on-line* fashion:

that is, the result of every action one takes during one's lifetime contributes to one's fitness, rather than one's fitness being judged only after one has completed learning about one's environment. Littman explores these questions using an extremely simple scenario: organisms that have to discover—via evolution, learning, or both—to distinguish food from poison using a variety of sensory inputs. The organisms are controlled by neural networks, evolution is modeled by a genetic algorithm that evolves neural network topologies, and learning consists of modifying network weights in response to environmental feedback.

Clearly in this scenario there are problems that cannot be solved by learning alone. Without the right architecture, there are some things a neural network cannot learn (e.g., a neural network without internal ("hidden") units cannot learn the XOR function). This carries over to the real world as well. For example, it is likely that dogs' brains are simply not the right architecture for learning complex language; even if dogs had humanlike vocal cords there is probably no way they could learn complex grammatical structure no matter how hard one tried to teach them. It is less clear whether the brains of nonhuman primates have the right architecture for learning language; this is currently a raging controversy in the psychology, linguistics, and animal-behavior communities.

In Littman's scenario there also are problems that cannot be solved by evolution alone. For example, if the environment is changing too quickly for slow genetic evolution to track, or is too complex for evolution to encode all possible responses genetically, or, as in Hinton and Nowlan's example (cf. Chapter 25), the necessary "tuning" for optimal behavior is in a very narrow range, then some form of learning seems necessary. The problem of acquiring language is a good example of all of these cases—it would be difficult to imagine having the ability to speak a particular language—say, Chinese—be hard-coded into one's genes rather than learned. Since the cultural environment changes rapidly, both over time and across distance, language has to adapt, but these changes are too fast for genetic evolution to keep up. A baby's local environment is also too uncertain—for example the parents might move from their original linguistic environment and thus force the baby to be brought up in a language different from the genetically encoded one. In addition, language is necessarily highly complex since it has to express complex ideas; it is hard to imagine how its full complexity could be encoded in the genome. Finally, languages are highly tuned for optimal communication—such fine tuning would likely be extremely difficult for an evolutionary process to discover. There are no doubt many other reasons as well that language has to be learned. (The degree to which complex linguistic structure is hard-wired in the brain via genetics is a controversy in linguistics—cf. Chomsky's notion of a "universal grammar.") But while particular languages must be learned, the ability to learn language requires the correct brain and body architecture, which must be produced via evolution.

I use the example of the relationship between language and genetics here because it is so vivid and fraught with controversy; a discussion of it could, of course, on its own fill an entire book. Littman's paper looks at much simpler scenarios in which learning and evolution interact. These scenarios (inspired by work of Todd and Miller) form a progression in which individuals can use increasingly sophisticated sensory information to decide what actions to take in various situations. Littman uses these examples to illustrate the standard types of learning identified in the machine learning community—supervised, reinforcement, and unsupervised learning—and how these can be combined with an evolutionary process such as a genetic algorithm. This leads into a discussion of one such combination Littman developed with David Ackley, "evolutionary reinforcement learning," in which the mapping of certain actions to "pleasure" or "pain" is evolved along with the network architecture for the actions. This is a novel approach to machine learning that, by combining evolution and learning in a particularly useful way, has the potential to be more powerful than other, more standard techniques.

Michael Littman

Chapter 26:
Simulations Combining Evolution and Learning

1. INTRODUCTION

Nature has devised innumerable ways to endow individual organisms with adaptive behavior. Many of these processes can be classified as being part of *evolution*, changes in the genome that alter inborn behavior; or *learning*, changes in an individual's behavior as a result of interactions with its environment. Researchers in the fields of genetic algorithms and artificial neural networks use caricatures of evolution and learning to solve difficult optimization problems.

The behavior of most real organisms is a consequence of learning and evolution in concert, so it is natural to assume that there is computational advantage in their combination. However, the formidable complexity of the two processes is multiplied when we consider their interactions.

This paper addresses the issue of how computational versions of learning and evolution have been made to interact in simulated systems. It examines various benefits of such combinations and details how supervised learning, reinforcement learning, and unsupervised learning can be adapted to fit into an evolutionary framework.

Adaptive Individuals in Evolving Populations, Ed. R. K. Belew & M. Mitchell,
SFI Studies in the Sciences of Complexity, Vol. XXVI, Addison-Wesley, 1996   **465**

2. EVOLUTION AND LEARNING

We will use a simple simulation framework for examining computational analogues of evolution and learning. At the top level is a genetic algorithm (GA) (Goldberg, 1989; Holland, 1975) that maintains a population of individuals specified by their genomes (fixed-length bit strings). The genetic algorithm collects *fitness* values for individual genomes by evaluating them according to some prespecified function and then selectively replicates and recombines genomes on the basis of this fitness. This constitutes the "evolutionary" level.

"Learning" takes place during the fitness evaluation of the genomes. The bits of the gene string are used to generate a neural network, which then undergoes a series of interactions with a simulated environment. These interactions may result in changes to the weights of the individual's network according to a prespecified learning algorithm. The interactions constitute the *lifetime* of the individual and last some fixed number of steps. The network learns insofar as the behavioral mapping it specifies changes over time. The fitness of the genome is then determined by some measure of performance of the individual over its entire lifetime. Although an individual's network might change a great deal during its lifetime, these changes are *not* incorporated back into the genome but instead "die" with the individual.

As a running example, we consider a family of environments based on Todd and Miller's associative eater environment (Todd and Miller, 1991). The scenario is depicted in Figure 1 and can be imagined as an underwater realm in which individual offspring are born into feeding patches, where they live out their entire lives. Feeding consists of deciding whether to consume substances that float past; these substances come in two forms, food and poison, and two colors, red and green. The association between color and substance is a function of the individual's feeding patch and cannot be known directly. On each step, each individual is presented with a substance (food or poison, red or green) and must decide, on the basis of sensory cues and experience alone, whether to consume the substance or to let it pass. Fitness is computed as the amount of food consumed minus the amount of poison consumed over a fixed lifetime of ℓ steps. Since half of the substances are food, and half are poison, fitness scores vary between $-\ell/2$ and $+\ell/2$.

As described here, no feedback is given to individuals that could be used to disambiguate between food and poison. In later sections, within-lifetime feedback about the consequences of eating food and poison will be the main way that the problem is manipulated to illustrate different types of learning.

The motivation for describing this model is that it is one of the simplest models that has been proposed that includes both evolution and learning components. Throughout the paper it is used to illustrate different approaches to studying how learning and evolution can interact.

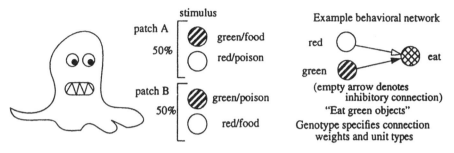

FIGURE 1   The generic associative eater environment.

2.1 EVOLUTION FINDS OPTIMAL BEHAVIOR

The behavior of each individual is controlled by a simple neural network encoded in the individual's ͵ ͺ If it were the case that poison was red and food green in all feeding patches, it ɪs easy to imagine that individuals would evolve with a simple network such as the one shown in the right half of Figure 1. The behavior of this network is to consume any green objects that float by (the filled-in arrow denotes a positive connection) and to ignore all red objects (the empty arrow denotes a negative connection: the eating unit will not activate).

Given the assumption that the substance to color mapping is constant, evolution can find networks that display optimal behavior. An experiment supporting this claim is provided in Todd and Miller's paper (their noise-free case).

2.2 LEARNING FINDS OPTIMAL BEHAVIOR

In a slightly different scena·˙   ˙dividuals are born into one of two feeding patches uniformly at random. In pa.   , food is green and poison red. In patch B it is the reverse. Individuals remain in the patch in which they were born, so for a single individual, the colors associated with food and poison are constant for its lifetime. A population of individuals with networks that exhibit fixed behavior such as the one shown in the right half of Figure 1 will not be optimal, since half of all individuals will be born into patch A and receive a fitness score of $+\ell/2$ and the other half will be born into patch B and receive $-\ell/2$ for an average fitness of 0.

Individuals that learn have the possibility of recognizing and exploiting structure in their surroundings and thus receiving fitness scores above 0 regardless of which feeding patch they are born into. However, the environment as described in Section 1 offers no opportunity for learning since it is impossible for an individual to distinguish food from poison. The environment must provide additional input for learning to be possible.

The machine learning community has identified three broad classes of learning: supervised, reinforcement, and unsupervised. They are distinguished by the amount and type of feedback information given to the learner. In supervised learning, the environment is assumed to provide the individual with direct and perfect information as to how it ought to behave in the situations it experiences (i.e., the right answer). The feedback in reinforcement learning is somewhat weaker in that the environment provides a single number, called a reinforcement signal, indicating the fitness consequence of performing the most recent action. Unlike supervised learning, the individual must experiment to find high fitness actions. In unsupervised learning, the feedback is weaker still and consists of latent structure in the input patterns themselves. Unsupervised learning algorithms can identify certain classes of structure which can be used to make decisions.

Given any of these kinds of feedback, learning algorithms exist that can alter the weights of a neural network to find optimal behavior (e.g., backpropagation [Rumelhart et al., 1986] for supervised learning, CRBP [Ackley, 1990] for reinforcement learning, and Kohonen networks [Kohonen, 1982] for unsupervised learning) independent of the individual's feeding patch. These references indicate that, given the correct network structure and feedback, learning is sufficient for finding optimal behavior.

2.3 COMBINATION SPEEDS ADAPTATION

Although either process is capable of solving difficult behavioral optimization problems, combinations of evolution and learning can lead to faster adaptation than either alone.

As an extreme example, Hinton and Nowlan (1987; reprinted in Chapter 25) showed that a problem that was extremely difficult for a simulated evolution system to solve was solved fairly easily by a system that contained elements of both evolution and learning. The mechanism they invoked to explain this result is known as the Baldwin effect (Baldwin, 1896; Schull, 1990), the essence of which is that learning alters the search space in which evolution operates by smoothing spikes in the fitness landscape, thus speeding evolution.

The acceleration of the evolutionary process by learning hinges on the structure of Hinton and Nowlan's simulation. There, some aspects of the individual are genetically fixed while others are filled in by a random search process they call "learning." An individual lives for a fixed number of steps, and learning ceases whenever the optimal behavior is discovered. An individual's fitness is the amount of time during which it displays optimal behavior. In this framework, learning assists evolution by making near-perfect individuals receive high fitness scores since such individuals quickly adapt, via learning, to exhibit optimal behavior. Evolution accelerates learning by making individuals nearly perfect to start with.

Fitness is measured "online"; that is, what matters is not just the individual's final behavior but also how quickly this behavior is discovered; thus, faster learning means higher fitness. This means that any genetically specified aspects of the learning process have the potential of serving as a source of guidance between the two processes, e.g., the initial weights of the neural network, the number of hidden units, and settings for learning rate and momentum parameters (Belew et al., 1991).

2.4 COMBINATION SOLVES MORE GENERAL PROBLEMS

The preceding section described how the combination of evolution and learning can make for quicker adaptation. However, the sort of straightforward combination used in these cases can only be applied when either evolution or learning alone is sufficient to solve the problem. It is also possible to combine evolution and learning in ways that make it possible to relax this requirement and thus solve problems that neither process can solve alone.

For evolution alone to find optimal behavior, the objective should not change radically from generation to generation. For learning alone to find optimal behavior, individuals need a feedback signal. Both of these assumptions are violated in Todd and Miller's original associative eater problem, as each individual is born into one of two feeding patches at random and no direct feedback is given to individuals during their lifetimes.

Todd and Miller show how to structure the simulation so that evolution can identify and take advantage of slowly changing, base-level properties of the environment while learning fills in the individual-specific details. This permits a hybrid system to solve problems that neither evolution nor learning could solve alone. Their example, and several others, are described in the following section.

3. ARCHITECTURES FOR LEARNING

One fundamental question must be answered before evolution and learning can be combined in a single system: from where does the feedback for learning come? In supervised learning and reinforcement learning, the feedback must be provided by the environment. In unsupervised learning, there is no direct feedback from the environment and instead the learner must exploit regularities in the *structure* of the environment to improve its behavior. Heuristic feedback methods, such as evolutionary reinforcement learning (Ackley and Littman, 1991), and auto-teaching (Nolfi and Parisi, 1993) are somewhat different as they involve learning via feedback signals, but the signals themselves are tailored by evolution.

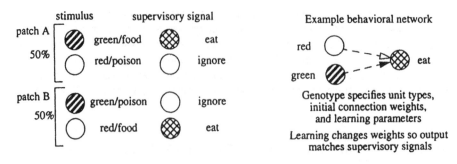

FIGURE 2 A supervised learning associative eater environment.

3.1 SUPERVISED LEARNING

Supervisory signals have been used in simulations that combine evolution and learning (Belew et al., 1991). In supervised learning, the environment provides a training signal that informs the individual of the precise action that it *should* have performed in each circumstance. In the associative eater environment, a supervised learner must discover which attributes of the environment are needed to implement the correct mapping from color to action by providing the correct action output (eat/ignore) for each color input (green/red). In Figure 2, we show a network architecture for this type of simulation (dashed lines indicate connections that are changed by learning).

This kind of information-rich feedback ought to be sufficient for learning alone to find optimal behavior. As described in Section 2.3, however, evolution can speed the process along in principle by generating networks that can learn quickly.

3.2 REINFORCEMENT LEARNING

Reinforcement signals provide less guidance than supervisory signals. Instead of giving the individual information as to the best action for each situation, in reinforcement learning only a single number is given which indicates how good the individual's last action was from the standpoint of achieving optimal fitness. By choosing actions that maximize the reinforcement signal, individuals can learn to behave optimally. In Figure 3, we summarize an associative eater environment in which a reinforcement signal is available.

For a behavioral repertoire consisting of two actions, reinforcement signals are practically identical to supervisory signals: being told that one of two actions is bad is the same as being told the correct action. When more actions are available, this type of feedback becomes weaker and weaker with respect to supervision. Nevertheless, in principle it is still sufficient for allowing any individual to learn optimal behavior given enough time (see, e.g., Ackley and Littman, 1990). ▪

3.3 UNSUPERVISED LEARNING

The learning methods described above use feedback signals that can be directly associated with optimal behavior. That is, if the feedback signal says you did something wrong, you did something wrong. This sort of learning is implausible in natural systems since the only truly unambiguous signal is death. Other signals are simply correlated with good or bad behavior to some degree. For instance, because of regularities in the natural environment, the smell of food might be considered a hint that an individual's recent actions constitute successful behavior. Unsupervised learning methods allow the individual to learn and exploit this type of association.

In an unsupervised learning framework, individuals receive no direct guidance from the environment as to which actions are better than others. Instead, they must distinguish between good and bad behavior using associations among sensory cues.

Todd and Miller (1991) described a simple scenario for the evolution of unsupervised learning, which they simulated to demonstrate its computational plausibility. Consider a version of the associative eater environment in which a second sensory input, smell, distinguishes food from poison consistently across food patches but with some probability of being misperceived. In particular, imagine food always smells sweet and poison sour but on any given step, "turbulence in the water" might cause a smell to be improperly recognized with 25% probability. This information is summarized in Figure 4.

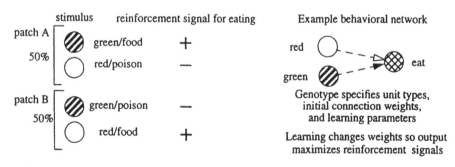

FIGURE 3   A reinforcement learning associative eater environment.

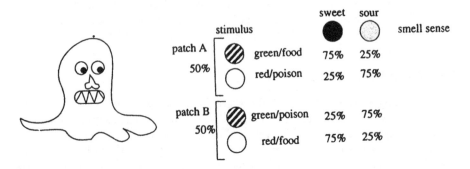

FIGURE 4   An unsupervised learning associative eater environment indicating correlations between color, smells, and object types.

How can an individual maximize its fitness in this environment? Eating anything that smells sweet is a reasonable strategy and an individual with this fixed strategy can expect a fitness score of:

$$\frac{3}{4}\left(\frac{\ell}{2}\right) - \frac{1}{4}\left(\frac{\ell}{2}\right) = \frac{\ell}{4}.$$

Recall that ℓ represents the number of decisions an individual makes during its lifetime. Such an individual does not use its color sensors at all and makes every decision based on its smell sense.

However, if an individual can keep statistics on which color tends to be identified with which smell (in patch A, green substances smell sweet three quarters of the time, for instance), then it can reliably identify a substance as food or poison by its color and earn a fitness score of $(\ell - k)/2$ where k is the number of trials before the individual has learned the association between color and substance. If k is small with respect to ℓ (less than $\ell/2$) this strategy is an improvement.

Todd and Miller (1991) describe an elegant network architecture that is capable of supporting this form of learning. The genome specifies a set of connections from sensor units to an eating unit. The left half of Figure 5 illustrates a simple network that eats anything that smells sweet. The right half of Figure 5 illustrates a feedforward network that can learn to eat substances of the proper color. The dashed line is a "changeable" connection that is strengthened if the green unit is correlated with eating and weakened if it is anticorrelated (i.e., it is a Hebbian

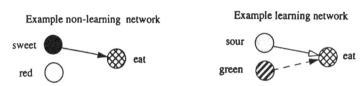

Example non-learning network   Example learning network

Genotype specifies unit types, initial connection weights, learning parameters,
and which connections are altered by learning

Learning changes selected (dashed) weights to correlate their inputs and outputs

FIGURE 5   Networks for an unsupervised learning associative eater environment.

connection). Consider the network's behavior in patch A (the argument is symmetrical for patch B). Since the eating unit has a direct connection from the sweet smell unit, the activity in these units will be correlated 75% of the time initially. After some number of trials that depends on the initial network weights, the dashed connection is strengthened to the point that the eating unit will fire on the basis of the green unit alone. Thus, the network learns the association between color and behavior and then begins to behave optimally using only sensory cues and no direct feedback.

Since the smell unit is positively correlated with whichever color predicts food, an unsupervised learning method can be used. However, it is the responsibility of evolution to recognize this relationship and to construct the proper network for exploiting it. Without the learning mechanism, individuals could not adapt to their food patch. Without evolution, the proper adaptation scheme could not be recognized. This is why, in certain environments, a system that combines evolution and learning is able to achieve greater levels of fitness than one using either learning or evolution alone.

3.4 EVOLUTIONARY REINFORCEMENT LEARNING

The unsupervised learning method described in the previous section uses a cross-generationally consistent input, namely sweet or sour smell, to determine which other input, namely red or green color, can be reliably mapped to appropriate behavior. As such, the smell units are being interpreted as a source of heuristic, though noisy, feedback. It is up to evolution to select how the sensory units will be used to generate these feedback signals.

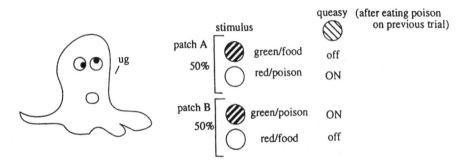

FIGURE 6   An evolutionary reinforcement learning associative eater environment.

Evolutionary reinforcement learning (Ackley and Littman , 1991) uses the same fundamental insight. To apply this model of learning, we need to assume that an individual has an extended interaction with its environment and that it can perceive the effects of its own actions. Further, assume there is a reliable way of mapping the resulting sensations to the *utility* of the previous action. An environment that satisfies these assumptions can be called a "constant utility" environment (Littman and Ackley, 1991) and evolutionary reinforcement learning can be used to find optimal behavior.

To illustrate this idea, here is another example modeled after the associative eater (see Figure 6). An additional input, labeled "queasy," is activated each time the individual consumes poison. The mapping from sensations and actions to resulting sensations is *not* constant between individuals since eating something red produces queasiness in patch A but not patch B. However, the mapping from resulting sensations to utility or fitness *is* constant since queasiness at time t always means that the action at time $t - 1$ was inappropriate. If the individual could map sensations to a feedback signal for the previous action, this could be used as a guide to learning.

A network architecture for this problem is illustrated in Figure 7. The behavior network (right) maps sensations to action, and the evaluation network (left) maps sensations to heuristic feedback signals. When the evaluation network gives a positive signal, a reinforcement learning algorithm (Ackley and Littman, 1990) is used to modify the behavior network to make the *previous* action more likely given the previous inputs. Similarly, negative signals from the evaluation network make the previous action less likely. Littman and Ackley (1991) studied a conceptually similar problem using this architecture and found that it worked quite effectively.

The primary influence of evolution here is in the specification of the fixed evaluation network, which generates the individual's heuristic feedback. Since individuals

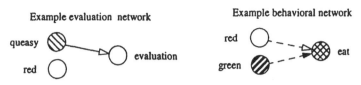

Genotype specifies unit types and connection weights of evaluation network,
and unit types, and initial connection weights of behavioral network

Learning changes weights of behavioral network to maximize evaluation output

FIGURE 7 Evaluation and behavior networks for an evolutionary reinforcement learning associative eater environment.

with good evaluation networks will tend to learn appropriate behaviors and earn high fitness, natural selection will create evaluation functions that do a good job of extracting the fitness consequences of actions. Evolution alone cannot solve this problem since individuals need to adapt to their own feeding patch. Learning alone is also inadequate because the relationship between the "queasy" unit and fitness can only be detected across generations. Once again, evolution and learning can be made to work together to solve a problem that neither could solve alone.

4. CONCLUSION

From a computer science perspective, evolution and learning are two approaches to the problem of generating adaptive behavior. Nature gives us an indication that there is some advantage to combining the two and, indeed, simulation results have been consistent with this observation.

This paper provides two computational arguments that the combination of evolution and learning is superior to either alone. For problems where evolution and learning are both capable of finding solutions independently, the combination can speed adaptation through the Baldwin Effect. For other problems, it is sometimes possible to combine evolution with unsupervised or heuristic learning methods to solve problems that neither learning nor evolution could be applied to alone.

Although the simulations reviewed in this paper suggest that there has been a great variety of work in the last few years combining evolution and learning in the form of genetic algorithms and neural networks, it seems we have barely scratched the surface. For example, Dennett (1981) describes a powerful combination of evolution and learning in which an evolved "inner environment" is used to help generate individual behavior by *imagined* trial-and-error learning. Implementations of this idea have begun to appear in the last few years (Riolo, 1991; Sutton, 1991) but not yet in an evolutionary setting.

Animals successfully employ forms of learning such as imprinting and imitation to develop survival skills. Individuals in our simulated worlds might also benefit from this form of learning, if honed properly by evolution. An even more exciting prospect is that, with the large number of types of learning that are possible, perhaps we should take a hint from natural evolution and let evolution choose the learning architecture as well (see, e.g., Chalmers, 1990).

One thing is certain. Each simulation extends our understanding of how intelligent behavior can arise from the interaction of "unintelligent" processes and also reinforces the idea that we have only begun to explore the question of how evolution and learning can be combined.

REFERENCES

Ackley, D. H., and M. L. Littman. "Generalization and Scaling in Reinforcement Learning." In *Advances in Neural Information Processing Systems 2*, edited by D. S. Touretzky, Vol. 2, 550–557. San Mateo, CA: Morgan Kaufmann, 1990.

Ackley, D. H., and M. L. Littman. "Interactions Between Learning and Evolution." In *Artificial Life II*, edited by C. G. Langton, C. Taylor, J. D. Farmer, and S. Rasmussen. Santa Fe Institute Studies in the Sciences of Complexity, Proc. Vol. X, 487–509. Redwood City, CA: Addison-Wesley, 1991.

Baldwin, J. M. "A New Factor in Evolution." *Am. Natural.* 30 (1896): 536–553.

Belew, R., J. McInerney, and N. N. Schraudolph. "Evolving Networks: Using the Genetic Algorithm with Connectionist Learning." In *Artificial Life II*, edited by C. G. Langton, C. Taylor, J. D. Farmer, and S. Rasmussen. Santa Fe Institute Studies in the Sciences of Complexity, Proc. Vol. X, 511–547. Redwood City, CA: Addison-Wesley, 1991.

Chalmers, D. J. "The Evolution of Learning: An Experiment in Genetic Connectionism." In *Proceedings of the 1990 Connectionist Models Summer School*, edited by D. S. Touretzky, J. L. Elman, T. J. Sejnowski, and G. E. Hinton. San Mateo, CA: Morgan Kaufmann, 1990.

Dennett, D. C. "Why the Law of Effect Won't Go Away." In *Brainstorms*, edited by D. C. Dennett, Ch. 5, 71–89. Cambridge, MA: Bradford Books/MIT Press, 1981.

Goldberg, D. *Genetic Algorithms in Search, Optimization, and Machine Learning.* Reading, MA: Addison-Wesley, 1989.

Hinton, G. E., and S. J. Nowlan. "How Learning Can Guide Evolution." *Complex Systems* 1 (1987): 495–502.

Holland, J. H. *Adaptation in Natural and Artificial Systems.* Ann Arbor, MI: University of Michigan Press, 1975.

Kohonen, T. "Clustering, Taxonomy, and Topological Maps of Patterns." In *Proceedings of the Sixth International Conference on Pattern Recognition*, edited by M. Lang, 114–125. Silver Spring, MD: IEEE Computer Society Press, 1982.

Littman, M. L., and D. H. Ackley. "Adaptation in Constant Utility Non-stationary Environments." In *Proceedings of the Fourth International Conference on Genetic Algorithms*, edited by R. K. Belew and L. Booker, 136–142. San Mateo, CA: Morgan Kaufmann, 1991.

Nolfi, S., and D. Parisi. "Auto-Teaching: Networks that Develop Their Own Teaching Input." In *Pre-proceedings of the Second European Conference on Artificial Life*, 845–862. Brussels, 1993.

Riolo, R. "Lookahead Planning and Latent Learning in a Classifier System." In *From Animals to Animats: Proceedings of the First International Conference on Simulation of Adaptive Behavior*, edited by J. A. Meyer and S. W. Wilson, 316–326. Cambridge, MA: MIT Press, 1991.

Rumelhart, D. E., G. E. Hinton, and R. J. Williams. "Learning Internal Representations by Error Backpropagation." In *Parallel Distributed Processing: Explorations in the Microstructures of Cognition. Volume 1: Foundations*, edited by D. E. Rumelhart and J. L. McClelland, Ch. 8. 1986.

Schull, J. "Are Species Intelligent?" *Behav. & Brain Sci.* **13** (1990): 61–73.

Sutton, R. S. "Planning by Incremental Dynamic Programming." In *Proceedings of the 1991 Machine Learning Workshop*, 1991.

Todd, P. M., and G. F. Miller. "Exploring Adaptive Agency II: Simulating the Evolution of Associative Learning." In *From Animals to Animats: Proceedings of the First International Conference on Simulation of Adaptive Behavior*, edited by J.-A. Meyer and S. W. Wilson, 306–315. Cambridge, MA: Bradford Books/MIT Press, 1991.

Charles Taylor

Preface to Chapter 27

In "Optimization with Genetic Algorithm Hybrids that Use Local Search," William Hart and Richard Belew address one of the foundation issues of evolutionary biology—how is it that organisms achieve their remarkable fit to the world in which they live? Based on general optimization theory, the authors conclude that the best fit will typically be found through a combination of global and local optimization techniques.

Evolutionary biologists have long found optimization theory to be as a useful and powerful tool (see, e.g., Pyke et al., 1977; Maynard Smith, 1978; Mangel and Clark, 1988; Parker and Maynard Smith, 1990). The usual approach has been to describe the state of some character, identify the constraints that are likely to be operating on it, and then use simple control theory or game theory to derive the theoretical optimal value that character should take. The theoretical optimum is then compared to the value actually observed.

For example, Shafir and Roughgarden observed foraging by Anolis lizards that live in the Caribbean (see Chapter 12). These lizards typically rest in trees, scanning the ground about them for food. On some occasions they will come down from their perch and attempt to capture and eat a passing insect. The problem is to understand when they will come down and attempt capturing the prey, and when they will simply "pass" and wait for the next prey to come by. There appears to be a tradeoff—it is necessary to leave the perch and chase the prey if the lizard is to eat, but if one expends time and energy chasing small and distant insects, then the occasional large and profitable insect that happens to meander near the perch may be missed. How should the lizard balance its wait and chase time? The optimal strategy must balance the expected energy intake from the food item, its distance, ease of catch, and distribution of alternative prey items, among other things. Roughgarden constructed a model of the process, and, with his student Sharoni Shafir, measured the distances that the lizards moved in their natural habitats. The fit was very close to the theoretical optimum.

While optimization is a useful heuristic for thinking about adaptation, there is debate about just how far and how strong that should be pushed (Gould and Lewontin, 1979; Orzack and Sober, 1994a,b). The roots to this debate go back even to the 1920s, in the conflicts among Fisher, Wright, and Haldane (Provine, 1986), and underlay the sometimes acrimonious debate between Dobzhansky and Muller and their students (described in Lewontin, 1974). Much of the dialectic in evolutionary biology today can be framed in those terms.

Largely missing from all this discussion, however, has been an informed analysis of *just how should optimization be achieved.* Irrespective of whether the fit between organism and environment is perfect or merely remarkable, there remains the question about how the hunt for an optimum is conducted: e.g., through natural selection, environmentally through learning, or interactively through development? Certainly there are issues regarding temporal and spatial heterogeneity (Levins, 1968), and issues regarding the amount of gene-gene interaction (Kauffman, 1993). But those are only a start.

The search space through which evolution must optimize is enormous. Reflect that there are approximately 10^9 nucleotides in the human genome. (Some animals contain less DNA than humans, e.g., bacteria, and some contain more, e.g., African toads.) Each nucleotide position is capable of taking any of four states. Ignoring some symmetries, there are thus some 4^{10^9} different genotypes to be searched. Beyond this, the fitness of each of genotype will depend on the environment and on the other genotypes in its biotic surroundings. The problem of searching this space to find an optimum is daunting, to say the least.

How is this search for an optimum conducted? Is there an optimal way to find the optimum—what might be termed meta-optimization? If so, is natural selection likely to find and favor that meta-optimum?

One way to understand how organisms search for optima has been through the study of genetic systems, particularly with the study of how and why sex has evolved. Sex, with its concomitant genetic recombination, will cause large changes in the set of genomes being explored, so would be better for global search, perhaps as required in a rapidly changing environment. Asexual reproduction would be best when fine-tuned, local search is desired.

A second way to understand how animals adapt to their environment has been to analyze traits in terms of heritability or genetic plasticity. This has been especially the case for human behavior. One extreme has been to view behavior as primarily the result of our genes (biological determinism), the outcome of genetic search by natural selection. The other extreme has been to describe our behavior as the result primarily of our environment (environmentalism), with learning and epigenesis emphasized. More likely, the phenotype cannot be viewed so simply, and must be viewed as the result of nonlinear and complex interactions—the outcome of both genetic and environmental search. But even here, the emphasis has typically been on reactions to an uncertain environment, with organisms employing what are sometimes termed "open" or "closed" programs, depending on their responsiveness to a changing environment (Mayr, 1976).

In the paper that follows, Hart and Belew suggest that there are other, more universal reasons to think that genotype and environment should both be engaged in determining how organisms respond to their environment. Search in a nonlinear and complex field, they suggest, is best accomplished by combining global and local search into hybrid strategies. In support of this they cite literature on the mathematics of search and on experiments they have performed and report here. They compare global searches by genetic algorithms with variants of local search (conjugant gradient and Solis-Wets), and find that hybrid methods typically do as well or better than either global or local search alone.

The lesson for biology is clear. We should expect neither all environmental determinism nor genetic determinism, but some combination of the two that will depend *on the nature of the problem* as well as environment predictability. Their argument is based on meta-optimization—the best way to solve the problem, irrespective of the constraints that nature might otherwise impose.

This is an intriguing idea, shedding light on how adaptation is itself best achieved. Their paper invites further investigation from theoreticians, for heuristics on when to trade off global vs. local search, and from empiricists, for insights about when the various tradeoffs actually do occur. It is a nice example of how the study of complex systems helps us to understand problems in evolutionary biology.

REFERENCES

Gould, S. J., and R. C. Lewontin. "The Spandrels of San Marco and the Panglossian Paradigm: A Critique of the Adaptationist Programme." *Proc. Roy. Soc. B* **205** (1979): 581–598.

Kauffman, S. *Origins of Order: Self-Organization and Selection in Evolution.* New York: Oxford University Press, 1993.

Levins, R. *Evolution in Changing Environments.* Princeton, NJ: Princeton University Press, 1968.

Lewontin, R. C. *The Genetic Basis of Evolutionary Change.* New York: Columbia University Press, 1974.

Mangel, M., and C. W. Clark. *Dynamic Modeling and Behavioral Ecology.* Princeton, NJ: Princeton University Press, 1988.

Maynard Smith, J. "Optimization Theory in Evolution." *Ann. Rev. Ecol. Syst.* **9** (1978): 31–56.

Mayr, E. *Evolution and the Diversity of Life: Selected Essays.* Cambridge, MA: Harvard University Press, 1976.

Orzack, S. H., and E. Sober. "How (not) to Test an Optimality Model." *Trends Ecol. & Evolution* **9** (1994a): 265–267.

Orzack, S. H., and E. Sober. "Optimality Models and the Test of Adaptationism." *Am. Naturalist* **143** (1994b): 361–380.

Parker, G. A., and J. Maynard Smith. "Optimality Theory in Evolutionary Biology." *Nature* **348** (1990): 27–33.

Provine, W. *Sewall Wright and Evolutionary Biology.* Chicago: University of Chicago Press, 1986.

Pyke, G. H., H. R. Pulliam, and E. L. Charnov. "Optimal Foraging: A Selective Review of Theory and Test." *Qtr. Rev. Biol.* **52** (1977): 137–154.

William E. Hart and Richard K. Belew

Chapter 27:
Optimization with Genetic Algorithm Hybrids that Use Local Search

INTRODUCTION

Computational models of evolution and learning offer to theoretical biology new tools for studying the interaction between these two elements in natural systems. For example, the combination of genetic algorithms (GAs) (Holland, 1975) with neural networks (Rumelhart et al., 1986) has received attention from a number of authors (Belew et al., 1991; Keesing and Stork, 1991; Montana and Davis, 1989; Nolfi et al., 1990), including authors in this volume (see Chapters 13, 21–27). An advantage of this particular combination is that the algorithmic properties of GAs and neural networks in isolation have already been studied extensively. In evolutionary models that combine GAs with neural networks, the GA performs a global, adaptive search of the space of possible neural networks, and the neural network learning algorithm (e.g., back-propagation) locally refines an initial estimate provided by the GA.

This dichotomy between a global search and a local search is a recurring theme in computational models of evolution and biology. In computational contexts, the hybridization of global and local search is known to produce more efficient algorithms. For example, most successful techniques for global optimization employ a local search algorithm to refine solutions generated by a global search (Törn and Žilinnskas, 1989). Thus there are computational reasons for augmenting a GA's standard operators to include the ability to perform a local search on members of its population. In fact, GAs have been combined with general local search methods for a number of different applications, such as combinatorial graph problems like the traveling salesman problem (Braun, 1990; Mühlenbein, 1991; Ulder et al., 1990) and the graph partitioning problem (von Laszewski, 1991). In most of these applications, the performance of the GA is substantially improved when the local search technique is employed. Mühlenbein et al. (1988, 1991a, 1991b), Ackley (1987), and McInerney (1992) have also developed application-independent versions of the GA for optimization with local search.

The computational advantage of such GA hybrids suggests that there may be abstract properties of hybridized global and local search algorithms that are common to evolution and learning in natural systems. Further, a computational analysis of models of evolution and local search may clarify different types of interactions between evolution and learning. In the following, we describe the relationship between a local search method and the GA's search operators. We then investigate the interaction between the global and local search performed by GA-local search (GA-LS) hybrids by measuring the efficiency of GA-LS hybrids on a class of difficult global optimization problems. Our analysis of these experiments allows us to make predictions concerning the relative efficiency of GA-LS hybrids on different optimization functions.

GENETIC ALGORITHM HYBRIDS WITH LOCAL SEARCH

When using local search with the GA, it is possible to perform the local search on a representation that differs from that used by the GA. In analogy to biological systems, we will call the representation used by the GA the *genotype*, and the representation used by the local search the *phenotype*. Let $\mathcal{G}$ be the space of potential genotypes and $\mathcal{P}h$ be the space of phenotypes. Genotypes are mapped to phenotypes by a maturation map $\delta(\cdot)$; if the local search uses the same representation as the GA, $\delta(\cdot)$ is the identity map. Let the function $f(\cdot)$ be the fitness function, $f : \mathcal{P}h \rightarrow \mathbf{R}$.

A local search algorithm iteratively improves its estimate of the optimum by searching for better points in some local neighborhood of the current solution.[1] The neighborhood of a local search algorithm is the set of solutions that can be

---

[1] These have also been called "hill climbing" algorithms (Mühlenbein et al., 1991a, 1991b).

reached from the current solution in a single "iteration" of the algorithm. A local search algorithm employs information about the fitness landscape when performing a search. For example, the fitness values of individuals from previous iterations are typically used. Because information about the fitness function is employed, a local search algorithm can be modeled as a mapping $\lambda_f : \mathcal{P}h \to \mathcal{P}h$. If iterative moves of the local search are defined using a local search operator $\mathbf{L}_f$, then

$$\lambda_f(ph) = \mathbf{L}_f^{(k)}(ph), \quad ph \in \mathcal{P}h, \tag{1}$$

where k denotes the number of applications of the local search operator.

The fact that a local search operator uses fitness information to select among the solutions in its neighborhood differentiates it from mutation, which depends only on information contained in the genotype. Mutation can be modeled as $M : \mathcal{G} \to \mathcal{G}$. Because mutation does not use fitness information, it is forced to make a random selection among a set of possible genotypes. Further, the set of genotypes that can be generated by mutation is also invariant to the fitness information of the genotypes. While "mutation" is sometimes used to refer to any and all genotypic modifications, we reserve the term "mutation" for completely random, "blind" modifications.[2]

Figure 1 shows how mutation and local search take a genotype g and generate new genotypes g' and g''. The mutation operator simply generates another genotype. Local search uses the maturation map δ to generate a phenotype, which is modified with a sequence of applications of a local search operator. When local search is used, the initial genotype g may or may not be modified. *Lamarckian local search* replaces g with $\delta^{-1}(\lambda_f(\delta(g)))$. The name is an allusion to Jean Baptiste de Lamarck's contention that (some) phenotypic characteristics acquired during a lifetime can become heritable traits. In our model, acquired characteristics correspond to phenotypic modifications due to the local search operator, and their heritability corresponds to the replacement of g with $\delta^{-1}(\lambda_f(\delta(g)))$.

EXPERIMENTAL DESIGN

We now examine factors that affect the way that local search is used by the GA. Our experiments examine the efficiency of GA-LS hybrids on a class of difficult optimization problems. The important dimensions of our experiments concern the frequency at which local search is used, the "intelligence" of the local search, and whether Lamarckian or non-Lamarckian local search is used.

---

[2]See Hart, Kammeyer, and Belew (1995) for a further discussion of the relationship between local search and mutation.

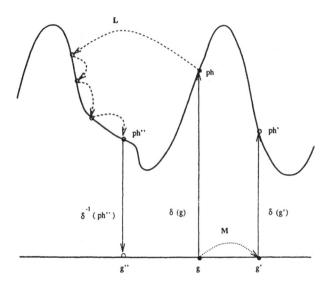

FIGURE 1   Illustration of genotypic and phenotypic local search. The bottom line represents the space of genotypes, and the top, bold curve indicates the fitness landscape. The individual g represents an initial genotype, whose phenotype is ph. The individual g' represents an individual generated by mutation (dotted line), whose phenotype is ph'. The phenotype ph'' represents the phenotype generated by performing a local search beginning with ph (dashed lines), and g'' is the genotype that would be generated by a Lamarckian local search.

FLOATING POINT GA

We use a GA with floating point encoding in our experiments. GAs have tradition-ally used binary encodings of real numbers to perform optimization on $\mathbf{R}^n$ (De Jong, 1975). While binary encodings have been used to successfully solve optimization problems, special manipulation of this encoding is often necessary to increase the efficiency of the algorithm (Schraudolph and Belew, 1992; Whitley et al., 1991). There is evidence that optimization on $\mathbf{R}^n$ can and should be performed with real parameters. Goldberg (1990) provides formal arguments that floating point GAs manipulate virtual alphabets, a type of schema that is appropriate in $\mathbf{R}^n$. The work by Wright (1991) and Janikow and Michalewicz (1991) suggests that floating point GAs can be more efficient, provide increased precision, and allow for genetic operators that are more appropriate for a continuous domain.

Proportional selection was used in our experiments with populations of size 50. Our floating point GA uses a two-point crossover that swaps the floating point values between two individuals. This is analogous to the two-point binary crossover when crossover points are only allowed at the boundaries between the binary-encoded real numbers. A crossover rate of 0.8 was used in the experiments, so an average of 80% of the individuals in each generated were generated by the crossover operator followed by mutation and the remaining individuals were generated by the mutation operator alone.

The mutation operator used with the floating point GA is Cauchy mutation, which adds a Cauchy deviate to one dimension of an individual. A random variable X has a Cauchy distribution denoted by $C(\alpha, \beta)$ if its density function is

$$f(x) = \frac{\beta}{\pi \{\beta^2 + (x - \alpha)^2\}} \quad \alpha \geq 0, \beta > 0, -\infty < x < \infty. \tag{2}$$

We used mutation with $C(0, 1)$ deviates in our experiments. In preliminary experiments (Hart, 1994), the Cauchy mutation operator was compared with the normal mutation operator, which uses normal deviates, and an interval mutation operator, which replaces one dimension of an individual with a value uniformly selected over the domain of that dimension. The Cauchy mutation operator was a good compromise between the local deviates of the normal mutation operator and the global deviates of the interval mutation operator. The mutation rate was computed with a formula derived from the analysis in Schaffer et al. (1989); when optimizing in $\mathbf{R}^n$ with a population size P, the mutation rate used is $\sqrt{e/n}/P$.

LOCAL SEARCH ALGORITHMS

The performance of the GA-LS hybrids is compared to the *multistart* (MS) method, which takes n samples from the search space (using a fixed distribution) and performs a complete local search on each of these points. A *complete* local search does not terminate until it reaches a local minimum. In our experiments, we selected the n samples from a uniform distribution.

Two types of local search algorithms were used with the GA and MS algorithms. The first was the conjugate gradient method, which uses gradient information to iteratively perform line searches (Press et al., 1990). The conjugate gradient method was terminated using gradient information to determine when the algorithm had reached a critical point on the function.

The second local search algorithm was a random search technique proposed by Solis and Wets (1981) ("Algorithm 1" with normally distributed steps). This algorithm does not use the function's gradient information. Local search is performed by adding a normal deviate in every dimension of the current solution, and accepting the new solution if it is better. The algorithm specifies parameters that reduce and increase the variance of the normal deviates in response to the rate at which better solutions are found. The control parameters for this algorithm were chosen so it would reliably converge to the local minima in the basin of attraction in which the search is started. We terminated the search after 200 function evaluations to avoid performing very long local searches on regions of the test functions that do not have local minima nearby.

TEST PROBLEM

Our experiments examine the optimization performance of these algorithms on a class of optimization test functions. This class of functions contains modified versions of Griewank's test problem (Törn and Žilinnskas, 1989). The test functions vary the weight of the quadratic term used in Griewank's function:

$$f_\sigma(x) = \sigma \sum_{i=1}^{n} x_i^2/4000 + \left(1 - \prod_{i=1}^{n} \cos(x_i/\sqrt{i})\right). \tag{3}$$

This function is a bumpy quadratic when σ is one and is a product of cosines when σ is zero. For all σ, the minimum value of the function is zero. We minimize this function over the domain $[-600.0, 600.0]^{10}$, so $n = 10$. This is the range and dimension proposed by Griewank (1989).

All of the functions in this class are difficult optimization problems. The basin of attraction of the global minimum is very small and there are a large number of other local minima. Further, the parameters of the function are nonseparable; the optimal value of each dimension depends on the values of the other parameters along the other dimensions. As we shall see later, this particular class of functions is interesting because it allows us to control the distance between the bottoms of the basins of attraction; as σ approaches zero, the values of the bottoms of the basins of attraction become similar.

EXPERIMENTAL FACTORS

The two main factors in our experiments were the test function and the type of optimization algorithm. Experiments were performed with two values of σ: 0.1 and 1.0. The GA was used without local search and with local search stochastically applied to the individuals generated by crossover and mutation. Local search was applied with three frequencies: 0.0625, 0.25, and 1.0. Both Lamarckian and non-Lamarckian local search was used with the GA using the two local search algorithms. Every combination of these factors was run with 16 different random seeds. The optimization algorithms were run until a solution was found whose value was less than 10^{-16} or until 150000 function evaluations were performed. A calculation of the gradient was counted as a single function evaluation in these experiments.

Differences between the experimental results are evaluated using the method of multiple statistical comparisons. Multiple comparisons are performed with the GH procedure, which compares samples with unequal variances (Toothaker, 1991). This method tests multiple null hypotheses which state that the means of each pair of samples are identical. The statistical comparisons reported have a confidence of $(p < 0.05)$.

RESULTS

Figures 2 and 3 summarize the results of these experiments. Table 1 defines the abbreviations used in these figures to denote the optimization methods. These figures plot the average value of the best solution after a given number of function evaluations. Figure 2 shows the average performance of the optimization algorithms that use conjugate gradient on the two test functions. Figure 3 shows the average performance of the optimization algorithms that use the randomized algorithm by Solis-Wets. The graphs describing the GA-LS hybrids using local search with a 0.25 probability were omitted to clarify the presentation. The efficiency of these algorithms was intermediate between the GA-LS hybrids with probabilities 0.0625 and 1.0.

TABLE 1 Abbreviations of optimization methods.

Abbreviation	Optimization Method
MS-CG	Multistart local search using conjugate gradient
GA	GA without local search
GA-nL-CG 0.0625	GA with non-Lamarckian local search using conjugate gradient at a frequency of 0.0625
GA/nL-CG 1.0	GA with non-Lamarckian local search using conjugate gradient at a frequency of 1.0
GA-L-CG 0.0625	GA with Lamarckian local search using conjugate gradient at a frequency of 0.0625
GA/L-CG 1.0	GA with Lamarckian local search using conjugate gradient at a frequency of 1.0
MS-SW	Multistart local search using Solis-Wets
GA	GA without local search
GA-nL-SW 0.0625	GA with non-Lamarckian local search using Solis-Wets at a frequency of 0.0625
GA/nL-SW 1.0	GA with non-Lamarckian local search using Solis-Wets at a frequency of 1.0
GA-L-SW 0.0625	GA with Lamarckian local search using Solis-Wets at a frequency of 0.0625
GA/L-SW 1.0	GA with Lamarckian local search using Solis-Wets at a frequency of 1.0

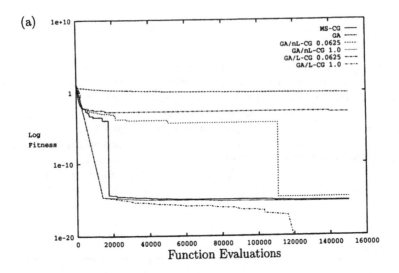

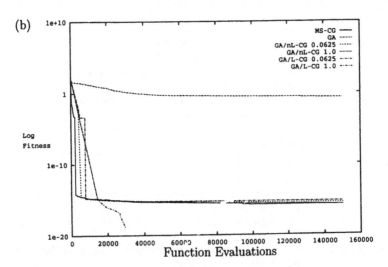

FIGURE 2   The two graphs show the average log-performance for the optimization algorithms using Conjugate Gradient (described in Table 1) for (a) $f_{1.0}$ and (b) $f_{0.1}$.

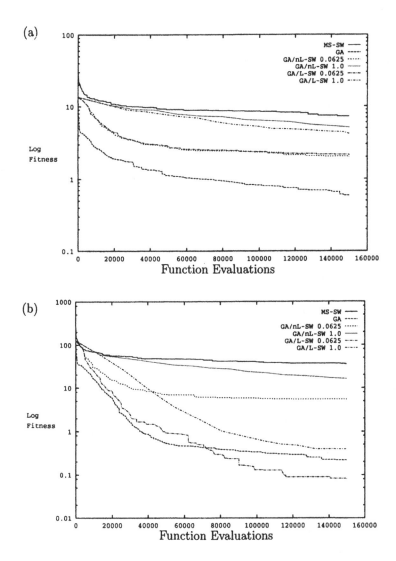

FIGURE 3 The two graphs show the average log-performance for the optimization algorithms using Solis-Wets (described in Table 1) for (a) $f_{1.0}$ and (b) $f_{0.1}$.

The statistical analysis of the results in Figure 2(a) indicates that the GA was significantly worse than the GA-LS hybrids, and that the Lamarckian GA-CG hybrid with frequency 1.0 was significantly better than multistart conjugate gradient and non-Lamarckian GA-CG hybrids. The statistical analysis of the results in Figure 2(b) was the same, except that the Lamarckian GA-CG hybrid with frequency 1.0 was not significantly better than the non-Lamarckian GA-CG hybrid with frequency 0.0625. All differences between the algorithms shown in Figure 3(a) were significant except for the difference between the GA and Lamarckian GA-SW hybrid with frequency 0.0625. All differences between the algorithms shown in Figure 3(b) were significant except for the differences between the Lamarckian and non-Lamarckian GA-SW hybrids at frequencies 0.0625 and 1.0 respectively.

The relative performance of the GA-LS hybrids was roughly the same for both test functions. The Lamarckian GAs that used local search infrequently were more efficient at the beginning, but eventually the Lamarckian GAs that used local search with frequency 1.0 became more efficient. This is clearly true for the GAs that use conjugate gradient. Extended simulations of the GAs that use Solis-Wets confirmed the same results. The Lamarckian GA-LS hybrids were more efficient than the non-Lamarckian GA-LS hybrids, except for the GA-CG hybrids that used local search with a frequency of 0.0625 on $f_{0.1}$.

The Lamarckian GAs using conjugate gradient quickly became more efficient than the GA. The Lamarckian GAs using Solis-Wets on the function $f_{1.0}$ were also more efficient than the GA, but required many more function evaluations to do so. On the function $f_{0.1}$, the GA was more efficient than the Lamarckian GA-SW hybrid, even after 500,000 function evaluations.

DISCUSSION

The different behaviors of the Lamarckian GA-LS hybrids provide insight into the way local search is used by the GA to guide the global search. In particular, these results indicate that the performance of a GA-LS hybrid is affected by the degree to which the population's fitnesses accurately reflect domain-wide characteristics of the fitness function. Let x^* be the optimum and x' be an initial estimate of the optimum. In the experiments, GA-LS hybrids using local search infrequently were more efficient at finding solutions for which $\epsilon = |f(x') - f(x^*)|$ is relatively large. This can be attributed to the fact that the GA's initial populations are quite diverse, so the population's fitnesses accurately reflect the domain-wide characteristics of the fitness function. Thus, the GA can reliably identify regions that are likely to contain optimal solutions, and applying local search infrequently avoids numerous local searches on individuals located in bad regions of the search domain.

The experimental results also indicate that GA-LS hybrids using local search frequently are more efficient at finding solutions with small ϵ accuracy. This can be attributed to the fact that the GA's populations typically lose diversity after several generations, and new individuals generated by the genetic operators become focused on some subset of the search space. In this case, the population's fitnesses do not reflect the domain-wide characteristics of the fitness function. Consequently, the GA cannot reliably identify regions that are likely to contain optimal solutions.

Refining individuals with local search can improve the efficiency of the GA-LS hybrids in two ways. First, the local searches may generate better solutions more efficiently than the GA. Second, the fitness of the refined solutions may reflect the domain-wide characteristics of the fitness function more accurately, especially when complete searches are performed.

Note that the population diversity is not the only factor that affects the degree to which a population's fitnesses reflect the domain-wide characteristics of the objective function. The overlap in ranges among the local minima of the function affects the degree to which fitness information can be used to discriminate between local minima. This factor certainly affects the reliability of the GA's competitive selection, thereby affecting the optimal local search frequency. This suggests that frequent local searches will be more efficient sooner when σ is small, because as σ becomes smaller a larger portion of the search space contains local minima close in value to neighboring local minima. A careful inspection of our experimental results with the Griewank and modified Griewank functions confirms this prediction. Extended simulations with the GA-SW hybrids confirmed this prediction when comparing the performance of GA-SW hybrids with frequencies 0.0625 and 0.25, but not when comparing GA-SW hybrids with frequencies 0.0625 and 1.0.

This analysis may also explain why the non-Lamarckian GA-CG hybrid was more efficient than the Lamarckian GA-CG hybrid when using a frequency of 0.0625 on $f_{0.1}$. In this case, non-Lamarckian local search may help retain some diversity in the population, since the individuals are not being replaced with the results of the local search. This additional diversity is important when optimizing $f_{0.1}$ because the values of neighboring local optima are very close, and the local search is not used often enough to help the GA find better local minima. The Lamarckian GA-CG hybrid quickly loses diversity in its population, after which it has a very difficult time finding better local minima. This same results may not have occurred with the GA-SW hybrids because the Solis-Wets algorithm is much less powerful than conjugate gradient.

Finally, the fact that the GA-SW hybrids are efficient with low local search for an extended period may be attributed to a number of factors. The stochastic search performed by the Solis-Wets algorithm proceeds downhill slowly, so this local search algorithm may not be useful until the local minima become very similar. The conjugate gradient algorithm is more efficient, and therefore more useful at an earlier stage of the optimization. Another factor may be the lengths of the local search; the conjugate gradient algorithm was allowed to perform much longer searches than Solis-Wets because it had well-defined stopping conditions. Longer local searchs for Solis-Wets may improve the precision of the estimated value of individuals in the population.

CONCLUSIONS

Our model of GAs with local search provides a well-defined description of the different factors involved with GA-LS hybrids. For example, it provides a clear distinction between Lamarckian and non-Lamarckian local search, and enables a direct comparison between the roles of mutation and local search.

Our analysis of GA-LS hybrids can help explain the performance that other researchers have observed in their GA-LS hybrids. In particular, it explains the observations by Davis (1989) and Mühlenbein (1991a) that local search is not needed in the initial stages of the optimization. Our analysis suggests that local search is useful in these initial stages, but is best used with a low frequency.

This analysis also suggests a number of other factors that could influence the behavior of GA-LS hybrids. We have seen that population diversity can affect the efficiency of GA-LS hybrids. The population diversity is affected by the size of the GA's population, as well as the method used to perform selection. The length of the local search is another influential factor, since it affects the level of refinement that is performed by the local search. Experiments by Hart (1994) examine the effects of these factors. The results of these experiments are consistent with the interpretation of our results.

These experimental results suggest that factors like population diversity and the structure of the fitness landscape in natural systems can influence the role that learning plays. The results predict that learning will be less important in environments for which the fitness distribution of the population is an accurate predictor of the evolutionary potential of members in the population. While our model of GA-LS hybrids simplifies many of the elements of evolution and learning, we expect that general properties like these may be common to evolution and learning in natural systems.

ACKNOWLEDGMENTS

Our thanks to Chuck Taylor and Filipo Menczer for their careful reviews of this paper.

REFERENCES

Ackley, D. H. *A Connectionist Machine for Genetic Hillclimbing.* Boston, MA: Kluwer, 1987.

Belew, R. K., J. McInerney, and N. N. Schraudolph. "Evolving Networks: Using the Genetic Algorithm with Connectionist Learning." In *Artificial Life II*, edited by C. G. Langton, C. Taylor, J. D. Farmer, and S. Rasmussen, 511–548. Santa Fe Institute Studies in the Sciences of Complexity, Proc. Vol. X. Reading, MA: Addison-Wesley, 1991.

Braun, H. "On Solving the Travelling Salesman Problems by Genetic Algorithms." In *Parallel Problem Solving from Nature*, edited by H.-P. Schwefel and R. Männer, 129–133. New York: Springer-Verlag, 1990.

De Jong, K. A. *An Analysis of the Behavior of a Class of Genetic Adaptive Systems.* Ph.D. Thesis, University of Michigan, Ann Arbor, 1975.

Goldberg, D. "The Theory of Virtual Alphabets." In *Parallel Problem Solving from Nature*, edited by H.-P. Schwefel and R. Männer, 13–22. New York: Springer-Verlag, 1990.

Hart, W. E. *Adaptive Global Optimization with Local Search.* Ph.D. Thesis, University of California, San Diego, 1994.

Hart, W. E., T. E. Kammeyer, and R. K. Belew. "The Role of Development in Genetic Algorithms." In *Foundations of Genetic Algorithms*, edited by L. D. Whitley and M. Vose, 315–332. San Francisco, CA: Morgan-Kaufmann, 1995.

Holland, J. H. *Adaptation in Natural and Artificial Systems.* Ann Arbor, MI: The University of Michigan Press, 1975.

Janikow, C. Z., and Z. Michalewicz. "An Experimental Comparison of Binary and Floating Point Representations in Genetic Algorithms." In *Proceedings of the Fourth International Conference on Genetic Algorithms*, edited by R. K. Belew and L. B. Booker, 31–36. San Mateo, CA: Morgan-Kaufmann, 1991.

Keesing, R., and D. G. Stork. "Evolution and Learning in Neural networks: The Number and Distribution of Learning Trials Affect the Rate of Evolution." In *NIPS 3*, edited by R. P. Lippmann, J. E. Moody, and D. S. Touretzky, 804–810. San Mateo, CA: Morgan-Kaufmann, 1991.

McInerney, J. M. N. *Biologically Influenced Algorithms and Parallelism in Non-Linear Optimization.* Ph.D. Thesis, University of California, San Diego, 1992.

Montana, D. J., and L. Davis. "Training Feedforward Neural Networks Using Genetic Algorithms." In *IJCAI 1989*, 762–767. San Mateo, CA: Morgan-Kaufmann, 1989.

Mühlenbein, H. "Evolution in Time and Space—The Parallel Genetic Algorithm." In *Foundations of Genetic Algorithms*, edited by G. J. E. Rawlins, 316–337. San Mateo, CA: Morgan-Kaufmann, 1991.

Mühlenbein, H., M. Gorges-Schleuter, and O. Krämer. "Evolution Algorithms in Combinatorial Optimization." *Parallel Computing* **7** (1988): 65–85.

Mühlenbein, H., M. Schomisch, and J. Born. "The Parallel Genetic Algorithm as Function Optimizer." In *Proceedings of the Fourth International Conference on Genetic Algorithms*, edited by R. K. Belew and L. B. Booker, 271–278, San Mateo, CA: Morgan-Kaufmann, 1991a.

Mühlenbein, H., M. Schomisch, and J. Born. "The Parallel Genetic Algorithm as Function Optimizer." *Parallel Computing* **17** (1991b): 619–632.

Nolfi, S., J. L. Elman, and D. Parisi. "Learning and Evolution in Neural Networks." Technical Report CRL 9019, Center for Research in Language, University of California, San Diego, 1990.

Press, W. H., B. P. Flannery, S. A. Teukolsky, and W. T. Vetterling. *Numerical Recipies in C—The Art of Scientific Computing*. Cambridge University Press, 1990.

Rumelhart, D. E., G. E. Hinton, and R. J. Williams. "Learning Internal Representations by Error Propagation." In *Parallel Distributed Processing*, edited by D. E. Rumelhart and J. L. McClelland, Vol 1, 318–362. Cambridge, MA: MIT Press, 1986.

Schaffer, J. D., R. A. Caruana, L. J. Eshelman, and R. Das. "A Study of Control Parameters Affecting online Performance of Genetic Algorithms for Function Optimization." In *Proceedings of the Third International Conference on Genetic Algorithms*, edited by J. David Schaffer, 51–60. San Mateo, CA: Morgan-Kaufmann, 1989.

Schraudolph, N. N., and R. K. Belew. "Dynamic Parameter Encoding for Genetic Algorithms." *Machine Learning* **9** (1992): 9–21.

Solis, F. J., and R. J.-B. Wets. "Minimization by Random Search Techniques." *Mathematical Operations Research* **6** (1981): 19–30.

Toothaker, L. E. *Multiple Comparisons for Researchers*. Sages Publications, 1991.

Törn, A., and A. Žilinnskas. *Global Optimization*, Vol. 350. Lecture Notes in Computer Science. Springer-Verlag, 1989.

Ulder, N. L. J., E. H. L. Aarts, H.-J. Bandelt, P. J. M. van Laarhoven, and Erwin Pesch. "Genetic Local Search Algorithms for the Traveling Salesman Problem." In *Parallel Problem Solving from Nature*, edited by Hans-Paul Schwefel and Reinhard Männer, 109–116. New York: Springer-Verlag, 1990.

von Laszewski, G. "Intelligent Structural Operators for the K-way Graph Partitioning Problem." In *Proceedings of the Fourth International Conference on Genetic Algorithms*, edited by R. K. Belew and L. B. Booker, 45–52. San Mateo, CA: Morgan-Kaufmann, 1991.

Whitley, D., K. Mathias, and P. Fitzhorn. "Delta Coding: An Iterative Dearch Strategy for Genetic Algorithms." In *Proceedings of the Fourth International Conference on Genetic Algorithms*, edited by R. K. Belew and L. B. Booker, 77–84. San Mateo, CA: Morgan-Kaufmann, 1991.

Wright, A. H. "Genetic Algorithms for Real Parameter Optimization." In *Foundations of Genetic Algorithms*, edited by G. J. E. Rawlins, 205–218. San Mateo, CA: Morgan-Kaufmann, 1991.

Glossary

SELECTED KEYWORDS

As the biologists, psychologists, and computer scientists contributing to this volume have attempted to understand one another, it has become clear that much of the confusion can be traced back to a number of "keywords" that are central to our conversation. In many cases, these terms have been quite precisely defined within one discipline or another but also retain more colloquial, "folk" connotations from general parlance. The challenge is to retain the hard-won precision of one field's technical term while extending its use to the interdisciplinary phenomena experienced by others. Our current effort to provide working definitions for some of the most critical keywords is modeled after Fox Keller and Lloyd's (1992) *Keywords In Evolutionary Biology*. Fox Keller and Lloyd make an eloquent argument for the importance of such a vocabulary to scientific progress, and for the insight these words can give into the scientific process they mediate. (The introduction to this volume discusses this connection in more depth.)

In our case, the following "definitions" have come from our authors and workshop participants, sources recommend by them, a group of students at the University of California at San Diego who were early readers of this manuscript, dictionaries, and encyclopedias. We are most grateful to all these contributors for their willingness to participate in this unusual intellectual amalgam. The terms are often defined in the context of readings in this volume. In some cases this large collection of authors has generated multiple, sometimes inconsistent and sometimes fully contradictory definitions of the same term. Rather than attempting to reconcile the various definitions into a single, authoritative one (as a logical positivist would recommend), we present the variations to the reader as an indication of the rich texture and scientific questions remaining to be answered.

accommodation   Adaptation on the individual level, via interaction with the environment; nonhereditary adaptation. See ASSIMILATION

- According to Piaget (Chapter 19), the infant's process of changing an existing schema to fit an external stimulus.

action   A primitive unit of behavior. In simulations, often one of a discrete set of choices the individual is faced with at each (discrete) moment in time.

adaptation   To make fit (as for a specific or new use or situation) often by modification; modification of an organism or its parts that makes it more fit for existence under the conditions of its environment (On-line Webster program). See also Fox Keller and Lloyd (1992).

- The process of changing the parameters of a system from their initial values to the values needed to solve a particular problem optimally.

• Genotypic change which results in an individual with fitness superior to that of its parent.

• Adaptation is also a description of a feature, or state, of adaptedness. For example, the gill is an adaptation to the marine environment.

adaptationist The tendency to believe that the evolutionary process unfailingly optimizes species to perfectly fit their niche.

• A school of thought in evolutionary theory that views most or all organismal traits as adaptive, and, more specifically, as having arisen as a direct result of natural selection acting on the trait in question.

adaptation, co- Mutually adapted, especially by natural selection (On-line Webster program).

• Used by Hinton and Nowlan (Chapter 25) to describe the specific situation in which a set of genetic alleles is not adaptive unless they occur in conjunction.

adaptations, pre- Genetic changes which result in no increase in fitness, but which have been analyzed *post hoc* as having laid the groundwork for subsequent adaptive changes.

• A trait or traits that lay the groundwork for future adaptive changes. These traits may be random with respect the subsequent adaptive changes, or they may have come into being through selection pressures which are irrelevant to the subsequent ones. For example, it has been speculated that the birds' proto-wings evolved first as "sweepers" by which running bipedal reptiles swept their prey into their mouths. The trait of swinging feathered arms while running at top speed may have been a preadaptation for flight.

adaptive computation See COMPUTATION, ADAPTIVE

allele One of a group of genes that occur alternatively at a given locus (On-line Webster program). See GENE

• In genetic algorithms terminology (e.g., as used by Hinton and Nowlan, Chapter 25), it is one of the forms that a gene can take (e.g., fixed present, fixed absent, or modifiable).

artificial intelligence The capability of a machine to imitate intelligent human behavior (On-line Webster program).

• The capacity of a digital computer or computer-controlled robotic device to perform tasks commonly associated with the higher intellectual processes characteristic of humans, such as the ability to reason, discover meanings, generalize, or learn from past experience. The term is also frequently applied to that branch of computer science concerned with the development of systems endowed with such capabilities (Brittanica On-line).

artificial life   A field of study devoted to understanding life by attempting to abstract the fundamental dynamical principles underlying biological phenomena, and recreating these dynamics in other physical media—such as computers—making them accessible to new kinds of experimental manipulation and testing (Langton et al., 1992).

assimilation   According to Piaget, the infant's process of incorporating external stimuli to an existing schema. See ACCOMMODATION

assimilation, genetic   The "genetic fixation" (via selective processes) of "acquired" characters (phenotypes). Prior to fixation, these characters seem to appear solely as a response to specific environmental stimuli, i.e., they are modifications whose range of variability is genetically controlled; after fixation (assimilation) they are also produced in the absence of the particular environmental condition ("pseudo-exogenous adaptations").

The phenotypes assumed to be somatically "acquired," stabilized, and genetically assimilated may in fact be a consequence of "threshold effects" (Stern, 1958). In the case of "threshold selection," the special environmental conditions may reveal which individuals (from among a number) already carry polygenes and modifiers of the phenotype in question. The character differences may be subthreshold and thus not discernible in the original environment and above threshold in the new environment where they are endowed with positive selective value. Under these circumstances genotypes whose action is above threshold in both the original and the new environment may arise by crossing (Reiger, Michaelis, and Green, 1991).

• Almost 50 years after Baldwin and his contemporaries, Waddington (1942) proposed a similar but more plausible and specific mechanism that has been called "genetic assimilation." Waddington reasoned that certain sweeping environmental changes require phenotypic adaptations that are not necessary in a normal environment. If organisms are subjected to such environmental changes, they can sometimes adapt during their lifetimes because of their inherent plasticity, thereby acquiring new physical or behavioral traits. If the genes for these traits are already in the population, although not expressed or frequent in normal environments, they can fairly quickly be expressed in the changed environments, especially if the acquired (learned) phenotypic adaptations have kept the species from dying off. The previously acquired traits can thus become genetically expressed, and these genes will spread in the population. Waddington demonstrated that this had indeed happened in several experiments on fruit flies (Mitchell, 1996).

- Waddington's version of the Baldwin effect, but with his perspective of mutation as a back-filling process, versus Baldwin's exploratory one.

- Schull (Chapter 3) sees genetic assimilation as a discipline-specific special case of Bateson's formulation. It takes the Baldwinian scenario and transposes it from the level of intelligent behavior to the level of physiological adaptation.

attainability The problem of providing a model that is not only accurate as far as fitting data, but also biologically or cognitively plausible. For Shafir and Roughgarden (Chapter 12), attainability motivates a solution to optimal foraging that is based on abstract operations available to the individual (lizard).

Baldwin effect The tendency of organisms that learn or acquire useful characteristics to be successful, leading to a higher probability of their reproduction and a fixation of their useful characteristics in the population despite the absence of direct inheritance of these characteristics.

- The tendency for ontogenetically acquired adaptations (e.g., acquired through learning) to produce evolutionarily mediated changes in a species—population-genetic adaptation in response to selection pressures produced in part by the prior ontogenetic adaptations. (N.B., This definition avoids the concepts of genes and population genetics, which post-date Baldwin's formulation.)

- According to Schull (Chapter 3), the "Baldwin effect" was a term Baldwin managed to attach to a family of processes about which he and several of his contemporaries (Lloyd Morgan and Poulton) theorized. Baldwin battled for the credit, and in doing so managed to narrow the scope and to obscure the generality of the concept. In his scenario, which was intrinsically self-limiting, the effect of intelligent plasticity was to usher in nonintelligent, nonplastic instincts, thereby causing evolution to cease.

- According to Hart and Belew (Chapter 27), the increase of evolutionary performance (e.g., average population fitness) in non-Lamarckian evolution when the frequency of learning is sufficiently small.

- Definitions vary in the scope of evolutionary situations that they encompass, but most seem to involve the evolution of genetically determined characteristics similar in function to those that arise through phenotypic plasticity. Todd (Chapter 21) suggests an even broader definition. In his simulation, when the sexual imprinting gene is on, the organism makes no reference to its mate preference genes. So, it seems that there would be little pressure for learned mate preferences to become innate. However, individual learning affects mate choice, which certainly affects the genetic traits of offspring. The article seems to say that the term "Baldwin effect" could encompass this.

behavior Actions taken by an organism that change the environment and/or the organism's relationship to the environment.

- Phenotypic traits exhibited in the interaction between an organism and its environment.

behavior, optimal In a simulation, the behavior of an individual with the largest possible fitness score.

behaviorism A school of psychology that takes the objective evidence of behavior (as measured responses to stimuli) as the only concern of its research and the only basis of its theory without reference to conscious experience (On-line ,Webster program).

- A school of psychology which posited that behavioral data and a behavioral theory of psychology must be restricted to measurable stimuli and measurable responses. Conscious experiences, purposive intentions, or any other psychological or mental processes were specifically considered beyond any science of behavior.

- According to Schull (Chapter 3), materialistic and behavioralistic interpretations of biology (e.g., those that rejected Lamarckism not only for its account of inheritance but also for its granting of a causal role to the individual organisms, needs, wants, and intentions) created a climate that was antipathetic to the broader implications of organic selection, even though they were actually quite compatible with them. Skinner (Chapter 18) saw that learned behavior could change selection pressures and thereby influence the course of biological evolution. Skinner's theories received little attention within biology, however.

benefit Selective advantage. Not to be confused with "utility"; traits useful to an individual may still not confer a reproductive advantage, and vice versa.

biology 1: A branch of knowledge that deals with living organisms and vital processes; 2a: the plant and animal life of a region or environment; b: the life processes of an organism or group; *broadly*: See ECOLOGY (On-line Webster program).

- According to Schull (Chapter 3), Lamarck was among the first to use this word. Lamarck thought it would be an appropriate new discipline that his philosophy of science could nurture.

biosphere Living beings together with their environment (On-line Webster program).

- What James referred to as the "earth" and "environment." (See Schull, Chapter 17.)

- What many now refer to as Gaia (Lovelock, 1987).

caloric The invisible fluid that was believed to cause heat. Throughout Lamarck's article (Chapter 4) fluids play an important causal role in his explanations. For example, Larmarck suggests that animals that butt heads together cause a flow of fluids to their heads which eventually causes horns to form (over many generations).

canalization   1a: To provide with a canal or channel; b: to make into or similar to a canal; 2: to provide with an outlet; *esp.*: to direct into preferred channels (On-line Webster program).

• Waddington demonstrated that plasticity (in morphological development) could influence selection of genes which facilitate, elaborate or "canalize" the particular adaptive response. Over several generations adaptive response could become so reliable that plasticity seems irrelevant even though it played a crucial role at the outset.

• According to Waddington (Chapter 7), canalization forces evolution on a particular path, on which the course of evolution will remain regardless of minor variations. His most explicit invocation of the metaphor is in his description of a ball rolling on the developmental landscape; the grooves worn are the canals.

• Reduced variance; tendency of a maturational system to remain constant in the face of (external, environmental) perturbation.

Central Dogma of biology   Information flows from DNA to RNA to proteins.

complexity   An imprecise term used to describe characteristics of systems including a large number of parts and interactions between parts, nonlinearity of interactions, and difficulty of analysis and prediction of behavior. The term is used precisely in some contexts, e.g., "computational complexity," "environmental complexity," etc.

complexity, environmental   A more complex environment has a larger number of more varied survival strategies. Complexity is a function of the "chemistry," "physics," the topology of the environment, the modes of interaction available to the simulated environment, and similar considerations.

computation, adaptive   Roughgarden et al. (Chapter 2) refer to an unattributed quote, "all reality is computation," and describe "adaptive computation" as an agent-oriented (or individual-oriented) method of computing (and maximizing) the fitness of discrete individuals engaging in discrete interactions with their environments. It is contrasted with differential equations, which treat the world as being continuous. The implication is that the discrete method is better at describing the complexity of small populations than the continuous method. See INDIVIDUAL-BASED MODELING

• The theory and practice of building computational systems that are able to adapt—i.e., improve their performance in response to their environment, continue to perform satisfactorily in changing environments, and cope well with wholly new environments. Adaptive computation researchers often take inspiration from natural adaptive systems in developing adaptive computer systems. Such nature-inspired systems include genetic algorithms, neural networks, economic approaches to resource allocation, and immune-system-based pattern recognition and computer security systems. Adaptive computation is also often taken to include computational modeling of natural adaptive systems, such as that done by Hinton and Nowlan (Chapter 25) and Menczer and Belew (Chapter 13).

computer science   The theory and practice of computation, typically taken to refer to computation on human-constructed electronic digital or analog devices. In this book the computer scientists focus primarily on using such devices to simulate natural systems and to perform experiments on these simulations. The word "computation" can also refer to information-processing activities in natural systems, such as learning in organisms. In this context, computer scientists can study the computational properties of these behaviors as a way of understanding aspects of nature.

contingency   A condition under which a behavior and its consequences occur.

correlation, dynamic   Two surfaces are dynamically correlated in a region in which movement that results in increased values on one surface in that region results in increased values on the other surface.

correlation, statical   Two surfaces are statically correlated if they tend to have peaks and valleys in similar places, or, at the least, if those areas with high values for one surface have high values for the other, and those areas with low values for one surface have low values for the other.

cost   Selective disadvantage. As with "benefit," this does not refer to a hardship or inconvenience but rather to a trait that decreases the organism's reproductive success.

development   All changes occurring to an individual during its lifetime, whether due to genetic causes or interaction with the environment. "Maturation" is used to refer to changes associated primarily with the former, and "learning" with the latter. See MATURATION, LEARNING, ONTOGENY

ecological   "...given in terms of the relationship between the animal and its environment, rather than in terms of the animal alone" (Johnston, Chapter 20). Johnston wants to distinguish ecological learning—that to which the cost/benefit analysis is applicable—from "surplus" (selectively irrelevant) learning.

ecology of mind Bateson's (Chapter 9) evocative term (and book title) for a science which accommodates mental phenomena in a science spanning both psychology and biology. In these matters, Bateson drew inspiration from Lamarck (a pre-Darwinian), Samuel Butler (an anti-Darwinian), and Alfred North Whitehead (an admirer of the trans-Darwinian William James). Schull's chapters are a post-Batesonian attempt to show where the ecology of mind comes from, to show where it might be leading, and to show how it is consistent with, but broader than, traditional Darwinism. Bateson's general formulation of the Baldwin effect is one example of the kind of discipline-spanning evolutionary theory that Bateson advocated. See SELECTION, ORGANIC

ecology 1: A branch of science concerned with the interrelationship of organisms and their environments; 2: The totality or pattern of relationships between organisms and their environment (On-line Webster program).

• Roughgarden et al. use the term in a pretty standard way, differing from Webster's only in that they choose to emphasize the role of the environment in determining individual development.

energy A measure of capability to perform work. Energy can be stored (potential energy) but in a closed system, according to the Second Law of Thermodynamics, it tends to be transformed into motion (kinetic energy). Energy in a closed system (potential + kinetic) is conserved.

energy, latent If performing work W upon a system enables it to perform work $W' > W$, we say that the system stores latent energy $W' - W$. From an ethological point of view, this means that behavior must supply work in order to realize (use or store) energy latent in an environment.

environment Used in reference to the physical environment but also seems to include habits. See also Fox Keller and Lloyd (1992).

• The complex of physical, chemical, and biotic factors that acts upon an organism or an ecological community and that ultimately determines its form and survival.

• Everything outside the structure (skin) of the individual organism.

• In Zhivotovsky et al.'s model (Chapter 10) a set of "external factors" that can be described by an mth order Markovian process. The transition probabilities are unchanging over time. The environment is assumed to undergo a very large number of transitions in the course of an individual's lifetime. The authors seem to have in mind something like whether it is raining, or where the individual happens to be today. This is in contrast to a slowly changing environment, such as whether the organism is living in a rain forest or a desert.

• In Shafir and Roughgarden's model (Chapter 12), environment is completely characterized by the abundance of prey per unit area-time. In the simulations this is a constant parameter. This greatly simplified model of environment seems appropriate for an "idea" model in which fitting actual data is not a primary concern, compared to tractability.

environmental information For Zhivotovsky et al. (Chapter 10), this is *a priori* information available to individual about an environment prior to its experience of it. It is encoded as a function giving the probability of the next environmental state, based on the last l states. This information may be complete (if $l = m$, the order of environmental process itself) or incomplete. This information is assumed to be absolutely correct. The case where it is not correct is not considered. This seems to be an unrealistic notion of information. Organisms probably have many ways of implicitly predicting future environmental states, but they almost certainly lack perfect information about conditional probabilities.

epigenesis The concept that an organism develops by the new appearance of structures and functions, as opposed to the hypothesis that an organism develops by the unfolding and growth of entities already present in the egg at the beginning of development (preformation) (King, 1990).

• Development in which differentiation of an individual's cells into organs and systems arises primarily through their interaction with the environment and each other.

epistasis A form of gene interaction whereby one gene interferes with the phenotypic expression of another nonallelic gene (or genes), so that the phenotype is determined effectively by the former and not by the latter when both genes occur together in the genotype. ... In population genetics and quantitative genetics, the term epistasis is sometimes used to refer to *all* nonallelic gene interactions (Reiger, Michaelis, and Green, 1991, p. 170; see also Fox Keller and Lloyd, 1992).

• Nonlinear interactions among genes due to their linkage on the chromosome under recombination events.

• The phenomenon of a phenotypic trait being affected by nonadditive interactions among multiple genes.

equilibration The process whereby an organism adapts to its environment, forming a correspondence. The process by which internal and external variables interact to produce a certain set of behaviors.

evolution A process of change in a certain direction; a process of continuous change from a lower, simpler, or worse state to a higher, more complex, or better state; a theory that the various types of animals and plants have their origin in other preexisting types and that the distinguishable differences are due to modifications in successive generations (On-line Webster program). See also Fox Keller and Lloyd (1992).

• Evolution is both process and result. The *process* of evolution includes all mechanisms of genetic change that occur in organisms through time, with special emphasis on those mechanisms that promote the adaptation of organisms to their environment or that lead to the formation of new reproductively isolated species. The *result* of evolution is observed in the evolutionary history of life on Earth and in the genetic relationships among species that exist today (Hartl, 1988, p. 143).

• Time course of any system. Physicists often describe dynamical systems in this way.

• The process of change from a lower (simpler) to a higher (more complex) form of biological organization through accumulation of variations and selective reproduction. However, few would disagree that parasites typically "devolve" from higher forms, and yet are simpler in significant ways from their progenitors.

• Changes in the genome that alter inborn behavior; in contrast to "learning."

exaptation   Selectively neutral traits that are subsequently adapted towards some function of selective consequence. See PREADAPTATION

feedback signal   Any information available to an adaptive process that might alter its state. In a learning framework, input that an individual can use to improve itself relative to some objective. Some examples include supervisory signals, reinforcement signals, unsupervised signals, and heuristic signals.

fitness   The relative success an individual achieves at reproduction, relative to other members of the same generation. See also Fox Keller and Lloyd (1992).

• The relative ability of an organism to survive and transmit its genes to the next generation (King, 1990).

• In population genetics, a quantitative measure of reproductive success of a given genotype, i.e., the average number of progeny left by this genotype as compared to the average number of progeny of other, competing genotypes (a.k.a. adaptive value, selective value) (Reiger, Michaelis, and Green, 1991).

• In the genetic optimization framework, a measure of how well an individual achieves the goal set by the designer of the simulation (e.g., predicting an environment); that which the adaptive system optimizes.

- The relationship between a reproducing pattern (gene, genotype, or phenotype) and its environment which determines the differential reproductive success of that entity relative to competing alternatives.

- The way to determine the reproductive success of an individual (candidate solution) with respect to the rest of the population. It can be exogenous (defined *a priori*) or endogenous (defined by interactions between individuals and an environment that may not be completely specified *a priori*).

- To be distinguished from "adaptedness." For example, imagine a hypothetical individual that produces ten times as many offspring as others in the population, but imagine also that all the offspring are sterile. The adult parent is more *fit*, but not well adapted.

fitness, inclusive Going beyond the fitness of an individual to consider the collective fitness of an individual and all genetically close relatives.

- The sum of an individual's fitness, quantified as the reproductive success of the individual and its relatives, with the relatives devalued in proportion to their genetic distance (Reiger, Michaelis, and Green, 1991).

function Baldwin (Chapter 5) speaks of "the functions which an organism performs in the course of his life history," "new or modified functions," "normal congenital functions," etc. Baldwin's central definition of "Organic Selection" involves the word "function": "the organism's behavior in acquiring new modes or modifications of adaptive functions with its influence of structure." Function is also, according to Baldwin, a central part of the Lamarckian argument: "the evidence drawn from function, 'use and disuse,' is discredited."

- In mathematics, a function is a relation between values in a set D and values in a set R such that for all values d in D, there is associated a unique value $r = f(d)$ in R. The value r can be viewed as a consequence (or result) of the value d. For example, a "fitness function" maps individuals to fitness scores (real numbers).

gene The basic unit of Mendelian inheritance which represents a contiguous region of DNA (or RNA in some viruses) corresponding to usually one (less often two or more) transcription unit or transcription. The particular sequence of nucleotides along the nucleic acid molecule represents a function unit of inheritance (defined operationally by the cis-trans test), a cistron.... Genes can be divided into those that code for polypeptides (structural genes), those that are transcribed into RNA...but not translated into proteins, and possibly those in which functional significance does not demand that they are transcribed at all (DNA sequences that may govern the punctuation or regulation of genetic transcription).

A gene consists of a linear array of potentially mutable units...between which intragenic genetic recombination can occur and permits the genetic mapping of mutational sites....

As a result of gen⟨   ⟩.⟨   ⟩.tion, alternative forms, referred to as alleles, of a particular gene are pr⟨ ⟩.⟨ ⟩.ed. The existence of a gene is extrapolated from these alleles which generally influence the same phenotypic character or trait (being the product of gene expression)....

Genes are the basis of both continuous (qualitative) and discontinuous (quantitative) characters and usually produce effects on a wide variety of biochemical and morphological characters. So-called "pleiotropic" "effects" [sic] are the result of a variety of effects originating from a single primary action, i.e., coding for a definite polypeptide ("one cistron-one polypeptide model").

The collection of genes contained within one chromosome in linear sequence (genetic map) and in definition positions constitute a linkage group. There is evidence, in some microorganisms at least, that genes which control a series of related biochemical reactions are often adjacent to each other in the linkage structure (gene cluster). The same is true for the case of the mutational sites within one gene. Sites with similar properties show a tendency to cluster and are not distributed randomly within the nucleotide sequence defined functionally as a gene (Reiger, Michaelis, and Green, 1991, pp. 189–191; see also Fox Keller and Lloyd, 1992).

genetic algorithm (GA) An evolutionary model that performs global adaptive search with respect to an objective function. A population of candidate solutions are encoded as gene vectors and evaluated relative to the objective function. Some vectors are selected to differentially reproduce and are subjected to genetic variation (e.g., mutation, crossover). The resulting set of gene vectors forms the next generation, and the cycle repeats itself until some termination criterion is reached.

genome See GENOTYPE

genotype 1: The genetic constitution of an individual or group; 2: a class or group of individuals sharing a specified genetic makeup (On-line Webster program; see also Fox Keller and Lloyd, 1992). See PHENOTYPE

habits The meaning of habits seems to be more general in usage than the modern use of the word and maps onto our use of the word "behavior."

immune system The system responsible for recognizing, and defending against, pathogens, toxins, and other foreign molecules.

individual 1a: A particular being or thing as distinguished from a class, species, or collection; b(1): a single human being as contrasted with a social group or institution; (2): a single organism as distinguished from a group; c: a particular person; 2: an indivisible entity; 3: the reference of a name or variable of the lowest logical type in a calculus (On-line Webster program).

individual-based modeling   See MODELING, INDIVIDUAL-BASED

Lamarckian inheritance   Inheritance in which characteristics acquired during the lives of one generation are passed on to the next.

- This notion of inheritance was required by (but not unique to) Lamarck's theory of biological evolution. Darwin (as well as Lamarck) implicitly assumed that acquired characteristics, like somatic characteristics, were transmitted physiologically through the germ line, an assumption that was refuted by August Weismann.

learning   Knowledge or skill acquired by instruction or study. Modification of a behavioral tendency by experience. Coming to know (On-line Webster program).

- "...any process in which, during normal, species-typical ontogeny, the organization of an animal's behavior is in part determined by some specific prior experience." (Johnston, Chapter 20.) Learning, for Johnston, excludes both conditioning *per se* and "nonspecific" experience such as inadequate nutrition.

- Often used synonymously with modification, independent of effect on resulting performance at some skill or knowledge.

- Refers to types of within-lifetime developmental growth that are primarily due to the individual's interaction with the environment. See MATURATION

learning algorithm   A set of rules that determine how an individual's experiences affect its future behavior.

mastery effect   The diminution of evolutionary improvement when learning alone is capable of attaining maximal performance at some task. Especially dependent on the non-Lamarckian character of a given simulation, as well as a lack of a fitness cost based on the amount of learning performed by a given individual. Both of these dependencies are met by the GA used by Hightower et al. (Chapter 11).

maturation   1a: The process of becoming mature; b: the emergence of personal characteristics and behavioral phenomena through growth processes; c: the final stages of differentiation of cells, tissues, or organs (On-line Webster program).

- Refers to types of within-lifetime developmental change that are primarily due to genetically determined growth processes. See LEARNING

meme   A hypothesized analog in systems of cultural change to the gene in biological evolution.

memory The retention and retrieval in the human mind of past experiences. . . . The function of remembering and its converse, forgetting, are normally adaptive. Learning, thought, and reasoning could not occur without remembering. On the other hand, forgetting has many functions, including time orientation by virtue of the tendency of memories to fade over time; adaptation to new learning by the loss or suppression of old patterns; and relief from the anxiety of painful experiences (Brittanica On-line).

• In Zhivotovsky et al. (Chapter 10), a record of the last k environmental states, which is to be used in making a decision about the next state. Nothing else counts as memory, in particular no information about states before the last k. It is not necessary that the organism 'recall' these states, only that they have an effect on the decision.

• Storage of past events. In Shafir and Roughgarden (Chapter 12), memory refers to the accumulation of prey appearances across a temporal window.

model A miniature representation of something; an example for imitation or emulation; a description or analogy used to help visualize something (as an atom) that cannot be directly observed; a system of postulates, data, and inferences presented as a mathematical description of an entity or state of affairs. (See On-line Webster program.) Roughgarden et al. (Chapter 2) follow the last definition very closely.

• A set of assumptions, identifying the essential features of a system, presented as a mathematical or algorithmic description of the system. This simplified description can be used as a theoretical laboratory in order to test hypotheses about the system.

model, individual-based Rather than assuming all individuals in a population are the same, or that average values of individuals in a population can be used to describe population phenomena, . . . individual-based models have been developed by ecologists and evolutionary biologists who want to emphasize individual differences and local interactions.

There are two kinds of individual-based models: (1) distribution and (2) configuration. Distribution models consider individual differences by lumping together individuals with common characters. Configuration models track all individuals in a population to incorporate both individual differences and local interaction. . . . In a broad sense, individual-based configuration models can be regarded as artificial life (Kawata and Toquenaga, 1994).

model, minimal idea A model that explores some idea or concept without the inconvenience of specifying any particular species, environment, or much of anything else. The premise is that some phenomenon of interest—say, sexual reproduction—is a computational entity whose properties will be pretty much the same across a wide range of possible universes. See MODEL, MINIMAL SYSTEM

model, minimal system   An exploration of the dynamics of some greatly simplified subset of the features of a real environment. Basically a minimal idea model with some contact to the real world. See MODEL, MINIMAL IDEA

model, synthetic system   Also called a systems model, a synthetic model is supposed to be an expansion of a minimal system model in which (ideally) all the assumptions are treated formally. The implication is that it is so huge that it could only be the product of large-scale collaboration; interestingly, Roughgarden et al. (Chapter 2) make no mention of synthetic models in their discussion of how computation might be useful to biology.

modification   Changes to the "body structure" of a phenotype.

mutation   Any genetic difference between parent and offspring.

- *Random* modifications to the genome. Note that crossover (the model of sexual recombination) in the genetic algorithm is distinguished from mutation under this definition, while included under the first definition.

- In modern biology, mutation usually refers to a change in the genotype that may or may not result in a phenotypic change. Lamarck used this term to refer to a morphological change in the phenotype.

Neo-Darwinian   A theory that holds natural selection to be the chief factor in evolution and specifically denies the possibility of inheriting acquired characters (On-line Webster program).

- The "Neo-" qualifies the integration of Darwin's theory about the process of evolution with genetic models developed following the rediscovery of Mendel's work and subsequent mathematical models of population genetics near the turn of the twentieth century.

neural network, artificial   A computational model, usually simulated on a general-purpose computer, that maps vectors of numbers to different vectors of numbers through a sequence of mathematical transformations. It is intended to behave in a manner analogous to biological neurons and, hence, consists of components termed "connections," "weights," and "units." The transformations are usually chosen to be simple thresholdlike operations. The mapping implemented by a neural net can be altered to more closely match a given pattern using any of a number of "learning" mechanisms, such as "error backpropagation."

ontogeny   The trajectory of changes occurring to an individual during its lifetime. See PHYLOGENY

- Ernst Haeckel's "biogenetic law" held that an individual's ontogeny *recapitulates* the phylogeny of its species, but this facile connection between the two forms of change is now discredited (Gould, 1977).

phenocopy A phenotypic variation that is caused by unusual environmental conditions and resembles the normal expression of a genotype other than its own (On-line Webster program).

- Schull (Chapter 3) sees Piaget's use (Chapter 19) of this term as a discipline-specific special case of organic selection. Piaget gave credit to both Baldwin and Waddington, and applied his term to the more general case in which genetic adaptedness comes to supplant or replace phenotypic adaptability, with little change in the observable phenotype. Plasticity would persist or would be enhanced due to its effect upon natural selection. Piaget thought of behavior as the "motor of evolution."

- A product of the convergence between a phenotypic variation and a genotypic variation which comes to take its place. The idea is that changes in the behavior of a particular organism create changes in the genetic makeup of that organism's descendants.

phenotype 1: The detectable expression of the interaction of genotype and environment constituting the visible characters of an organism; 2: a group of organisms sharing a particular phenotype (On-line Webster program). See also Fox Keller and Lloyd (1992).

phylogeny 1: The racial history of a kind of organism; 2: the evolution of a genetically related group of organisms as distinguished from the development of the individual organism; 3: the history or course of the development of something (On-line Webster program). See ONTOGENY

plasticity 1: The quality or state of being plastic; *esp.*: capacity for being molded or altered; 2: the ability to retain a shape attained by pressure deformation; 3: the capacity of organisms with the same genotype to vary in developmental pattern, in phenotype, or in behavior according to varying environmental conditions (On-line Webster program). See also Fox Keller and Lloyd (1992).

- According to Schull (Chapter 3), plastic individuals are often purposive, goal-driven, and intelligent, and plasticity is a prerequisite for ontogenetic adaptation and learning. Further, if plastic individuals play a causal role in evolutionary processes, then the question arises whether the evolutionary process, itself is purposive, goal-driven, and/or intelligent.

population A group of individuals of a species living in a certain area. Because they are the same species they can interbreed, although these breeding patterns might show local structure.

- A local population or "Mendelian" population is the local panmictic unit, for example, in "island" models of structured populations. A collection of local populations makes up the species.

- 1: The whole number of people or inhabitants in a country or region; 2: the act or process of populating; 3a: a body of persons having a quality or characteristic in common; b(1): the organisms inhabiting a particular area or biotope; (2): a group of interbreeding biotypes that represents the level of organization at which speciation begins; 4: a group of individual persons, objects, or items from which samples are taken for statistical measurement (On-line Webster program).

preadaptation Traits initially evolved in response to one selective pressure but then adapted to a second. See EXAPTATION

preformation Alternative (now discredited) account of development. See EPIGENESIS

psychology 1: The science of mind and behavior; 2a: the mental or behavioral characteristics of an individual or group; b: the study of mind and behavior in relation to a particular field of knowledge or activity; 3: a treatise on psychology (On-line Webster program).

punishment Something "bad" that happens to an organism after a behavior has occurred. Punishment cannot be avoided as it occurs after the behavior has already happened. See REINFORCEMENT, NEGATIVE

purposiveness 1: Serving or effecting a useful function though not as a result of planning; 2: having or tending to fulfill a conscious purpose or design; purposeful (On-line Webster program).

- According to Schull (Chapter 3), Darwin's theory demonstrated that purely mechanistic processes (mutation and differential reproduction) could explain evolutionary adaptation without necessarily invoking "purposiveness." In this light, organismic behavior and psychological phenomena were seen as results rather than causes of evolutionary progress.

reinforcement, negative Any stimulus that, when removed following a response, increases the probability of the response (Hilgard, Atkinson, and Atkinson, 1979). See PUNISHMENT

- The modification of behavior based upon the omission of a stimulus (typically a punishment).

reinforcement, positive Any stimulus that, when removed following a response, decreases the probability of the response (Hilgard, Atkinson, and Atkinson, 1979). See REWARD

- The modification of behavior based upon the presentation of a stimulus (typically a reward).

reward Something "good" that happens to an organism after a behavior has occurred. Reward cannot be avoided as it occurs after the behavior has already happened. See REINFORCEMENT, POSITIVE.

search, global An algorithm that estimates extreme values of a function over a specified domain.

search, local An algorithm that refines an initial estimate of a solution to a function. Specifically, an algorithm that identifies a local optimum near the initial estimate. Examples include the back-propagation method for neural networks (also called stochastic approximation); the conjugate gradient method, which exploits gradient information in some local neighborhood; and Solis-Wets, which does not require explicit gradient information.

selection The process that chooses among variants, and that is itself shaped by those it selects for and against; in James' view, selection operates at nearly all levels, within and across individuals.

selection, intra- Plasticity of an individual organism, changes within an individual in response to environment.

selection, organic Baldwin's own term for any mechanism based on natural selection by which the adaptive achievements of individuals could influence the direction of evolution. Ontogenetic adaptation to local circumstances could prolong the race long enough for natural selection to accumulate genetic variations that would support the behaviors in question. Ontogenetic adjustments would be mediated by learning. The originally learned behavior would become innate. Credit for the origin of adaptations could thus be given to creative individuals, not to the heritable adaptations. A way of reconciling the Darwinian mechanism of evolution (natural selection) with a Lamarckian characterization of evolution (as purposive and individual-driven). See BALDWIN EFFECT, ECOLOGY OF MIND

- An individual's ability to adjust to its local environment in its own lifetime (ontogenetic adaptation) is valuable but limited: an organism's investment in adapting to one domain will tend to limit its capacity to adapt to other domains or in other directions. Since the time scale for ontogenetic adaptation is so much shorter than that of natural selection, adaptive strategies will often be "invented" first through individual plasticity rather than genetic mutation and population-genetic evolution. But once invented, the persistence of these adaptations may affect selection pressures and allow time for mutations such that population-genetic evolution occurs. Thus, the role of population-genetic evolution is often to "ratify" rather than to generate adaptations which set the course of evolution. Schull (Chapter 3) suggests that this general formulation developed by Bateson (Chapter 9) can subsume a number of discipline-specific special cases, such as "genetic assimilation," "canalization," "phenocopy," and the "Baldwin effect."

• A term originally associated (by Baldwin and his contemporaries) with what is now known as the "Baldwin effect." For example, Simpson (Chapter 8) essentially equates this term with his own characterization of "Baldwin effect."

selection, sexual   A particular form of selection. While natural selection refers to any differential reproductive success due to differences in survival ability, sexual selection refers to those effects due to differences in mate choice. See also Fox Keller and Lloyd (1992).

• Sexual selection occurs when females (typically) of a particular species tend to select mates according to some criterion (e.g., who has the biggest, most elaborate plumage or antlers), so males having those traits are more likely to be chosen by females as mates. The offspring of such matings tend to inherit the genes encoding the sexually selected trait and those encoding the preference for the sexually selected trait. The former will be expressed only in males, and the latter only in females. Fisher (1930) proposed that this process could result in a feedback loop between (typically) females' preference for a certain trait and the strength and frequency of that trait in males. As the frequency of females that prefer the trait increases, it becomes increasingly sexually advantageous for males to have it, which then causes the preference genes to increase further because of increased mating between females with the preference and males with the trait. Fisher (1930) termed this "runaway sexual selection." (Mitchell, 1996.)

• Some sociobiologists distinguish between "sexual selection" and "natural selection," as disjoint types. "We view *sexual selection* as arising from variance in mating success and *natural selection* as arising from variance in other components of fitness" (Arnold and Wade, 1984).

simulation   1: The act or process of simulating; 2: a sham object: counterfeit; 3a: the imitative representation of the functioning of one system or process by means of the functioning of another (a computer ∼ of an industrial process); b: examination of a problem often not subject to direct experimentation by means of a simulating device (Merriam-Webster's Collegiate Dictionary).

soma   All of an organism except the germ cells (On-line Webster program). See WEISMANN'S DOCTRINE

variation   Changes with a "germinal origin"; changes to the genotype.

• The existence of variants; the range over which natural selection can select.

Weismann's Doctrine The German biologist August Weismann distinguished between two substances that make up an organism: the soma, which comprises most body parts and organs, and the germ plasm, which contains the cells that give rise to the gametes and hence to progeny. Early in the development of an egg, the germ plasm becomes segregated from the soma, that is, from the cells that give rise to the rest of the body. This notion of a radical separation between germ and soma prompted Weismann to assert that inheritance of acquired characteristics was impossible, and it opened the way for his championship of natural selection as the only major process that would account for biological evolution (Brittanica On-line).

REFERENCES

Arnold, S., and M. Wade. *Evolution* **38** (1984): 720–734.

Brittanica On-line: http://www.eb.com:180/eb.html.

Fisher, R. A. *The Genetical Theory of Natural Selection.* Clarendon, 1930.

Fox Keller, E., and E. A. Lloyd, eds. *Keywords in Evolutionary Biology.* Cambridge, MA: Harvard University Press, 1992.

Gould, S. J. *Ontogeny and Phylogeny.* Cambridge, MA: Harvard University Press, 1977.

Hartl, D. L. *A Primer of Population Genetics.* 2nd ed. Sinauer Associates, 1988.

Hilgard, E. R., R. L. Atkinson, and R. C. Atkinson. *Introduction to Psychology.* 7th ed. New York: Harcourt Brace Jovanovich, 1979.

Kawata, M., and Y. Toquenaga. "From Artificial Individuals to Global Patterns." *Trends in Ecology and Evolution* **9(11)** (1994): 417–421.

King, R. C. *A Dictionary of Genetics.* 4th ed. Oxford University Press, 1990.

Langton, C. G., C. Taylor, J. D. Farmer, and S. Rasmussen, eds. *Artificial Life II.* Santa Fe Institute Studies in the Sciences of Complexity, Proc. Vol. XIII. Reading, MA: Addison-Wesley, 1992.

Lovelock, J. E. *Gaia: A New Look at Life on Earth.* Oxford University Press, 1987.

Merriam-Webster's Collegiate Dictionary. 10th ed. Springfield, MA: Merriam-Webster, Inc, 1993.

Mitchell, M. *An Introduction to Genetic Algorithms.* Cambridge, MA: MIT Press, 1996.

On-line Webster program. Based on *Webster's Seventh Dictionary.* Springfield, MA: Merriam-Webster Inc., 1965.

Reiger, R., A. Michaelis, and M. M. Green. *Glossary of Genetics.* 5th ed. Berlin: Springer-Verlag, 1991.

Stern, C. "Selection for Subthreshold Differences and the Origin of Pseudo Exogenous Adaptations." *Amer. Nat.* **92** (1958): 313.

Waddington, C. H. "Canalization of Development and the Inheritance of Acquired Characters." *Nature* **150** (1942): 563–565.

Index

Made in the USA
Coppell, TX
18 February 2021